39.50

THERMAL IMAGING SYSTEMS

OPTICAL PHYSICS AND ENGINEERING

Series Editor: **William L. Wolfe**
Optical Sciences Center, University of Arizona, Tucson, Arizona

1968:
M. A. Bramson
Infrared Radiation: A Handbook for Applications

1969:
Sol Nudelman and S. S. Mitra, Editors
Optical Properties of Solids

1970:
S. S. Mitra and Sol Nudelman, Editors
Far-Infrared Properties of Solids

1971:
Lucien M. Biberman and Sol Nudelman, Editors
Photoelectronic Imaging Devices
   Volume 1: Physical Processes and Methods of Analysis
   Volume 2: Devices and Their Evaluation

1972:
A. M. Ratner
Spectral, Spatial, and Temporal Properties of Lasers

1973:
Lucien M. Biberman, Editor
Perception of Displayed Information

W. B. Allan
Fibre Optics: Theory and Practice

Albert Rose
Vision: Human and Electronic

1975:
J. M. Lloyd
Thermal Imaging Systems

Winston E. Kock
Engineering Applications of Lasers and Holography

A Continuation Order Plan is available for this series. A continuation order will bring delivery of each new volume immediately upon publication. Volumes are billed only upon actual shipment. For further information please contact the publisher.

THERMAL IMAGING SYSTEMS

J. M. Lloyd

Honeywell Inc.
Radiation Center
Lexington, Massachusetts

PLENUM PRESS • NEW YORK AND LONDON

Library of Congress Cataloging in Publication Data

Lloyd, J   Michael, 1945-
  Thermal imaging systems.

  (Optical physics and engineering)
  Includes bibliographical references and index.
  1. Imaging systems. 2. Infra-red technology. I. Title.
TA1570.L55   621.36'72   75-9635
ISBN 0-306-30848-7

First Printing–May 1975
Second Printing–March 1979

© 1975 Plenum Press, New York
A Division of Plenum Publishing Corporation
227 West 17th Street, New York, N.Y. 10011

United Kingdom edition published by Plenum Press, London
A Division of Plenum Publishing Company, Ltd.
Davis House (4th Floor), 8 Scrubs Lane, Harlesden, London, NW10 6SE, England

Printed in the United States of America

PREFACE

This book is intended to serve as an introduction to the technology of thermal imaging, and as a compendium of the conventions which form the basis of current FLIR practice. Those topics in thermal imaging which are covered adequately elsewhere are not treated here, so there is no discussion of detectors, cryogenic coolers, circuit design, or video displays. Useful information which is not readily available because of obscure publication is referenced as originating from personal communications.

Virtually everyone with whom I have worked in the thermal imaging business has contributed to the book through the effects of conversations and ideas. I gratefully proffer blanket appreciation to all those who have helped in that way to make this book possible. The contributions of five people, however, bear special mention: Bob Sendall, Luke Biberman, Pete Laakmann, George Hopper, and Norm Stetson. They, more than any others, have positively influenced my thinking.

The many supervisors I have had in three years of writing this book have generously supported my efforts. Among them, I especially wish to thank Ed Sheehan and Don Looft of the Army's Night Vision Laboratory, who originally supported the work, and Bob Norling, Bob Rynearson, and Sheldon Busansky of Honeywell who continued that support. This book was ultimately made possible by its inclusion in the Optical Physics and Engineering Series. For that, I wish to thank Bill Wolfe, the Series editor, His confidence, patience, and criticisms are much appreciated. I also wish to recognize and to thank Miss Beth Whittemore of Honeywell who typed numerous drafts of the manuscript, and Miss Gail Logan of Frank Thompson Associates, who supervised the preparation of the final manuscript. Finally, my profound appreciation goes to LaVonne, my editor-at-home, whose editorial skills have clarified many passages.

Readers are encouraged to contact me at Honeywell to transmit corrections, alternate views, and new sources of information.

Mike Lloyd
Acton, Massachusetts

v

CONTENTS

Chapter Four

Chapter Five

Chapter Six

Optics 212

Chapter Seven

Scanning Mechanisms 283

Chapter Eight

Thermal Imaging System Types 329

SYMBOL TABLE

Latin Alphabet Symbols

Symbol	Meaning	Units
a	detector horizontal linear dimension	[cm]
A	system horizontal field of view	[angular degree]
A_d	detector area	[cm^2]
A_c	displayed noise correlation area	[cm^2]
A_t	displayed target area	[cm^2]
A_o	effective optical collecting area	[cm^2]
arg ( )	"argument of complex number" operator	dimensionless
b	detector vertical linear dimension	[cm]
B	system vertical field of view	[angular degree]
c	speed of light in vacuum	[cm/sec]
c_1, c_2, c_3	constants in Planck's laws	—
C	contrast	dimensionless
C	capacitance	[farad]
C	design complexity	[cm]
C_R	radiation contrast	dimensionless
$_0C_{40}$	spherical aberration coefficient	[cm^{-3}]
Circ ( )	circular function	dimensionless
Comb ( )	periodic delta function array	dimensionless
CFF	critical flicker-fusion frequency	[Hertz]
CTD	critical target dimension	[meter]
d	linear blur diameter	[cm]

Symbol	Meaning	Units
$D^*(\lambda)$	specific detectivity as a function of wavelength	$[\text{cm Hz}^{1/2}/\text{watt}]$
$D^*(\lambda_p)$	peak specific detectivity	$[\text{cm Hz}^{1/2}/\text{watt}]$
D_0	optical system clear aperture diameter	$[\text{cm}]$
$e(t)$	voltage as a function of time	
E	sensor efficiency	dimensionless
f	optical focal length	$[\text{cm}]$
f	temporal or spatial frequency (depending on context)	$[\text{Hz}]$ or $[\text{cy/mrad}]$
f_R	reference frequency	$[\text{Hertz}]$
f_T	target spatial frequency	$[\text{cy/mrad}]$
Δf	electrical noise equivalent bandwidth	$[\text{Hz}]$
f_x, f_y	Cartesian angular spatial frequencies	$[\text{cy/mrad}]$
$f_c, f_\tau, f_1, f_2, f_0$	characteristic frequencies	$[\text{Hz}]$ or $[\text{cy/mrad}]$
f_c	cutoff spatial frequency	$[\text{cy/mrad}]$
$\dot{F}$	frame rate	$[\text{Hz}]$
$F/\#$	optical focal ratio (F-number)	dimensionless
$F\{\ \}$	forward Fourier transform operator	dimensionless
$F^{-1}\{\ \}$	inverse Fourier transform operator	dimensionless
$g(f)$	noise voltage spectrum	$[\text{volts/Hz}^{1/2}]$
G	gain	dimensionless
h	Planck's constant	$[\text{watt sec}^2]$
H	irradiance	$[\text{w/cm}^2]$
H_a	absolute humidity	$[\text{gm/cm}^3]$
H_a'	absolute humidity at saturation	$[\text{gm/cm}^3]$
H_r	relative humidity	dimensionless
$I(x,y)$	image signal distribution as a function of angular or linear Cartesian coordinates	dimensionless
$i(t)$	current as a function of time	$[\text{amperes}]$
i	image distance	$[\text{cm}]$
I	scanner interlace factor	dimensionless

Symbol	Meaning	Units
I	relative coefficient of refractive index	$[^\circ K^{-1}]$
$I_m(\ )$	"imaginary part of complex number" operator	dimensionless
J	radiant intensity	[watt/sr]
K	Boltzmann's constant	[watt sec/$^\circ$K]
$\vec{K}$	wave vector	$[cm^{-1}]$
$k(\lambda)$	atmospheric absorption coefficient	$[km^{-1}]$
L	luminance	[foot-Lambert]
L_B	background luminance	[foot-Lambert]
L	scan lines per target height	[foot-Lambert]
LSF	line spread function	dimensionless
M	optical angular magnification	dimensionless
MDTD	minimum detectable temperature difference	$[^\circ K]$
MTF	modulation transfer function	dimensionless
MRTD	minimum resolvable temperature difference	$[^\circ C]$
n	refractive index	dimensionless
n	number of detectors	dimensionless
N	angle of surface normal	[radian]
N	radiance	[watt/cm^2 sr]
N_e	equivalent line number or equivalent system bandwidth	[cy/mrad]
NETD	noise equivalent temperature difference	$[^\circ K]$
O(x,y)	object signal distribution with angular or linear Cartesian coordinates	dimensionless
O	object distance	[cm]
O	overscan factor	dimensionless
O_h	hyperfocal distance	[cm]
OTF	optical transfer function	dimensionless
P	power of a lens or lens surface	$[cm^{-1}]$

Symbol	Meaning	Units
P	radiant power	[watt]
P	water vapor partial pressure	[mm of mercury]
$p(x,y)$	entrance pupil function	dimensionless
P	sensor performance	[sec$^{-1/2}$ mr^{-1} °K^{-1}]
P_C	probability of success as a function of a performance criterion C	dimensionless
Q	electric charge	[coulomb]
Q_λ	spectral radiant photon emittance	[photons/cm^2 sec μm]
Q_B	background photon flux	[photons/cm^2 sec]
$r(x)$	impulse response	dimensionless
$\tilde{r}(f)$	OTF or MTF depending on context	dimensionless
$\tilde{r}_o(f)$	MTF of optics	dimensionless
$\tilde{r}_d(f)$	MTF of detector	dimensionless
$\tilde{r}_e(f)$	MTF of processing electronics	dimensionless
$\tilde{r}_m(f)$	MTF of video monitor	dimensionless
$\tilde{r}_s(f)$	MTF of system	dimensionless
$\overset{\sqcap}{r}(f)$	square wave response	dimensionless
R	resistance	[ohm]
R	slant range to target	[m]
R	radius	[m]
$R(\lambda)$	responsivity	[volts/watt]
$R_e(\ )$	"real part of complex number" operator	dimensionless
R_p	Petzval radius	[cm]
Rect ( )	rectangular function	dimensionless
S	surface area	[cm^2]
S{ }	system operator	dimensionless
Sinc ( )	(Sin x)/x function	dimensionless
SiTF	signal transfer function	[ft-lambert/°K]
SNR	signal-to-noise ratio	dimensionless

Symbol	Meaning	Units
SNR_p	preceived signal-to-noise ratio	dimensionless
SNR_i	image point signal-to-noise ratio	dimensionless
SWR	eye sine wave response function	dimensionless
t	time	[second]
t	optical element thickness	[cm]
t	lens center thickness	[cm]
T	absolute temperature	[degree Kelvin]
T_e	effective eye integration time	[second]
T_f	frame time	[second]
T_T, T_B, T_A	target, background, and atmospheric absolute temperatures	[°K]
ΔT	temperature difference	[°K]
U	radiant energy	[joule]
V	voltage	[volts]
V	dispersion index	dimensionless
W	radiant emittance	[watt/cm^2]
W	precipitable water	[pr cm/km]
W'	precipitable water at saturation	[pr cm/km]
W_λ	spectral radiant emittance	[watt/cm^2 μm]
W(x,y)	aberration function	[cm]
W_B, W_T	background and target spectral radiant emittances	[watt/cm^2 μm]
$\dfrac{\partial W}{\partial T}$	differential change in radiant emittance with a differential change in temperature, within some specified spectral band	[watt/cm^2 °K]
x,y	angular or linear Cartesian coordinates, as implied by context	[m] or [mrad]

Greek Alphabet Symbols

Symbol	Meaning	Units
α	detector horizontal angular subtense	[mrad]
α	coefficient of thermal expansion	[cm/cm $^\circ$K]
β	detector vertical angular subtense	[mrad]
γ	scan mirror deflection angle	[radian]
$\gamma(\lambda)$	atmospheric extinction coefficient	[km^{-1}]
δ	angular blur diameter	[radian]
$\delta(\ )$	dirac delta function	dimensionless
ΔT	temperature difference	[$^\circ$K]
Δ_1, Δ_2	separations of principal planes and lens vertices	[cm]
ϵ	emissivity	dimensionless
ϵ	linear defocus	[cm]
ξ, η	linear or angular Cartesian coordinates	[m] or [mrad]
η	fractional obscuration	dimensionless
λ	wavelength	[μm]
μ	statistical mean	dimensionless
ω	optical frequency	[sec^{-1}]
Ω	solid angle	[sr]
Ω_{cs}	cold shield angle	[sr]
η_q	detective quantum efficiency	dimensionless
η_{cs}	cold-shielding efficiency	dimensionless
η_v	vertical scan duty cycle	dimensionless
η_H	horizontal scan duty cycle	dimensionless
η_{sc}	overall scan duty cycle	dimensionless
η_O	optical efficiency	dimensionless
σ	line spread function standard deviation	[mrad]
σ	Stefan-Boltzmann constant	[watt/cm^2 $^\circ$K^4]
$\sigma(\lambda)$	atmospheric scattering coefficient	[km^{-1}]

Symbol	Meaning	Units
θ	scan angle in object space	[radian]
θ_c	critical target angular subtense	[mrad]
τ	time constant	[second]
τ_d	detector dwelltime	[second]
$\tau_A(\lambda)$	atmospheric transmission	dimensionless
$\overline{\tau}_a$	average atmospheric transmission within some specified spectral band	dimensionless
$\tau_0(\lambda)$	optical transmission	dimensionless
$\overline{\tau}_0$	average optical transmission within some specified spectral band	dimensionless

CHAPTER ONE — INTRODUCTION

1.1 The Function of Thermal Imaging Systems

Thermal imaging systems extend our vision beyond the short-wavelength red into the far infrared by making visible the light naturally emitted by warm objects. Because of the eye's lack of response in the absence of 0.4 to 0.7 μm light, a device is needed which will image the dominant energy at night as the eye does during the day. This night eye must respond generally to the photons emitted at body temperature because these dominate when reflected solar radiation is absent. Specifically, it must have spectral response at wavelengths where significant emissivity, temperature, and reflectivity differences exist in the scene. This is necessary to assure that radiation patterns will exist which are sufficiently similar to the corresponding visual reflectivity patterns to make visual interpretation of the converted scene possible.* This spectral sensitivity must also coincide with an atmospheric transmission "window" which does not excessively absorb the desired radiation.

The difficulty of producing thermal images may be appreciated by considering the perfection with which the eye produces visible images. The eye is an optimized sensor of visible radiation in three respects. First, the eye's spectral response range of 0.4 to 0.7 μm coincides with the peak of the sun's spectral output. Approximately 38 percent of the sun's radiant energy is concentrated† in this band, and terrestrial materials tend to have good reflectivities there. Second, the eye is an ideal quantum-noise-limited device because the retinal radiation detectors have low noise at the quantum energy levels in this band. Third, the response of these retinal detectors to the photons emitted at body temperature is negligible, so that this long wavelength thermal energy does not mask the response to the desired wavelengths. This optimization allows the eye to perform its primary functions, which are

---

*We will be concerned solely with passive imaging systems which sense natural scene radiation, rather than with active systems which sense the reflected radiation from a thermal illuminator.

†Compare this with the 8 to 14 μm band where only about 0.08 percent of the sun's energy is found.

1

the detection of reflectivity differences in objects illuminated by 0.4 to 0.7 micrometer radiation, discernment of patterns in these reflectivity differences, and the association of these patterns with abstractions derived from previous visual and other sensory experiences. To be as effective as the human eye, the night thermal eye must also image the dominant light, be quantum-limited, and reject extraneous light.

Whereas natural visual spectrum images are produced primarily by reflection and by reflectivity differences, thermal images are produced primarily by self-emission and by emissivity differences. Thus in thermal imaging, we are interested in thermally self-generated energy patterns. Thermal system capabilities are usually described in terms of scene temperatures rather than in radiometric terms. Therefore some clarifying definitions are necessary. All of the scene temperature, reflectivity, and emissivity contributions taken together can be represented at any point in the scene by an effective temperature at that point. This is the temperature which would produce the measured irradiance near the point if the point were an ideal blackbody radiator; that is, if the point radiated the maximum theoretically possible power at the effective temperature. Similarly, the irradiance measured through an intervening attenuating atmosphere may be thought of as being produced by an apparent temperature less than the effective temperature.

This simplification is possible because most thermal imaging systems have broad spectral bandpasses and accept all polarizations. Consequently they are insensitive to the mechanisms which produce the effective temperature differences. The variations in the effective temperature of a scene tend to correspond to the details in the visual scene, so a thermal imaging system provides a visible analog of the thermal scene, hopefully with an efficient transfer of useful information from one spectral regime to the other. Wormser[1] discusses the desirability of thermal imaging and describes how conventional thermal imagers work.

This book is concerned primarily with those mechanically-scanning devices which convert radiation in the far infrared spectral region to visible radiation in real time and at an information update rate (or frame rate) comparable to that of television. For lack of a more descriptive abbreviation such as TV for television or I^2 for image intensification, thermal imaging users have adopted the term FLIR, the acronym for Forward Looking Infra-Red. This term distinguishes fast framing thermal imaging systems from downward looking single channel thermal mapping systems, and from single-framing thermographic cameras. Although the term FLIR originally implied an airborne system, it is now used to denote any fast-framing thermal imager.

Thermal imaging in a FLIR is produced in the following way. An optical system collects, spectrally filters, and focuses the infrared scene radiation onto an optically-scanned multi-element detector array. The detectors convert the optical signals into analog electrical signals which are then amplified and processed for display on a video monitor. FLIR is similar in external aspects to television in that the final image often appears on a TV-type monitor operating at TV-type frame rates. Contrast (video gain) and brightness (background level) controls are used, and spatial patterns in the displayed thermal scene strongly resemble those of the corresponding visible scene.

The major applications of real time thermal infrared imaging presently are for military and intelligence purposes. The armed forces have publicized[2] the fact that they have thermal rifle scopes, missile and gun sights, and airborne FLIRs. Their functions include target acquisition, fire control, aerial navigation, surveillance, and intelligence gathering. Non-military applications of thermal imaging are limited, but growing. They include thermal pollution surveys, breast cancer detection and other medical diagnoses, forest fire prevention, air-sea rescue, manufacturing quality control, detection of fissures and loose material in coal mines, preventive maintenance inspections of electrical power equipment, and earth resources surveys. As the utility of thermal imagery becomes more widely recognized, the list is certain to grow. Some likely additions are commercial aircraft landing and taxying aids, floating debris avoidance systems for surface-effect vessels, crime prevention, and smoke-penetration aids for fire fighters.

Thermal imagers are superior in performance to other types of passive-sensing electro-optical imaging devices when operability at any time of the day or night and under all weather conditions is the primary consideration. The major reasons for this superiority are the high contrast rendition which is achieved and the good atmospheric transmission windows which are available. Image intensifiers and low light level television systems rely largely on reflectance differences between the target and the background for detection and recognition. In the visible spectral region and when using non-color devices, broadband reflectance differences between interesting targets and their backgrounds tend to be moderate, especially when camouflage is employed.

Thermal imaging systems, however, usually suppress the average (dc or background) value of the scene radiance, so that only scene variations around the average are displayed. Thus high image contrasts are achieved, especially since the camouflaging of temperature differences is very difficult. We will see in Chapter Ten that an observer's image interpretation performance strongly depends on image contrast so thermal systems in this respect are superior to most visible wavelength devices.

A minor disadvantage is that thermal scenes usually have fewer shadows than visible scenes, so that some three-dimensional contour information is lost. This is usually offset by the fact that all surfaces in the thermal scene radiate energy so that all forward surfaces tend to be visible. Another minor disadvantage is that thermal emitters are not necessarily good visual reflectors, so that visually bright objects may be dark in thermal scenes, and *vice versa*. Nonetheless, considerable intelligence is usually available through thermal imagery. Some often-cited examples are storage tank fluid levels, ship wakes, evidence of recent operation of a vehicle, operation of an industrial plant revealed by hot gaseous emissions or heated areas, and object identifications by noting characteristic surface heating patterns. We will now briefly trace the history of thermal imaging systems.

1.2 A History of Thermal Imaging

The first thermal imaging system was the circa-1930 Evaporagraph, a rather insensitive non-scanning device described in Section 8.7. The Evaporagraph could not satisfy most thermal imaging tasks because of inherent contrast, sensitivity, and response time limitations. Two alternative approaches to thermal imaging were evident in the 1940's. One was to develop discrete-detector, mechanically-scanning analogues of television systems. The other was to develop an infrared vidicon, or some other non-mechanically-scanning device. The first approach has been immensely successful, while the second has had only modest success to date. Since non-scanning devices have not yet matched the performance of scanning systems, we will defer discussion of them until Chapter Eight.

The original scanning thermal imagers were called thermographs. They were single detector element, two-dimensional, slow framing scanners which recorded their imagery on photographic film and consequently were not real-time devices. In 1952 the Army built the first thermograph in this country using a 16-inch searchlight reflector, a dual axis scanner, and a bolometer detector. This event was followed in the period from 1956 to 1960 by rapid development of Army-supported military thermographs, which have since come to be used almost exclusively for civilian applications.

Up to the late 1950's, the development of fast-framing thermal imagers was infeasible because detectors with fast time response were not available. Electrical signal frequency bandwidths had to be limited to a few hundred Hertz because poor detector response above this range gave low image signal-to-noise ratios. The development of cooled short-time-constant indium antimonide (InSb) and mercury doped germanium (Ge:Hg) photo-detectors made possible the first fast framing sensors.

The first real-time FLIR sensor was an outgrowth of downward looking strip mapper technology. Strip mappers are essentially thermographs with the vertical scan motion generated by the aircraft motion relative to the ground, and were developed and used most extensively by the Army and the Air Force for reconnaissance.

The first operating long wavelength FLIR was built at the University of Chicago in 1956 with Air Force support. It was a modified AN/AAS-3 strip mapper designated the XA-1. The modification added a nodding elevation mirror to the counter-rotating wedge scanner of the mapper so that the single detector element traced out a two dimensional raster pattern. Due to the lack of a pressing military need following the Korean war, this development was not pursued further.

To the author's knowledge, the next real-time long wavelength device was a ground-based FLIR built by the Perkin-Elmer Corporation for the Army in 1960. This was called the Prism Scanner because it used two rotating refractive prisms to generate a spiral scan for its single element InSb detector. It had a 5-degree circular field with 1 milliradian detector subtense, a frame rate of 0.2 frame per second, a thermal sensitivity of about 1°C, and a long-persistence phosphor CRT display. The Prism Scanner launched the continuing development of ground based sensors for the Army and for civilian applications which still has not exhausted the possibilities for compactness and civil benefits.

The airborne FLIR concept was resurrected in two independent programs in the early 1960's by the Air Force and Texas Instruments Inc., and by the Navy and Hughes Aircraft Company. Prototypes were completed and flight tested in 1965, and were so successful that they spawned an amazing proliferation of airborne FLIRs and applications. From that point on, the FLIR business burgeoned, and between 1960 and 1974 at least sixty different FLIRs were developed and several hundred were produced. Surface and airborne FLIRs have come to be functionally similar and in many cases identical so that the term FLIR now properly connotes any real-time thermal imager.

The FLIR technology reached maturity with the development of detector arrays which perform at or near theoretical thermal sensitivity limits while providing signal response fast enough to sustain the wide bandwidths necessary for high ratios of field-of-view to resolution. Compact and efficient cryogenic coolers and detector dewars made possible FLIRs with reasonable power consumptions, and continuing advances in electronic component packaging have further reduced signal processing volume and power consumption. FLIR technology is rapidly evolving toward the small size, low power, and moderate cost necessary to make FLIR a common electro-optical imaging device.

When FLIR systems were in their infancy, the analysis and evaluation of thermal image quality were as tentative as the systems themselves. Performance was indicated by a spatial two-point resolving power, and by the irradiance input which produced a unity signal-to-noise ratio output from the sensor. These were usually measured at the pre-amplified detector output, not at the visual analog display. The system "resolution" and "thermal sensitivity" were dominated by then-poor detector performance, so image quality probably correlated reasonably well with the pre-amplifier measurements.

However, as the technology advanced and low-noise, high-resolution components were developed, image quality did not improve accordingly. The reason is that the lessons of similar electro-optical technologies were not applied. The subsequent efflorescence of interest in image quality and the application to FLIR of ideas from other technologies form much of the subject matter of this book. We shall see that there is much to be learned from modern optics, photography, television, and the psychophysics of vision. They suggest the fundamental image quality parameters which are applicable to FLIR, and which foster the analyses which made possible the television-like quality of current FLIR systems.

Thermal imaging is a fascinating but demanding technology. A great deal of the satisfaction of working in it derives from synthesizing one's own viewpoint from the seven basic disciplines involved. These disciplines are:

- radiation theory and target signatures,
- atmospheric transmission of thermal radiation,
- optical design,
- detector and detector cooler operation,
- electronic signal processing,
- video displays, and
- human search processes and visual perception using FLIR.

The following section introduces the minimum number of thermal imaging concepts necessary to provide an understanding of the more detailed information in later chapters.

1.3 Thermal Imaging Fundamentals

The basic elements of any thermal imaging problem are described in the blocks of the sequential process of Figure 1.1. These elements are obvious, but failure to adequately analyze their effects in a given problem can lead to serious design errors, so we will dwell on them briefly.

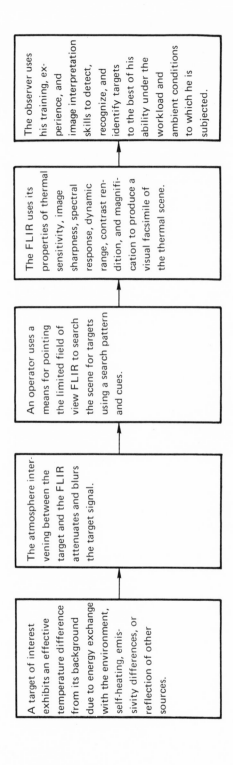

Figure 1.1 Sequence of events in the thermal imaging process.

To be detected, and subsequently recognized and identified, an object must produce an object-to-background apparent temperature difference of sufficient magnitude to distinguish it from other variations in the background. The intervening atmosphere must not excessively blur or attenuate this signal. The operator must use an effective search procedure, know what to look for, and be able to point his sensor in the proper direction. The sensor must collect the radiant signal with an optic, and convert it to an electrical signal with good signal-to-noise ratio in a detector operating in an appropriate spectral band. This electrical signal must then be reconverted to an optical signal on a video display.

Finally, the operator must be able to optimize his visual information extraction capability by using video gain and brightness controls. This entire process of conversion of an infrared scene to an analog visible scene must be performed so that contours, orientations, contrasts, and details are preserved or enhanced, and without introducing artifacts and excessive noise. The observer must be experienced and motivated, should have "good eyes", and should not be distracted from his visual search task.

A conventional scanning thermal imaging system is depicted in the block diagram of Figure 1.2, which shows the basic functions which are incorporated into thermal imagers as a class. Particular systems may combine some functions and eliminate others, and there are many different specific implementations of these functions. Figure 1.3 shows a simple schematic of one possible implementation of a scanning FLIR. The optical system collects, spectrally filters, spatially filters, and focusses the radiation pattern from the scene onto a focal plane containing a single small detector element. An optomechanical scanner consisting of a set of two scanning mirrors, one sweeping vertically and the other horizontally, is interposed between the optical system and the detector. The ray bundle reaching the detector from the object moves as the mirrors move, tracing out a TV-like raster in object space as shown in Figure 1.3. This process of detecting the scene sequentially is called scene dissection.

The incident electromagnetic field from the scene produces a disturbance within the detector which is usually proportional to the energy transported by the field. The type of disturbance depends on the type of detector, and may be an induced charge separation, an induced current, or a resistance change. Usually the incident radiation produces a voltage, and the two-dimensional object radiance pattern is converted into a one-dimensional analog voltage train by the scanning process. An infrared detector under a particular set of operating conditions is characterized by two performance

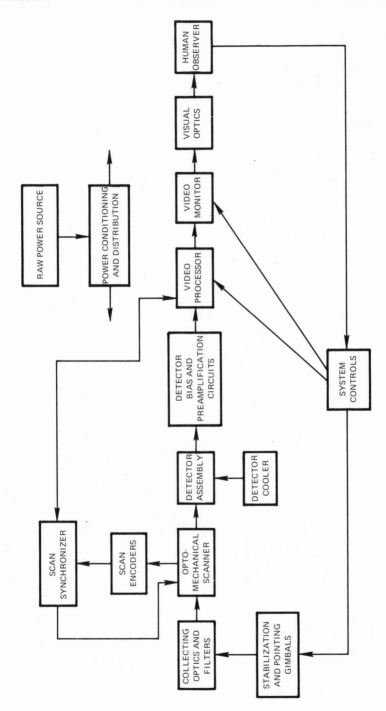

Figure 1.2 Elements of a scanning thermal imaging system.

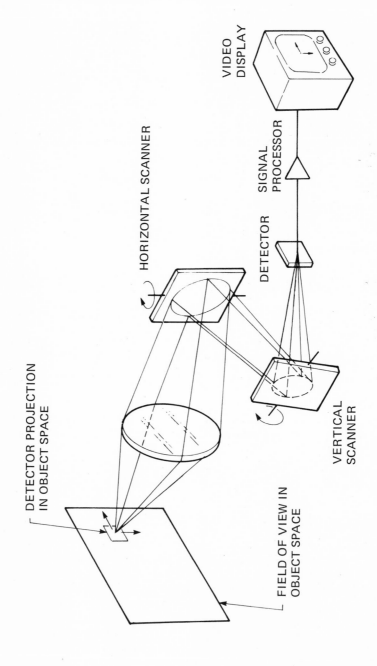

Figure 1.3 A simplified single-detector dual-axis scanner.

measures: the responsivity R and the specific detectivity D*. The responsivity is the gain of the detector expressed in volts of output signal per watt of input signal. The specific detectivity is the detector output signal-to-noise ratio for one watt of input signal, normalized to a unit sensitive detector area and a unit electrical bandwidth.

These two measures are functions of the following parameters:

electrical frequency: $f[Hz]$

optical wavelength: $\lambda[\mu m]$

detector area: $A_d[cm^2]$

detector noise measurement bandwidth: $\Delta f[Hz]$

detector rms noise voltage measured in
the bandwidth Δf: $V_n[volts]$

detector signal voltage as a function of
wavelength and electrical frequency: $V_s(\lambda,f)$ [volts]

distribution of irradiance on the detector
with wavelength: $H_\lambda[watt/cm^2\ \mu m]$.

Responsivity and specific detectivity are given[3] by:

$$R(\lambda,f) = \frac{V_s(\lambda,f)}{H_\lambda A_d},\qquad(1.1)$$

and

$$D^*(\lambda,f) = \frac{(A_d\ \Delta f)^{1/2}V_s(\lambda,f)}{H_\lambda A_d V_n} = \frac{R(\lambda,f)(A_d\ \Delta f)^{1/2}}{V_n}.\qquad(1.2)$$

If the device views objects against a terrestrial background, the radiation contrasts are very small, as will be demonstrated in Chapter Two. The corresponding analog voltage contrasts are thus also small, so that if these voltages are converted linearly into analog visual optical signals, observers will have difficulty recognizing and detecting objects because of the low image contrast. Contrast commonly is increased by ac coupling the detector signal to the pre-amplifier. This is an approximate method for removing the background signal. The simplest possible ac-coupling circuit is shown in Figure 1.4. This circuit blocks the dc component of the signal so that only variations from the scene average are passed.

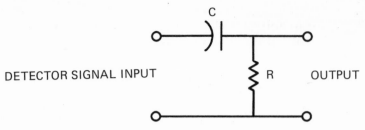

Figure 1.4 An ac coupling circuit.

The display scanning must be perfectly synchronized with the scene scanning, so both the display and the scan mirrors are driven by a scan synchronizer. For example, with a cathode ray tube display the scan-synchronization signals drive the vertical and horizontal beam deflection circuits.

Any scanning thermal imager may be defined in terms of a few basic parameters, as follows. An individual detector commonly has a sensitive area which is rectangular with horizontal dimension "a" and vertical dimension "b" in centimeters. The projection of the detector dimensions through an optic with effective focal length* f in cm gives detector angular subtenses, or projections, of α and β defined for small detectors by $\alpha = a/f$ and $\beta = b/f$ radians. Angular subtense is usually expressed in units of milliradians (10^{-3} radian). The angular dimensions in object space within which objects are imaged is called the field of view of the sensor. FLIRs conventionally have rectangular fields of view with dimensions A degrees wide and B degrees high.

The frame time T_f is the time in seconds between the instant the system scans the first point of the field of view, completes the scanning of every other point in the field, and returns to the first point. The rate at which complete pictures are produced by the system is called the frame rate $\dot{F}$, $\dot{F} = 1/T_f$.

The detector or detectors in the system do not produce informative video during the entire frame period T_f, because some time is required for performance of functions such as automatic gain adjustment, dc restoration, and scan mechanism retrace. The correspondence between active scan time and spatial coverage is demonstrated by Figure 1.5. The large rectangle in Figure 1.5 represents the spatial coverage which could be obtained by a scanner if all of the frame time T_f were used for scene coverage. The solidly-crosshatched areas represent the scan coverage lost if time is required for vertical scan retrace. In that case the system is said to have a vertical scan duty cycle η_v of less than unity. The dashed-crosshatched areas represent

---

*The distinction between f as focal length and f as frequency will be clear from the context.

coverage lost due to horizontal scan retrace or needed for automatic equalization functions. This is described by a horizontal scan duty cycle η_H. The inner blank rectangular area is the active scan area, and the time period corresponding to it is $\eta_v \eta_H T_f$. The overall scan duty cycle η_{sc} is the product $\eta_v \eta_H$.

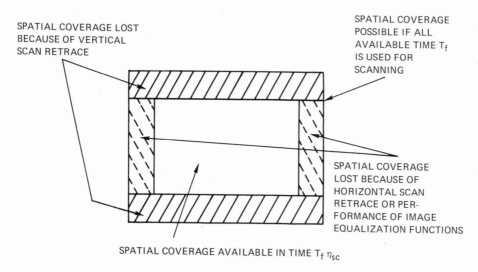

SPATIAL COVERAGE LOST BECAUSE OF VERTICAL SCAN RETRACE

SPATIAL COVERAGE POSSIBLE IF ALL AVAILABLE TIME T_f IS USED FOR SCANNING

SPATIAL COVERAGE LOST BECAUSE OF HORIZONTAL SCAN RETRACE OR PERFORMANCE OF IMAGE EQUALIZATION FUNCTIONS

SPATIAL COVERAGE AVAILABLE IN TIME $T_f \, \eta_{sc}$

Figure 1.5 Scan efficiency and spatial coverage.

Another useful number which characterizes the scan action is the detector dwelltime, τ_d, in seconds. The dwelltime may be understood by imagining that a geometrical point source of thermal energy is placed in the path of a detector scan. The time between the instant the leading edge of the detector intersects that point and the instant the trailing edge leaves the point is called the dwelltime. Alternatively, one may think of the dwelltime as the time required to scan through one detector angular subtense α. If the angular scan rate in object space is constant, the dwelltime will be the same throughout the field of view; otherwise the dwelltime will be a function of the position in the field of view.

For example, if the scan is linear, a single element system with unity scan duty cycle and contiguous scanning lines has a constant dwelltime of $\tau_d = \alpha\beta/AB\,\dot{F}$. If the scan duty cycle is not unity,

$$\tau_d = \frac{\alpha\beta\eta_{sc}}{AB\,\dot{F}}.$$

(1.3)

Figure 1.6 illustrates these concepts. The subsequent chapters of this book analyze the elements of the thermal imaging process of Figure 1.1, and present elaborations on the simple system discussed above.

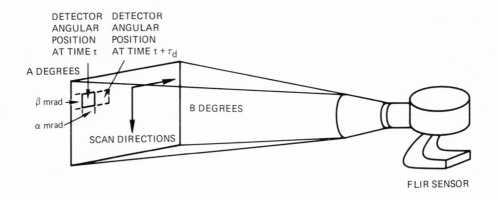

Figure 1.6 Dwelltime and detector angular subtense.

The state-of-the-art in component technology changes so rapidly that it would not be useful to discuss specific component implementations here. Rather the intent of this book is to serve as an introduction for the unini-tiated reader so that he may go forth armed with the answers to those embarrassingly simple questions no one likes to ask, such as "What defines the field of view?" or "Why is dc restoration desirable?" Detector, cooler, and video monitor data are readily available from manufacturers and have been covered adequately elsewhere, so we will concentrate on those aspects of thermal imaging not covered in the standard texts, the topics of optics, scanners, and system analysis.

1.4 Sources of Thermal Imaging Information

Most of the literature of thermal imaging systems relates to military developments and is classified or has limited accessibility. Articles in the open literature concerning thermal imaging, visual psychophysics, or ob-server performance using aided vision are usually found in the following journals:

Journal of the Optical Society of America,
Applied Optics,
Optica Acta,

Proceedings of the Institute of Electrical and Electronic Engineers (IEEE),

IEEE Journal of Quantum Electronics,

IEEE Transactions on Electron Devices,

Journal of the Society of Motion Picture and Television Engineers,

Human Factors,

Photographic Science and Engineering,

Vision Research,

Soviet Journal of Optical Technology,

Aviation Week and Space Technology,

Infrared Physics,

Optical Spectra,

Optical Engineering,

and Electro-Optical System Design.

The fundamentals of general infrared technology may be found in the texts by Hudson[3], Jamieson, et al.[4], Wolfe[5], Kruse, et al.[6], Hadni[7], Bramson[8], Smith, et al.[9], and Holter, et al.[10].

1.5 The Example System

For the sake of clarity, a single example system will be used in subsequent chapters to demonstrate the proper application of the principles derived. The example system will be a purely hypothetical device of a type which enjoys considerable attention and use, a FLIR consisting of a parallel-beam scanner which is adaptable by telescope changes to a variety of applications. This type of FLIR is represented by the "Universal Viewer" described by Daly, et al.[11], and by the "Thermal Imaging System," of Laakmann[12]. For our example we will hypothesize a FLIR operating in the 8 to 12 μm spectral band, and having a 4:3 aspect ratio field of view and a 30 frames per second, 60 fields per second scan format. The fundamental principles of the system are illustrated in Figure 1.7. For the present we will not specify the nature of the telescope, the scan mechanism, the detector lens, the cold shield, or the detector array. We will specify only the following minimum necessary parameters:

detector array individual element size: $a = b = 0.005$ cm
detector lens focal length: $f_d = 5$ cm
detector lens clear aperture: $D_d = 2$ cm
number of scan lines: 300
number of dwelltimes per scan line: 400
interlace: 2 to 1.

These parameters further imply a field of view of 400 mrad by 300 mrad (22.9 x 17.2 degrees) and a detector angular subtense of 1 x 1 mrad for the basic scanner and detector lens without a telescope. In subsequent chapters we will consider alternative implementations of the unspecified features of this example system, such as array configurations, telescope types, scan mechanisms, signal processing schemes, and performance tradeoffs.

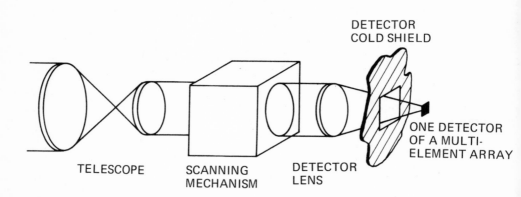

Figure 1.7 Basic features of the example system.

REFERENCES

[1] E.M. Wormser, "Sensing the Invisible World", Appl. Opt., *1*, pp 1667-1685, September 1968.

[2] B. Miller, "FLIR Gaining Wider Service Acceptance", Aviation Week and Space Technology, pp 42-49, 7 May 1973 and "Cost Reductions Key to Wider FLIR Use", AWAST, pp 48-53, 21 May 1973.

[3] R.D. Hudson, *Infrared System Engineering*, Wiley-Interscience, 1969.

[4] J.A. Jamieson, R.H. McFee, G.N. Plass, R.H. Grube, and R.G. Richards, *Infrared Physics and Engineering*, McGraw-Hill, 1963.

[5] W.L. Wolfe, Editor, *Handbook of Military Infrared Technology*, Superintendent of Documents, U.S. Government Printing Office, 1965.

[6] P.W. Kruse, L.D. McGlauchlin, and R.B. McQuistan, *Elements of Infrared Technology: Generation, Transmission and Detection*, Wiley, 1963.

[7] A. Hadni, *Essentials of Modern Physics Applied to the Study of the Infrared*, Pergamon, 1967.

[8] M.A. Bramson, *Infrared Radiation: A Handbook for Applications*, Translated by R.B. Rodman, Plenum, 1968.

[9] R.S. Smith, F.E. Jones, and R.P. Chasmar, *The Detection and Measurement of Infrared Radiation*, Oxford University Press, 1958.

[10] M.R. Holter, S. Nudelman, G.H. Suits, W.L. Wolfe, and G.J. Zissis, *Fundamentals of Infrared Technology*, MacMillan, 1962.

[11] P.J. Daly, et al., United States Patent Number 3760181, *Universal Viewer for Far Infrared*, 18 September 1973.

[12] P. Laakmann, United States Patent Number 3723642, *Thermal Imaging System*, 28 May 1971.

CHAPTER TWO – THERMAL RADIATION
THEORY AND ATMOSPHERIC TRANSMISSION

2.1 Introduction

Every object whose temperature is not absolute zero emits what is loosely called "thermal radiation". We will be concerned here only with that thermal radiation emitted in the 3 to 14 μm wavelength region for two reasons. First, most of the energy emitted by an object at terrestrial temperatures is emitted within this spectral wavelength band in an energy exchange between the electromagnetic field and the thermally-excited vibration-rotation electron energy level states of the body. Second, there are transmissive atmospheric windows in this band which permit signal detection over comparatively long ranges.

Thermal radiation at terrestrial temperatures consists primarily of self-emitted radiation from vibrational and rotational quantum energy level transitions in molecules, and secondarily from reflection of radiation from other heated sources. In many imaging applications, the actual mechanism is unimportant, and the only important factor is the existence of apparent temperature differences. The fundamental element in thermal radiation is the Planck blackbody radiation theorem. The basic premise of Planck's theorem is that thermal radiation is generated by linear atomic oscillators in simple harmonic motion which emit not continuously but in discrete quanta whose energy E is a function of the radiation frequency ν given by E = hν, where h is Planck's constant. Descriptions of Planck's theory are presented by Bramson[1], Jamieson, et al.[2], and Merritt and Hall[3].

Thermal radiators are characterized by their radiation emission efficiencies using three categories: blackbodies, greybodies, and selective radiators. This radiation efficiency property is described by the quantity spectral emissivity, $\epsilon(\lambda)$, which may be considered a radiation emission efficiency at a given wavelength λ. The blackbody is an idealization; it emits and absorbs the maximum theoretically available amount of thermal radiation at a given temperature. A blackbody has $\epsilon = 1$ within a given wavelength range of interest, and a greybody has ϵ = constant < 1 within a specified band. A selective radiator has $0 \leq \epsilon(\lambda) \leq 1$ and may exhibit virtually any kind of single-valued dependence on λ. These quantities are also dependent upon the viewing angle, but in analysis we usually deal with simple narrow-angle geometries and assume an average emissivity for all angles.

2.2 The Radiation Laws

The bare essentials of radiation geometry are depicted in Figure 2.1, where S is a two-dimensional radiating surface, dS is a differential surface area, and R(x,y) is the radius from the origin to a point (x,y) on the surface S. The cone defined by the movement of R around the periphery of S is described by a solid angle Ω in steradians:

$$\Omega = \int\int_S \frac{dS}{R^2(x,y)} = \int\int_S \frac{dx\,dy}{R^2(x,y)} . \tag{2.1}$$

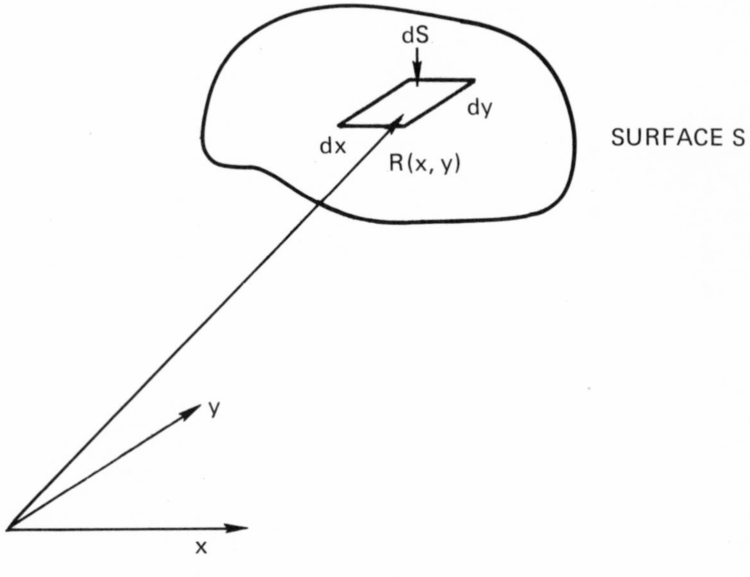

Figure 2.1 Radiation geometry.

Before summarizing the radiation laws, we have to consider notation. Several competitive notation systems are in use, including the very general phluometric system[4]. However, the quantities recommended by Hudson[5] suffice and are widely used in thermal imaging. The following basic notation is used: surface area S [cm^2], solid angle Ω [sr], time t [s], and the partial differentials with respect to these quantities: $\partial/\partial S$, $\partial/\partial\Omega$, and $\partial/\partial t$.

Letting U represent radiant energy, we define the following quantities:

radiant power = P[watts] $\overset{\Delta}{=} \partial U/\partial t$

radiant emittance = W[watt/cm²] $\overset{\Delta}{=} \partial P/\partial S = \dfrac{\partial^2 U}{\partial S \, \partial t}$

radiant intensity = J[watt/sr] $\overset{\Delta}{=} \partial P/\partial \Omega = \dfrac{\partial^2 U}{\partial S \, \partial t}$

radiance = N[watt/cm² sr] $\overset{\Delta}{=} \partial^2 P/(\partial S \, \partial \Omega) = \dfrac{\partial^3 U}{\partial S \, \partial \Omega \, \partial t}$.

The spectral distribution of each of these quantities is found by taking the partial derivative with respect to wavelength, and is denoted by a subscripted λ as in spectral radiant emittance W_λ.

One radiation law frequently assumed to be valid for analytical purposes is Lambert's cosine law. This states that the radiant intensity J[watt/sr] from an ideal perfectly diffuse source is proportional to the cosine of the angle between the normal to the surface and the viewing angle. This is a good approximation for many materials at a near-normal viewing angle. For a flat surface radiating W[watt/cm²], integration over 2π steradians of W weighted by Lambert's law yields

$$N = W/\pi. \tag{2.2}$$

Planck's blackbody radiation law (where the symbols have the values given in Table 2.1) is

$$W_\lambda (\lambda,T) = \left(\frac{2\pi h c^2}{\lambda^5}\right) \left(\frac{1}{e^{ch/\lambda kT} - 1}\right) \left[\frac{watt}{cm^3}\right] . \tag{2.3}$$

Equation 2.3 is more conveniently expressed in units of [watt/cm² μm] for λ in [μm] as:

$$W_\lambda (\lambda,T) = \frac{c_1}{\lambda^5 (e^{c_2/\lambda T} - 1)} , \tag{2.4}$$

where c_1 and c_2 have the values given in Table 2.1. Planck's law is the partial derivative with respect to wavelength of the total blackbody radiation W(T) in [watt/cm²]. This law must be expressed as a derivative with respect to λ rather than as a simple function of λ because it describes a distribution. That is, at a specific wavelength $W(\lambda) = 0$ whereas $W_\lambda \neq 0$. Integration of Planck's law yields the Stefan-Boltzmann law, originally empirical,

$$W(T) = \int_0^\infty W_\lambda (T,\lambda)\, d\lambda = \left(\frac{2\pi^5\, k^4}{15c^2 h^3}\right) T^4. \qquad (2.5)$$

Table 2.1
Radiometric Constants

c (speed of light in a vacuum) = 2.9979×10^{10} [cm/sec]

h (Planck's constant) = 6.6256×10^{-34} [watt $\sec^2$]

k (Boltzmann's constant) = 1.38054×10^{-23} [watt sec/$°K$]

σ (the Stefan-Boltzmann constant) = 5.6697×10^{-12} [watts/cm^2 $°K^4$]

c_1 = 3.7415×10^4 [watts $\mu m^4/cm^2$]

c_2 = 1.4388×10^4 [$\mu m\,°K$]

c_3 = 1.8837×10^{23} [$\mu m^3/sec\ cm^2$]

The constant term is called the Stefan-Boltzmann constant σ,

$$\sigma = \frac{2\pi^5 k^4}{15c^2 h^3}. \qquad (2.6)$$

For some analytical purposes it is useful to have radiation quantities expressed in units of [photons/second]. Any quantity distributed with wavelength and expressed in units of [watts] may be expressed in units of [photons/second] by dividing by the energy of one photon, hc/λ. For example, Planck's law converted to spectral radiant photon emittance Q_λ [photons/sec cm^3] becomes

$$Q_\lambda = \left(\frac{2\pi c}{\lambda^4}\right)\left(\frac{1}{e^{ch/\lambda kT}-1}\right). \qquad (2.7)$$

The expression for Q_λ in [photons/sec cm^2 μm] for λ in [μm] is

$$Q_\lambda = \frac{c_3}{\lambda^4 (e^{c_2/\lambda T}-1)} \qquad (2.8)$$

with the constants as given in Table 2.1.

The spectral radiant emittance W_λ is plotted as a function of wavelength at several interesting temperatures in Figure 2.2. The spectral radiant photon emittance Q_λ is plotted for $T = 300°K$ in Figure 2.3.

The observed radiance from a solid object is the sum of the radiation transmitted and emitted by the body itself, and the radiation reflected by the body from other thermal sources. As the self-emitted radiation is a function of the object's temperature, it is useful to inquire how W_λ changes with differential changes in T. The reader may easily verify that

$$\frac{\partial W_\lambda}{\partial T} = \frac{c_1\, c_2\, e^{c_2/\lambda T}}{\lambda^6 T^2\, (e^{c_2/\lambda T}-1)^2} = W_\lambda \frac{c_2\, e^{c_2/\lambda T}}{\lambda T^2 (e^{c_2/\lambda T}-1)} \cong W_\lambda \frac{c_2}{\lambda T^2} \qquad (2.9)$$

when $e^{c_2/\lambda T} \gg 1$. The exact expression for $\partial W_\lambda/\partial T$ is plotted for terrestrial temperatures in Figure 2.4.

Some useful defined quantities are: background radiant emittance (W_B), target radiant emittance (W_T), differential change in W with differential change in T $(\partial W/\partial T)$, and radiation contrast (C_R). These are defined for target and background absolute temperatures T_T and T_B, and for specific spectral bands, by

$$W_B = \int_{\lambda_1}^{\lambda_2} W_\lambda\, (T_B)\, d\lambda \qquad (2.10)$$

$$W_T = \int_{\lambda_1}^{\lambda_2} W_\lambda\, (T_T)\, d\lambda \qquad (2.11)$$

$$\frac{\partial W}{\partial T} = \int_{\lambda_1}^{\lambda_2} \frac{\partial W_\lambda\, (T_B)}{\partial T}\, d\lambda \qquad (2.12)$$

$$C_R = \frac{W_T - W_B}{W_T + W_B}^* . \qquad (2.13)$$

---

*If the definition of contrast as $C = (W_T - W_B)/W_B$ is preferred, $C_R \simeq C/2$ for $W_T \simeq W_B$. The term contrast has been applied to many different photometric and radiometric quantities. For a target level L_T and background L_B, contrast has been variously defined as: $(L_T - L_B)/(L_T + L_B)$, $(L_T - L_B)/L_T$, $(L_T - L_B)/L_B$, and L_T/L_B. As different authors use different definitions, the appropriate one will be noted when the term contrast is associated with specific numbers.

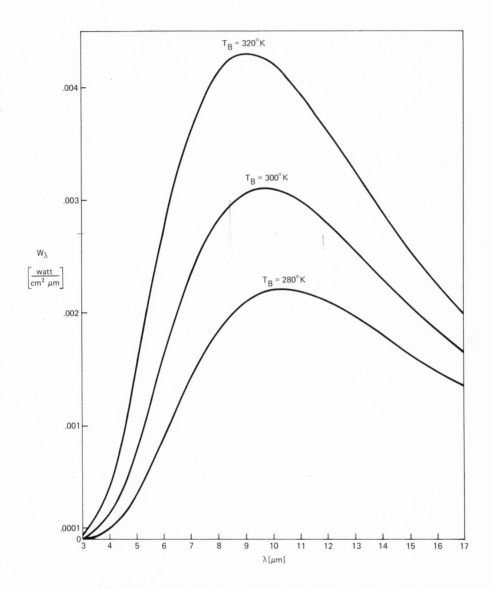

Figure 2.2 Planck's law for spectral radiant emittance at
three background temperatures T_B.

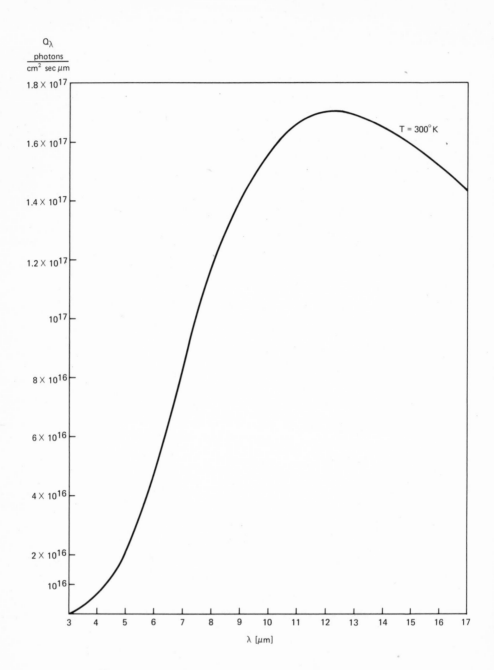

Figure 2.3 Planck's law for spectral photon emittance at a
background temperature of 300° K.

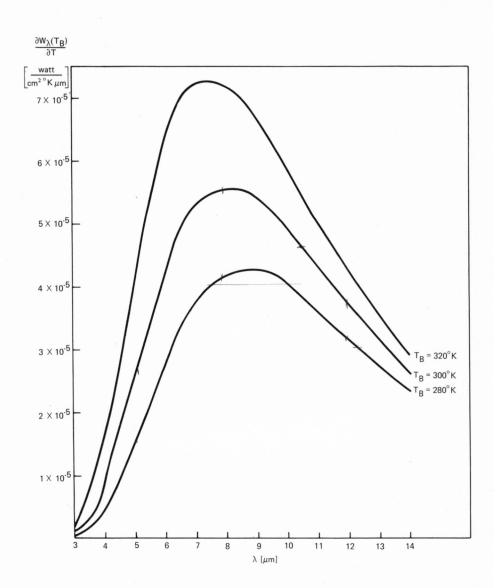

Figure 2.4 The thermal derivative of Planck's law evaluated
at three background temperatures.

To keep the notation simple, the spectral band is not explicitly noted in these quantities and is implied by the context. The majority of terrestrial scenes do not often have apparent temperature differences in excess of $\pm 20°K$ from ambient, so we will confine our attention to that range.

The quantities W_T, W_B, $\partial W/\partial T$, and C_R calculated for an ambient temperature of $300°K$ and a target ΔT of $10°K$ for two commonly used atmospheric windows are given below:

Spectral Band	W_B(watts/cm^2)	W_T(watts/cm^2)	$\dfrac{\partial W}{\partial T}\left(\dfrac{\text{watts}}{\text{cm}^2 °K}\right)$	C_R for $\Delta T = 10°C$
$3.5 - 5\,\mu m$	5.56×10^{-4}	7.87×10^{-4}	2×10^{-5}	$\dfrac{0.172}{0.074}$ = 2.35
$8 - 14\,\mu m$	1.72×10^{-2}	1.99×10^{-2}	2.62×10^{-4}	

Table 2.2 gives the $\partial W/\partial T$'s for other spectral bands and background temperatures, and Figures 2.5 and 2.6 plot the most often used results. Figure 2.7 compares the 3.5 to 5 μm and the 8 to 14 μm spectral bands.

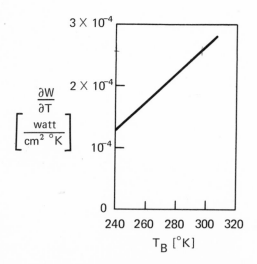

Figure 2.5 Values of the integrated thermal derivative of Planck's law for the 8 to 14 μm band as a function of background temperature.

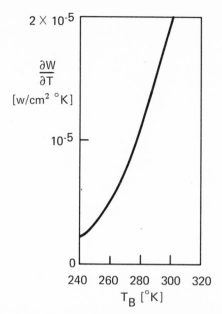

Figure 2.6 Values of the integrated thermal derivative of Planck's law for the 3.5 to 5 μm band as a function of background temperature.

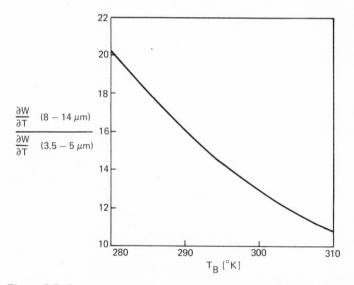

Figure 2.7 Comparison of the integral of the thermal derivative of Planck's law for the two spectral bands as a function of background temperature.

Table 2.2
Selected Values of the Spectral Integral of the Thermal Derivative
of Planck's law.

$$\frac{\partial W}{\partial T} = \int_{\lambda_1}^{\lambda_2} \frac{\partial W_\lambda(T_B)}{\partial T} \, d\lambda \quad \left[\frac{W}{cm^2 \, ^\circ K}\right] \text{ for}$$

$\lambda_1 \, [\mu m]$	$\lambda_2 \, [\mu m]$	$T_B = 280^\circ K$	$T_B = 290^\circ K$	$T_B = 300^\circ K$	$T_B = 310^\circ K$
3	5	1.1×10^{-5}	1.54×10^{-5}	2.1×10^{-5}	2.81×10^{-5}
3	5.5	2.01×10^{-5}	2.73×10^{-5}	3.62×10^{-5}	4.72×10^{-5}
3.5	5	1.06×10^{-5}	1.47×10^{-5}	2×10^{-5}	2.65×10^{-5}
3.5	5.5	1.97×10^{-5}	2.66×10^{-5}	3.52×10^{-5}	4.57×10^{-5}
4	5	9.18×10^{-6}	1.26×10^{-5}	1.69×10^{-5}	2.23×10^{-5}
4	5.5	1.83×10^{-5}	2.45×10^{-5}	3.22×10^{-5}	4.14×10^{-5}
8	10	8.47×10^{-5}	9.65×10^{-5}	1.09×10^{-4}	1.21×10^{-4}
8	12	1.58×10^{-4}	1.77×10^{-4}	1.97×10^{-4}	2.17×10^{-4}
8	14	2.15×10^{-4}	2.38×10^{-4}	2.62×10^{-4}	2.86×10^{-4}
10	12	7.34×10^{-5}	8.08×10^{-5}	8.81×10^{-5}	9.55×10^{-5}
10	14	1.3×10^{-4}	1.42×10^{-4}	1.53×10^{-4}	1.65×10^{-5}
12	14	5.67×10^{-5}	6.10×10^{-5}	6.52×10^{-5}	6.92×10^{-5}

The radiation contrast C_R varies with changes in target and background temperatures for the two spectral bands as shown in Figures 2.8 and 2.9. A useful number to remember is the differential change in radiation contrast with a differential temperature difference, $\partial C_R / \partial T$. From the above calculations, we see that $\partial C_R / \partial T$ is approximately $0.7\%/^\circ K$ for the 8 to 14-μm band and $1.7\%/^\circ K$ for the 3 to 5-μm band. This clearly shows that any system which linearly converts thermal energy to visible energy will not exhibit the high contrasts necessary for acceptable image quality. Thus it is necessary to have some sort of background radiance subtraction scheme in thermal imaging.

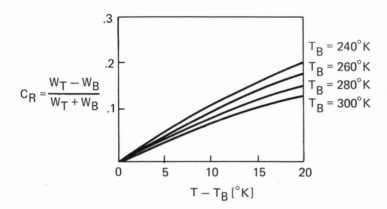

$$C_R = \frac{W_T - W_B}{W_T + W_B}$$

Figure 2.8 Radiation contrast for the 8 to 14 μm band as a function
of target-to-background temperature difference for
four background temperatures.

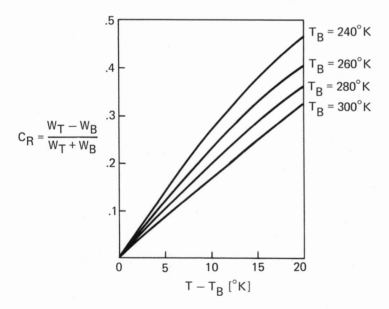

$$C_R = \frac{W_T - W_B}{W_T + W_B}$$

Figure 2.9 Radiation contrast for the 3.5 to 5 μm band as a function
of target-to-background temperature difference for
four background temperatures.

The radiation contrast is most simply evaluated for the case of black-body target emission over all wavelengths. In that case, the total radiant emittance W(T) is given by equation 2.5,

$$W(T) = \sigma T^4,$$

so

$$\frac{\partial W(T)}{\partial T} = 4\,\sigma T^3,$$

and

$$C_R = \frac{4\,\sigma T^3\,\Delta T}{2\,\sigma T^4 + 4\,\sigma T^3\,\Delta T}$$

$$= \frac{2\,\Delta T}{T + 2\,\Delta T}$$

$$\cong \frac{2\,\Delta T}{T} \tag{2.15}$$

for small ΔT.

Then $C_R/\Delta T \cong 2/T$, and for terrestrial temperatures poorer contrast is achieved broadband than is possible with the two narrow spectral bands we have considered. These are the basic concepts of thermal radiation theory which we will need here. Expanded treatments are given in any of references 1 through 9.

2.3 Atmospheric Transmission

Thermal radiation is attenuated in passage through terrestrial atmospheres by the processes of absorption and of scattering by gas molecules, molecular clusters (aerosols), rain, snow, and suspensions such as smoke, fog, haze, and smog. The following molecules (in order of importance) absorb radiation in the thermal infrared in broad bands centered on the wavelengths indicated: water (2.7, 3.2, 6.3 μm), carbon dioxide (2.7, 4.3, 15 μm), ozone (4.8, 9.6, 14.2 μm), nitrous oxide (4.7, 7.8 μm), carbon monoxide (4.8 μm), and methane (3.2, 7.8 μm). Molecular absorption is the greatest source of extinction except in dense suspensions, and water vapor, carbon dioxide, and ozone are the most significant absorbers. In the lower atmosphere, absorption by nitrous oxide and carbon monoxide may usually be neglected. The 6.3 μm water band and the 2.7 and 15 μm carbon dioxide bands effectively limit atmospheric transmission in the 2 to 20 μm thermal spectral range to two atmospheric windows. These are the 3.5 to 5 μm and the 8 to 14 μm bands.

At a specific wavelength λ for a specific atmospheric state, the atmospheric transmission $\tau_A(\lambda)$ is given by the Lambert-Beer law,

$$\tau_A(\lambda) = \exp(-\gamma(\lambda)R), \tag{2.16}$$

where R is the range or path length and in the notation of McClatchey, et al.[10], $\gamma(\lambda)$ is the extinction coefficient. The extinction coefficient is the sum of an absorption coefficient $\sigma(\lambda)$ and a scattering coefficient $k(\lambda)$,

$$\gamma(\lambda) = \sigma(\lambda) + k(\lambda). \tag{2.17}$$

The scattering and absorption coefficients are in turn sums of molecular and aerosol components

$$\sigma(\lambda) = \sigma_m(\lambda) + \sigma_a(\lambda) \tag{2.18}$$

and

$$k(\lambda) = k_m(\lambda) + k_a(\lambda). \tag{2.19}$$

The extinction coefficient is a complicated function of λ, so over a broad spectral band the atmospheric transmission prediction problem consists of an integration over wavelength and over an accumulation of absorbers. The average transmission for a specific band λ_1 to λ_2 is

$$\overline{\tau}_A = \frac{1}{\lambda_2 - \lambda_1} \int_{\lambda_1}^{\lambda_2} \exp(-\gamma(\lambda)R)d\lambda. \tag{2.20}$$

A broad survey of atmospheric transmission definitions and problems is given by Farrow and Gibson[11]. Reliable measurements of atmospheric transmission in the infrared are rare. The best are still those well known measurements of Taylor and Yates[12], reproduced in Figures 2.10, 2.11, and 2.12, and the measurements of Streete[13] shown in Figures 2.13 and 2.14. The detailed mechanisms of atmospheric effects have been presented elsewhere[2,7], so we will not consider them here; rather, we will identify and use a reliable analytical model.

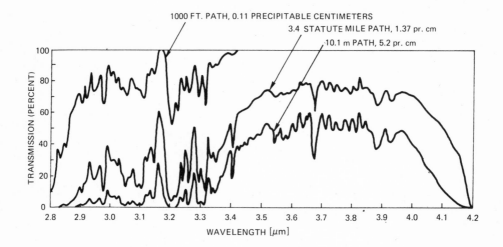

Figure 2.10 Taylor and Yates 2.8 to 4.2 μm data adapted from reference 12.

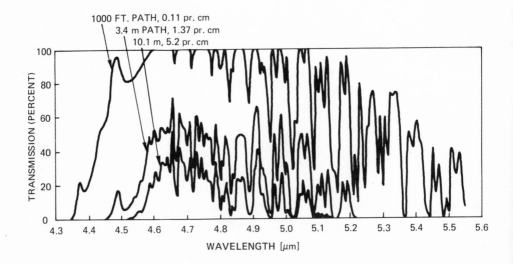

Figure 2.11 Taylor and Yates 4.3 to 5.6 μm data adapted from reference 12.

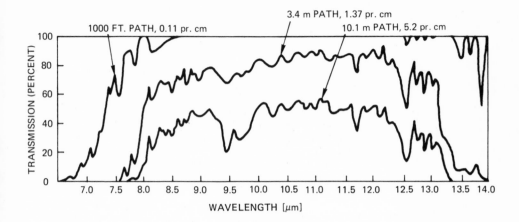

Figure 2.12 Taylor and Yates 6.5 to 14.0 μm data adapted from reference 12.

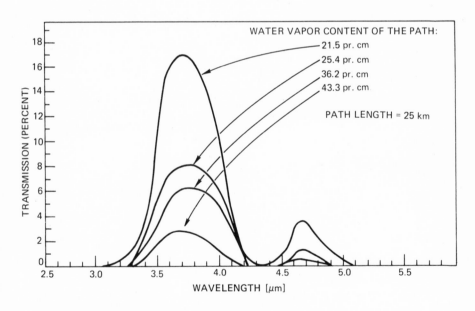

Figure 2.13 The 3 to 5 μm data of Streete adapted from reference 13.

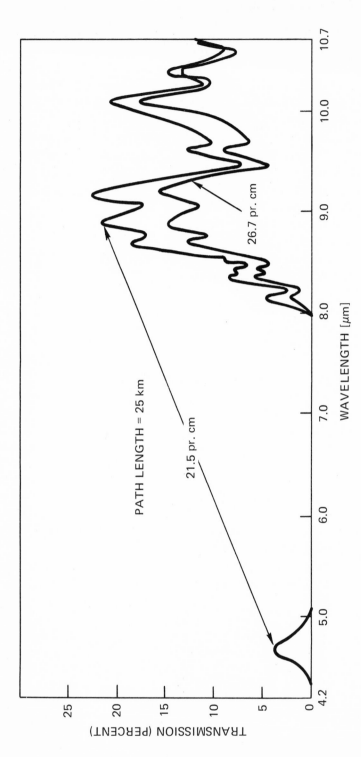

Figure 2.14 The 4.2 to 10.7 μm data of Streete adapted from reference 13.

Gaseous molecular absorption occurs when vibrational and rotational atomic motions in a molecule are allowed such that a net dipole moment change is produced. Then energy exchange with the electromagnetic field can result by resonant absorption and emission of quanta. Consequently, individual symmetric molecules do not exhibit vibration/rotation absorption because they cannot generate a net dipole moment. Also, molecular absorption by the electronic bands of atoms is insignificant. Thus for these two reasons individual molecules of the three main atmospheric species, diatomic nitrogen and oxygen, and monatomic argon do not absorb infrared. In a gas, however, additional states are introduced by the effect called pressure broadening, and appreciable absorption occurs.

Molecular absorption causes extinction of a beam of thermal radiation because the gas reradiates the absorbed energy in all directions rather than in the original direction of propagation. A good explication of molecular absorption is given by Anding[14], who presents the Elsasser, random, and random-Elsasser models of molecular absorption. Anding recommends appropriate models for the various molecular species, and compares predictions using these models with empirical data. McClatchey, et al.[10], is a good source for the analysis of the effects of molecular scattering and of aerosol extinction. The regular texts on infrared technology contain summaries of the various single line and band absorption models.

At high relative humidities, water vapor molecules tend to cluster and to produce an aerosol, a condition intermediate between a humid atmosphere and a fog or haze. Using reliable molecular absorption and scattering models, and a reliable aerosol scattering model, Hodges[15] compared predictions to empirical results and found that liquid aerosol absorption is quite significant at high relative humidity. For low altitudes and moderate ranges at which water vapor concentrations are low, Hodges asserts that molecular absorption alone reasonably accounts for extinction.

Scattering by molecules, aerosols, fogs, hazes, and clouds is describable by the mechanism known as Mie scattering. The Mie theory is valid for small particle, or Rayleigh, scattering where the particle is much smaller than a wavelength, and for large particle, or nonselective, scattering where the particle size p is much greater than a wavelength. Since Rayleigh scattering exhibits a λ^{-4} dependence, molecular scattering is insignificant beyond about $2\,\mu$m. The result is that scattering by molecules ($p \ll \lambda$) is insignificant compared to absorption, and only molecular aggregates produce scattering.

The best single source of models for all sources of atmospheric extinction is the report of McClatchey, et al.[10]. They detail methods for predicting the effects in the 0.25 to 25 μm region of the following mechanisms:

- Discrete line molecular absorption for all significant gases;
- Pressure-broadened line molecular absorption by water vapor;
- Molecular scattering for all species;
- Aerosol extinction for all species.

The paper also gives tables of characteristics for seven model atmospheres: tropical, mid-latitude summer, mid-latitude winter, subarctic summer, subarctic winter, clear (23 km visibility), and hazy (5 km visibility).

Figures 2.15 through 2.19 are taken from McClatchey, et. al. These figures show transmittance as a function of wavelength for the appropriate absorber density or for path length, for molecular absorption by non-pressure-broadened water vapor, for the pressure-broadened water vapor continum, for the uniformly mixed gases, for molecular scattering, and for aerosol extinction. The curves may be used directly for a path through a uniform atmosphere. Otherwise an equivalent path must be calculated per the original paper. The paper contains many other graphs for individual species. These curves are applied by tracing the appropriate transmittance scale onto a transparent sheet. The traced scale is then used by placing the desired absorber amount (or path length) on the horizontal scale line with the vertical scale over the wavelength of interest. The transmittance is read off where the curve crosses the scale.

The absorber concentration of water vapor may be described in several ways. The most fundamental measure is the absolute humidity, the water mass in grams of a one-cubic centimeter volume of air containing water vapor at some specified temperature and pressure. The absolute humidity H_a [gm/cm^3] is a function of water vapor partial pressure P[mm of mercury] and atmospheric temperature T_A[$°$K]. The absolute humidity $H_a{}'$ of a mass of air saturated with water vapor is given by

$$H_a' = 2.89 \times 10^{-4} \frac{P}{T_A}, \tag{2.21}$$

where the partial pressure P also depends on the temperature T_A. A more familiar measure of absorber concentration is the relative humidity. Relative humidity is the ratio of absolute humidity under a specified set of conditions to the absolute humidity when the volume is saturated under the same conditions.

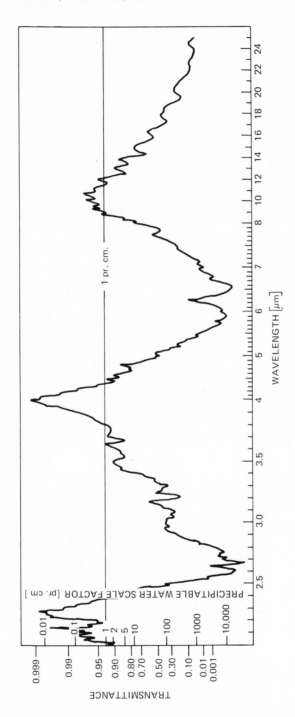

Figure 2.15 Discrete line molecular transmittance of water vapor, adapted from reference 10.

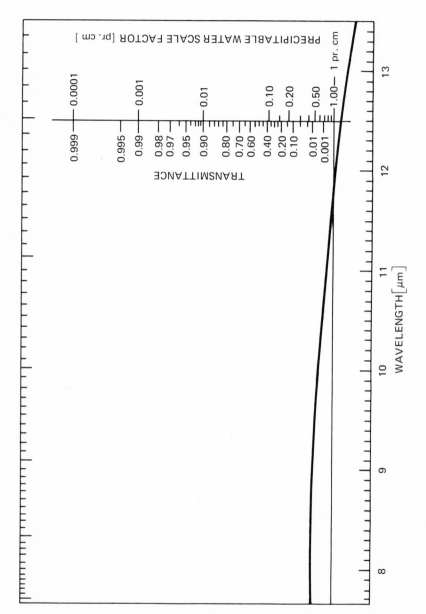

Figure 2.16 Pressure-broadened molecular transmittance of water vapor, adapted from reference 10.

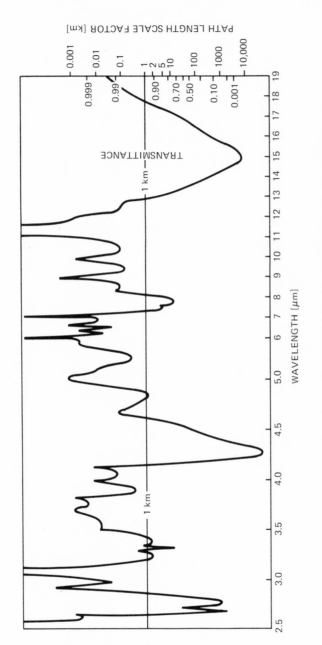

Figure 2.17 Molecular transmittance of uniformly mixed CO_2, N_2O, CO, and CH_4, adapted from reference 10.

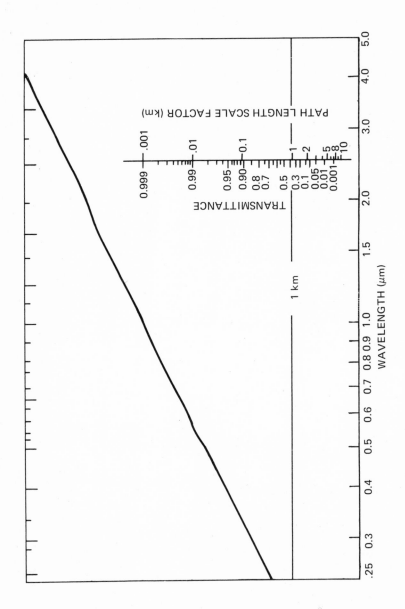

Figure 2.18 Transmittance for molecular scattering by all atmospheric species, adapted from reference 10.

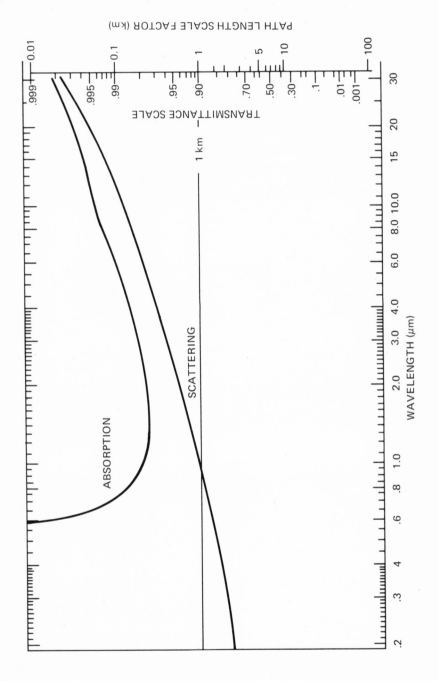

Figure 2.19 Transmission by aerosols, adapted from reference 10.

Absolute humidity and relative humidity do not directly indicate the total amount of absorber in a given path length. The precipitable water is the quantity which combines path length and relative humidity in a meaningful measure of total path absorption. Precipitable water W is defined in terms of depth of the column of liquid which would result from the condensation of all the vapor in a cylinder of air having a particular path length, and is usually expressed in units of precipitable centimeters of water per kilometer of path length [pr cm/Km].

Precipitable water is expressed in terms of liquid water density d, relative humidity H_r (expressed as a decimal fraction), and absolute humidity H_a as follows:

$$W \left[\frac{\text{pr cm}}{\text{Km}}\right] = \frac{H_r \, H_a \left[\frac{\text{gm}}{\text{cm}^3}\right]}{d \, [\text{gm/cm}^3]} = \frac{10^5 \, H_r \, H_a \left[\frac{\text{gm}}{\text{cm}^2 \, \text{Km}}\right]}{d \, [\text{gm/cm}^3]}$$

$$= 10^5 \frac{H_r \, H_a}{d} \left[\frac{\text{pr cm}}{\text{Km}}\right].$$

(2.22)

Since d is approximately constant at 1 gm/cm³,

$$W = 10^5 \, H_r \, H_a \left[\frac{\text{pr cm}}{\text{Km}}\right].$$

(2.23)

The total depth of precipitable water in a path is expressed in units of precipitable centimeters. For a path having uniform relative and absolute humidities, the total precipitable water is found by multiplying W by the path length R in kilometers. The precipitable water and the absolute humidity at saturation are shown in Figure 2.20.

Moser[17] has determined that at saturation and at sea level, the precipitable water W' is given by

$$W' \left[\frac{\text{pr cm}}{\text{Km}}\right] = 0.492 + 3.094 \times 10^{-2} \, T_A$$

$$+ 0.95 \times 10^{-3} \, T_A{}^2 + 2.888 \times 10^{-5} \, T_A{}^3$$

(2.24)

to an accuracy of ±1.6% for $0°C \leqslant T_A \leqslant 40°C$ and $0 \leqslant H_R \leqslant 1$, and by

$$W' \left[\frac{\text{pr cm}}{\text{Km}}\right] = 0.502 \, e^{\, 0.06 \, T_A}$$

(2.25)

to an accuracy of ±3.6% for $0 \leqslant T_A \leqslant 35°C$.

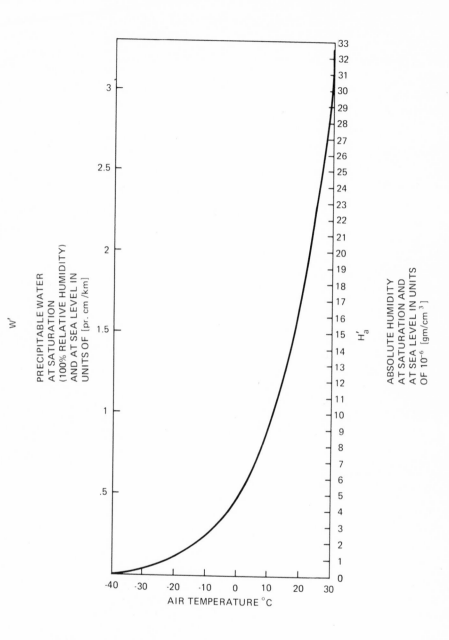

Figure 2.20 Precipitable centimeters of water vapor per kilometer
of path length and absolute humidity versus temperature at saturation.

Multiplication of the above expressions by the relative humidity H_r yields the precipitable water W at other than saturated conditions.

Hodges[15] calculated the average transmission in seven nominally "3-5 μm" and "8-14 μm" bands using the empirical Yates and Taylor data, found the contributions of each extinction process, and derived coefficients which permit predictions to agree reasonably well with Yates and Taylor. The plot in Figure 2.21 of Hodges' reduction of Yates and Taylor data in 8 to 14 μm bands shows an approximately linear relation between average band transmission $\bar{\tau}_A$ and total precipitable water vapor W_T [pr cm] in the path. This interesting result is fitted[16] well by:

$$\bar{\tau}_A = .8326 - .0277\,W_T. \tag{2.26}$$

The residual absorption loss of .167 at $W_T = 0$ presumably is due to extinction by other species. Plots of the 3 to 5 μm data do not show any clear dependence on W_T, reflecting the fact that CO_2 absorption is more important in the 3-5 μm band than in the 8-14 μm band.

The principles of the paper of McClatchey, et al.[10] are incorporated in a computer program written by Selby and McClatchey[18]. This program was used by MacDonald[19] to generate the curves shown in Figures 2.22 through 2.42. To provide a comparison of transmission loss at 3 to 5 μm and at 8 to 12 μm, MacDonald selected three slant ranges: 4, 8, and 16 nmi, and three altitudes: 300 feet, 4000 feet, and 12,000 feet. Two model atmospheres were used: the tropical and the 1962 U.S. standard, both with and without haze. Transmission variations with visibility at various altitudes and ranges are shown for the 8 to 12 μm band in Figures 2.24 through 2.27, and for the 3 to 5 μm band in Figures 2.28 through 2.33. The differences between tropical and U.S. standard atmospheres for various altitudes and slant ranges for the 8 to 12 μm band are shown in Figures 2.34 through 2.42.

Excluding source and detector characteristics, the results indicate that in mid-latitude dry air masses, transmission in the 8 to 12 μm band is superior because CO_2 heavily absorbs 3 to 5 μm radiation. However, in clear high humidity tropical air masses, the 3 to 5 μm range is clearly superior due to heavy absorption of 8 to 12 μm radiation by water vapor. The fact that haze attenuates the 3 to 5 more than the 8 to 12, however, tends to make the 8 to 12 μm range the preferred choice because long haze-free slant paths through humid atmospheres are not frequently encountered. The primary quantity of interest, however, is not absolute transmission but the ratio of thermal sensitivity to transmission. In this respect, the 8 to 12 μm region is usually superior to the 3 to 5 μm region because of an inherent sensitivity difference between the two regions even when no transmission losses are present, as discussed in Section 5.12.

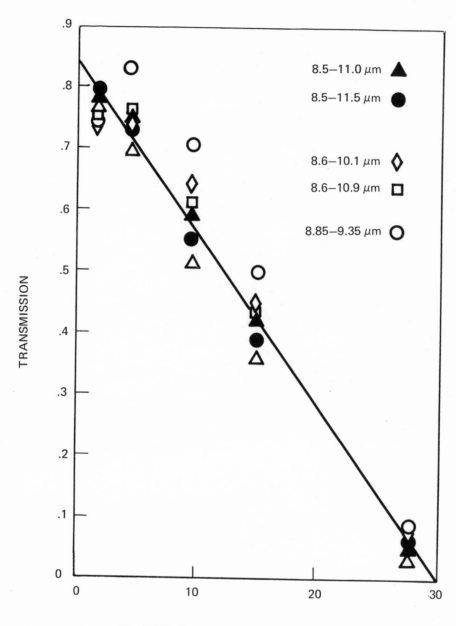

Figure 2.21 Average bandpass transmission versus total precipitable
water for five bands, adapted from reference 15.

A comparison of atmospheric transmission over long sea paths in the two spectral bands is given by Jaeger, et al.[20]. Using good InSb and (Hg,Cd)Te detectors in a radiometer, they showed that over 1 to 13 Km paths in the background-variation-limited noise (sea surface changes), experimental predictions of 8-14 μm signal transmission superiority are correct.

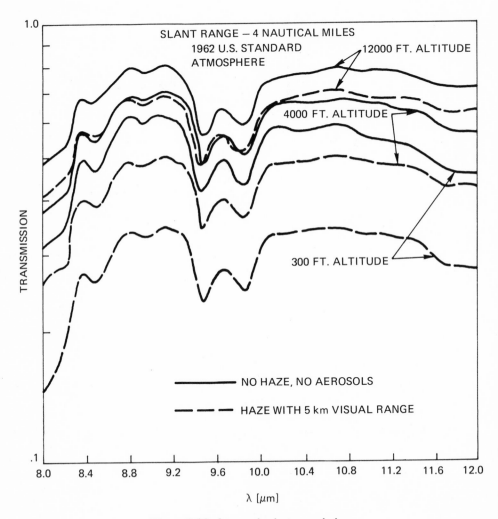

Figure 2.22 Atmospheric transmission.

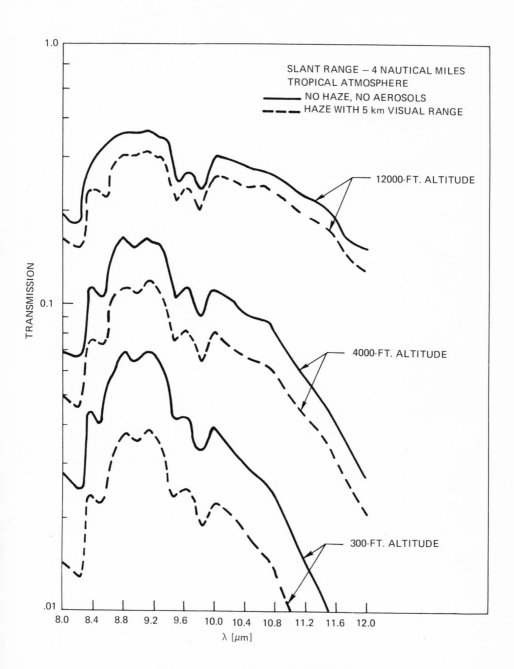

Figure 2.23 Atmospheric transmission.

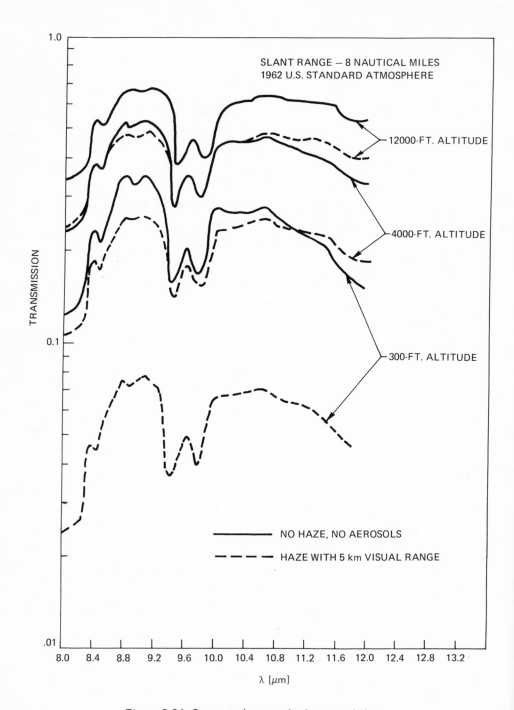

Figure 2.24 Computed atmospheric transmission.

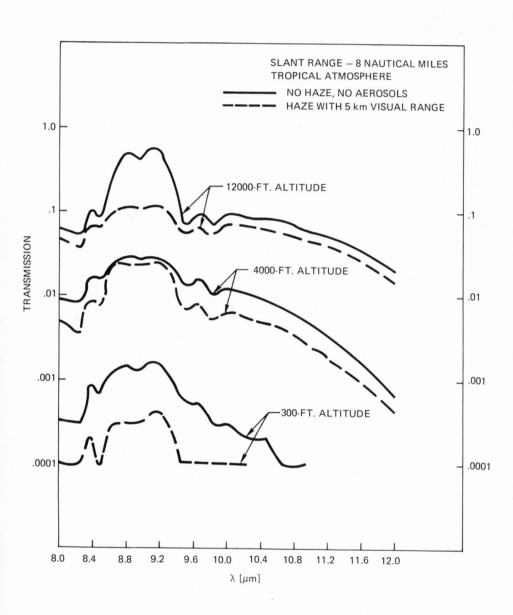

Figure 2.25 Atmospheric transmission.

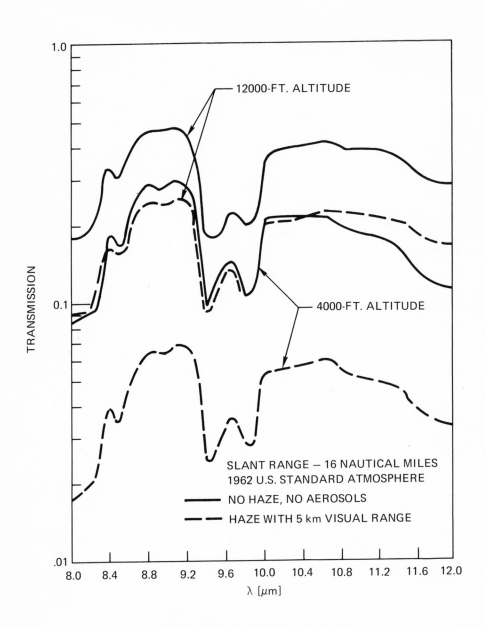

Figure 2.26 Atmospheric transmission.

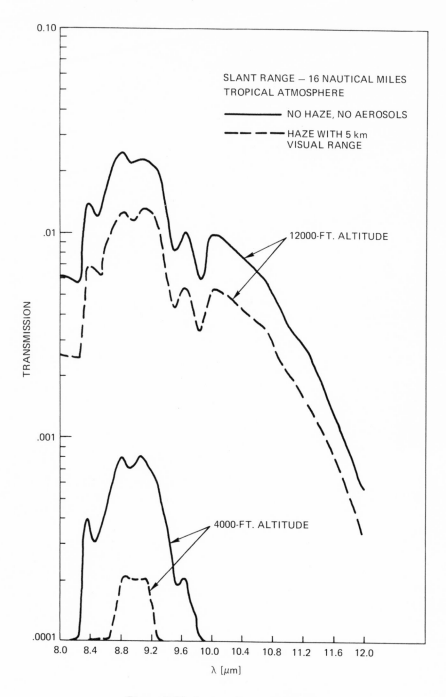

Figure 2.27 Atmospheric transmission.

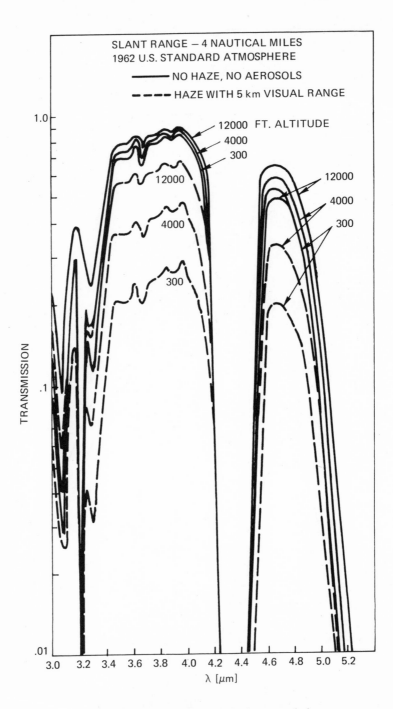

Figure 2.28 Atmospheric transmission.

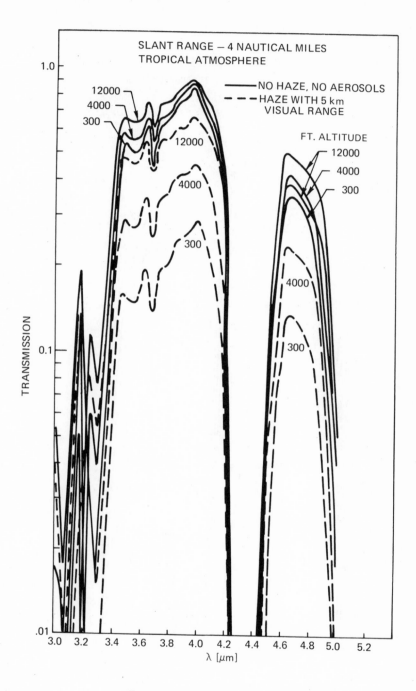

Figure 2.29 Atmospheric transmission.

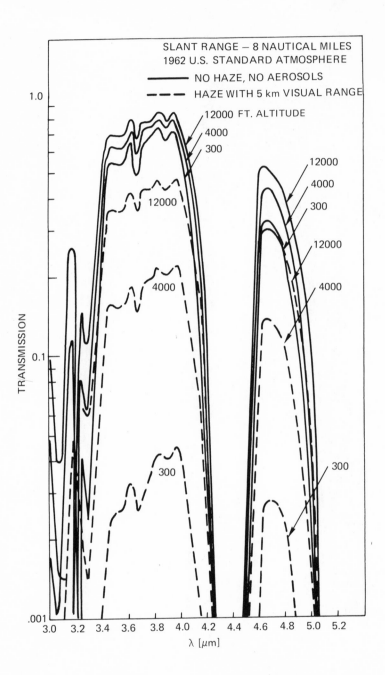

Figure 2.30 Atmospheric transmission.

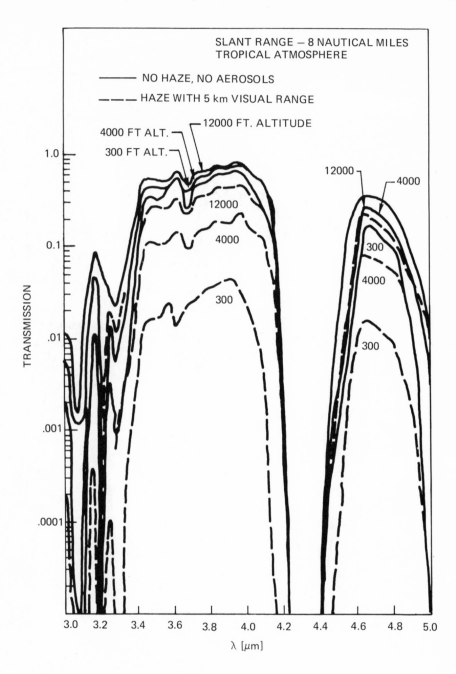

Figure 2.31 Atmospheric transmission.

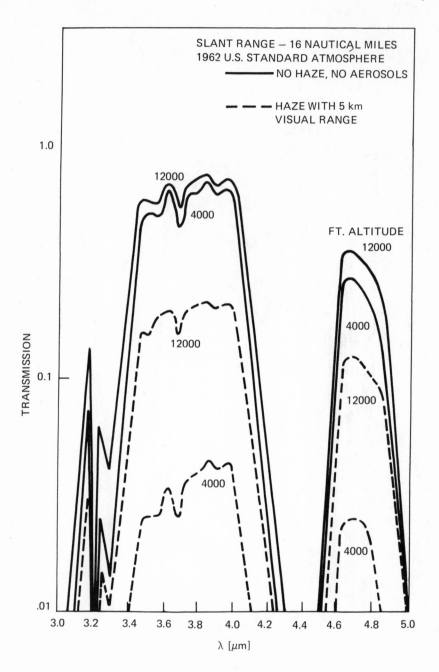

Figure 2.32 Atmospheric transmission.

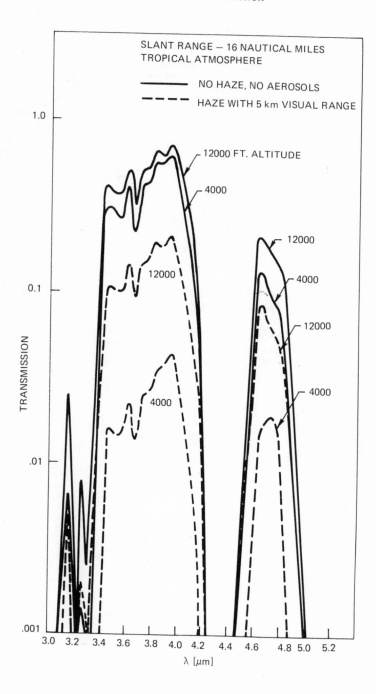

Figure 2.33 Atmospheric transmission.

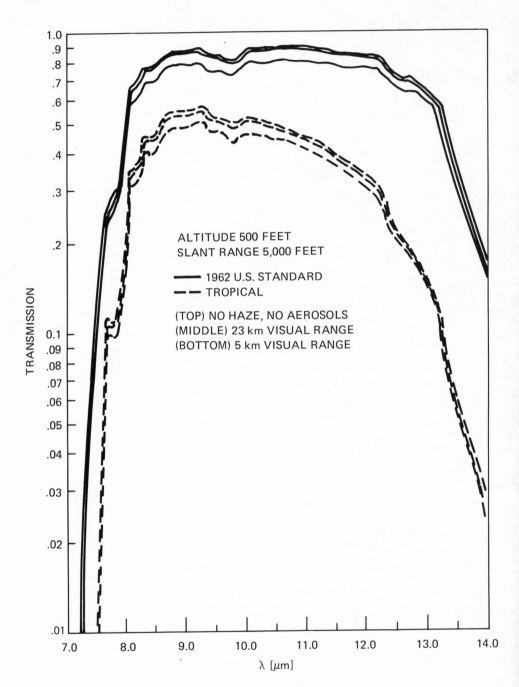

Figure 2.34 Atmospheric transmission.

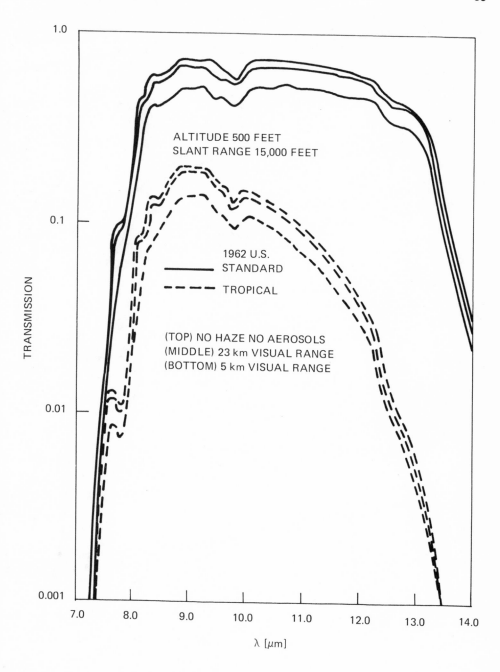

Figure 2.35 Atmospheric transmission.

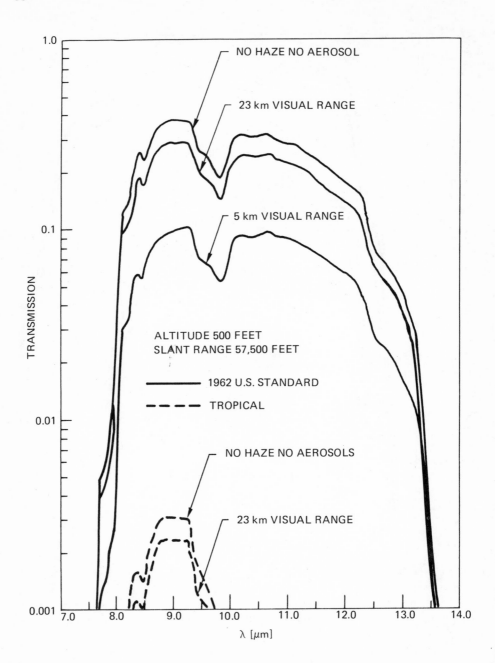

Figure 2.36 Atmospheric transmission.

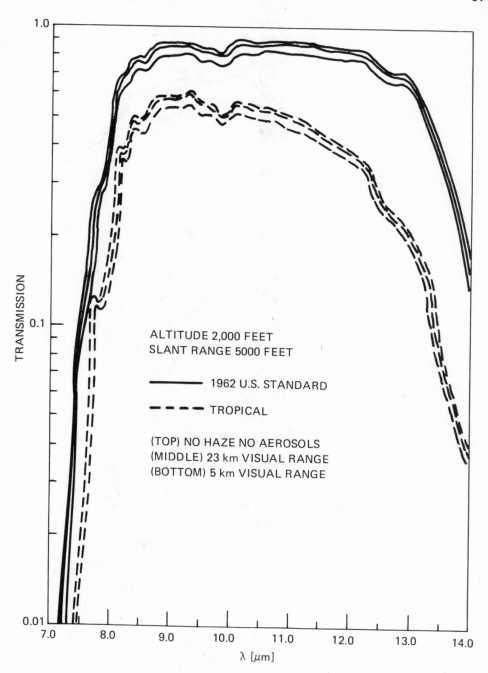

Figure 2.37 Atmospheric transmission.

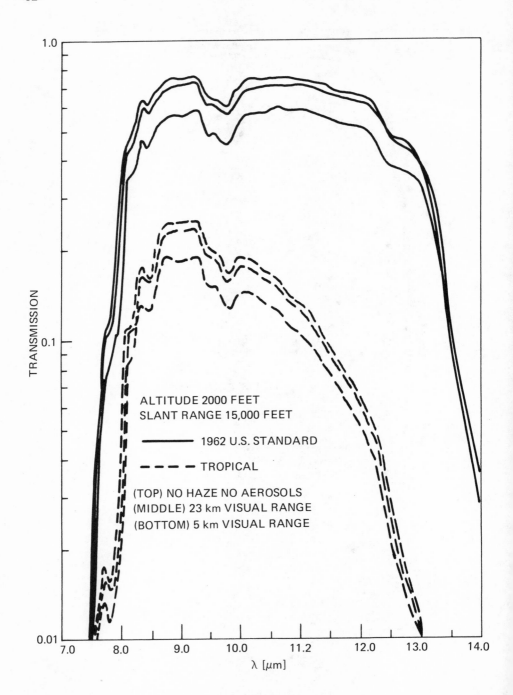

Figure 2.38 Atmospheric transmission.

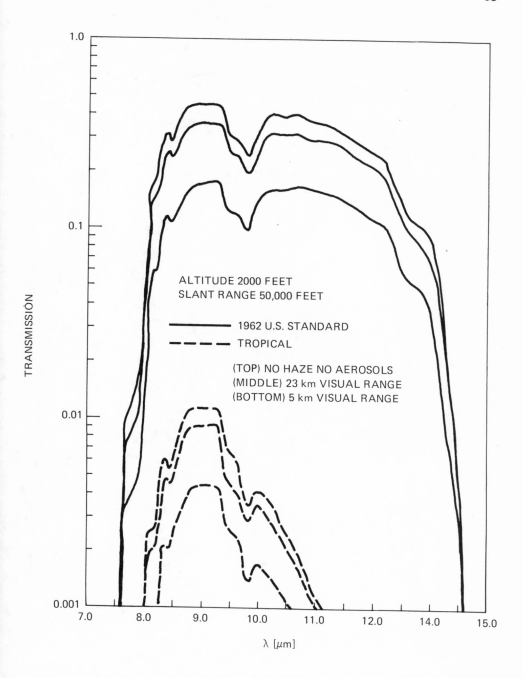

Figure 2.39 Atmospheric transmission.

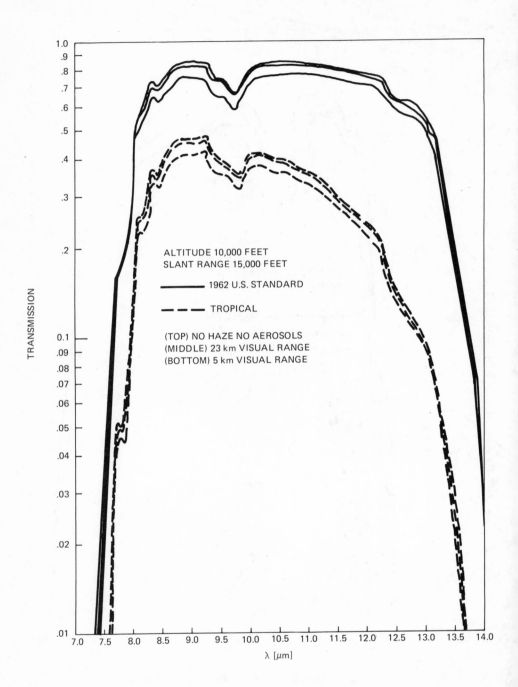

Figure 2.40 Atmospheric transmission.

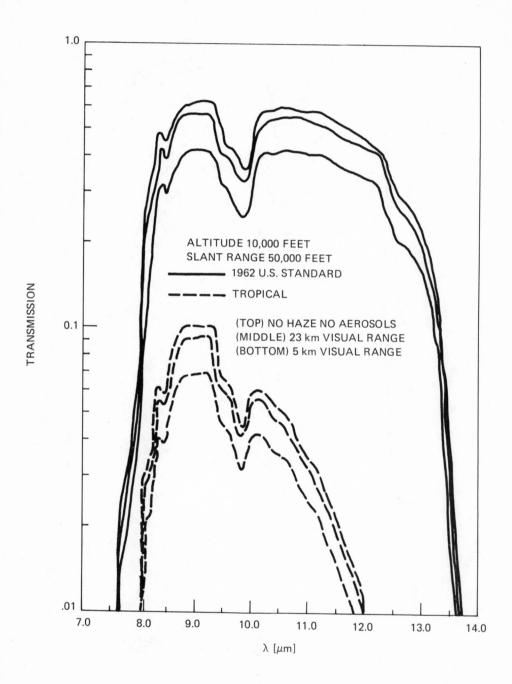

Figure 2.41 Atmospheric transmission.

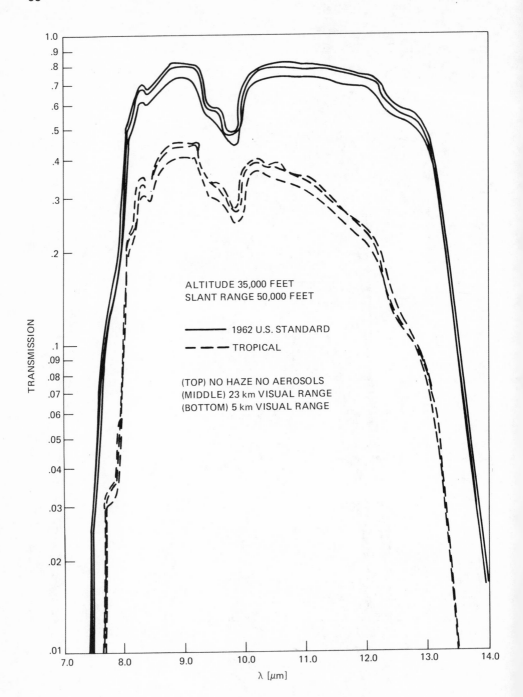

Figure 2.42 Atmospheric transmission.

REFERENCES

[1] M.A. Bramson, *Infrared Radiation: A Handbook for Applications*, translated by R.B. Rodman, Plenum Press, 1968.

[2] J.A. Jamieson, R.H. McFee, G.N. Plass, R.H. Grube, and R.G. Richards, *Infrared Physics and Engineering*, McGraw-Hill, 1963.

[3] T.P. Merritt and F.F. Hall, Jr., "Blackbody Radiation", Proc. IRE, *47*, pp 1435-1441, September 1959.

[4] R.C. Jones, "Terminology in Photometry and Radiometry", JOSA, *53*, pp 1314-1315, November 1963.

[5] R.D. Hudson, Jr., *Infrared System Engineering*, Wiley-Interscience, 1969.

[6] A. Hadni, *Essentials of Modern Physics Applied to the Study of the Infrared*, Pergamon, 1967.

[7] W.L. Wolfe, Editor, *Handbook of Military Infrared Technology*, Superintendent of Documents, U.S. Government Printing Office, 1965.

[8] P.W. Kruse, L.D. McGlauchlin, and R.B. McQuistan, *Elements of Infrared Technology: Generation, Transmission and Detection*, Wiley, 1963.

[9] K. Seyrafi, Editor, *Engineering Design Handbook: Infrared Military Systems*, Part One, U.S. Army Materiel Command Pamphlet 706-127.

[10] R.A. McClatchey, R.W. Fenn, J.E.A. Selby, F.E. Volz, and J.S. Garing, "Optical Properties of the Atmosphere", U.S. Air Force Cambridge Research Laboratories report AFCRL-72-0497, Hanscom Field, Massachusetts, 24 August 1972.

[11] J.B. Farrow and A.F. Gibson, "Influence of the Atmosphere on Optical Systems", Optica Acta, *17*, pp 317-336, May 1970.

[12] J.H. Taylor and H.W. Yates, "Atmospheric Transmission in the Infrared", JOSA., *47*, pp 223-226, March 1957.

[13] J.L. Streete, "Infrared Measurements of Atmospheric Transmission at Sea Level", Appl. Opt., *7*, pp 1545-1549, August 1968.

[14] D. Anding, "Band-Model Methods for Computing Atmospheric Slant-Path Molecular Absorption", IRIA State of the Art Report 7142-21-T, February 1967, reprinted January 1969.

[15] J.A. Hodges, "Aerosol Extinction Contribution to Atmospheric Absorption in Infrared Wavelengths", Appl. Opt., *11*, pp 2304-2310, October 1972.

[16] J.A. Hodges, Xerox Electro-Optical Systems, Pasadena, California, personal communication.

[17] P.A. Moser, Naval Air Development Center, Warminster, Pennsylvania, personal communication.

[18] J.E.A. Selby and R.M. McClatchey, "Atmospheric Transmittance from 0.25 to 28.5 μm: Computer code LOWTRAN 2", U.S.A.F. Cambridge Research Laboratories Report AFCRL-72-0745, 29 December 1972.

[19] D.A. MacDonald, Honeywell Radiation Center, Lexington, Massachusetts, personal communication.

[20] T. Jaeger, A. Nordbryhn, and P.A. Stockseth, "Detection of Low Contrast Targets at 5 μm and 10 μm: A Comparison", Appl. Opt., *11*, pp 1833-1835, August 1972.

CHAPTER 3 — LINEAR FILTER THEORY

3.1 Applicability

There is frequent need in thermal imaging practice to describe the responses of a system to various signals, to predict spatial resolution, and to devise image enhancement schemes. A tool which is useful for this purpose is linear filter theory[1,2], a branch of Fourier transform analysis used for analyzing the signal response of the class of linear, invariant, stable systems. Linear filter theory is applicable to optical, electro-optical, mechanical, and electronic devices. It originated in the description of electrical networks and servomechanisms and was extended to optical systems. Linear filter theory forms an indispensible part of image analysis and is necessary to every thermal imaging system design evaluation.

Goodman[1] defines a system as a mapping of a set of input functions into a set of output functions. We will consider here only single-valued system functions or one-to-one mappings, so that noisy systems are excluded. Figure 3.1 indicates the coordinate scheme and nomenclature we will use. A point in object space is located by referring to a set of Cartesian angular coordinates (x_o, y_o) measured in the x and y directions between the optical pointing axis and the line of sight from the aperture to the point. The image plane angular coordinates (x_i, y_i) are measured from the eye, and for a distortionless system they are given by $x_i = Mx_o$ and $y_i = My_o$, where M is the angular magnification.

Linear filter theory is applicable only to optical systems which map object distributions into image distributions by the process of convolution. Most imaging systems use convolutionary processes to the maximum extent possible because such processes generally produce the best imagery. Consequently it is important to understand convolution and to be able to recognize a non-convolutionary process.

3.2 The Convolution Integral

Consider a two-dimensional object whose radiant emittance is described by an object distribution $O(x_o, y_o)$ given as a function of object space angular

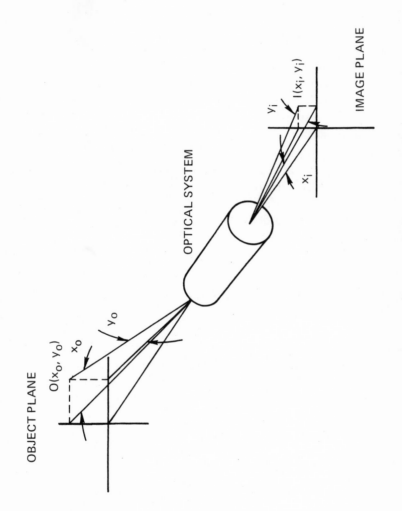

Figure 3.1 Object and image plane spatial coordinates.

Cartesian coordinates (x_0, y_0). Let the operation of the imaging system on any object be denoted by the operator $S\{\quad\}$. Then the image distribution will be $S\{O(x_0, y_0)\}$.

A system is defined to be linear if for arbitrary complex scalar weightings c and d of arbitrary object functions $O_1(x_0, y_0)$ and $O_2(x_0, y_0)$, the following equation is satisfied:

$$S\left\{cO_1(x_0, y_0) + dO_2(x_0, y_0)\right\} = cS\left\{O_1(x_0, y_0)\right\} + dS\left\{O_2(x_0, y_0)\right\}. \quad (3.1)$$

A consequence of system linearity is that any object function can be represented as a series, or in the limit as an integral, of simpler functions. To assure that this important simplification can be made, attention must be confined to systems which are, or can be approximated to be, linear operators in time and space on the signals of interest.

In general, object functions are four-dimensional functions of space and time, but for analytical purposes we usually consider that depth is not significant and therefore describe objects as functions of two angular spatial coordinates and one time coordinate. It is convenient to decompose an object function into a weighted integral of Dirac delta functions using the "sifting property" of the Dirac delta $\delta(x)$. The mathematical expression of the sifting property in two dimensions is:

$$O(x_0, y_0) = \int\limits_{-\infty}^{\infty}\!\!\!\int O(\xi, \eta)\, \delta\,(x_0 - \xi)\, \delta\,(y_0 - \eta)\, d\xi d\eta, \quad (3.2)$$

and its geometrical interpretation is shown in Figure 3.2. The meaning of this integral is that an arbitrary function is representable as an infinite sum of weighted and displaced delta functions.

The image distribution $I(x_i, y_i)$ is

$$I(x_i, y_i) = S\left\{O(x_0, y_0)\right\}$$

$$= S\left\{\int\limits_{-\infty}^{\infty}\!\!\!\int O(\xi, \eta)\, \delta\,(x_0 - \xi)\, \delta\,(y_0 - \eta)\, d\xi d\eta\right\}. \quad (3.3)$$

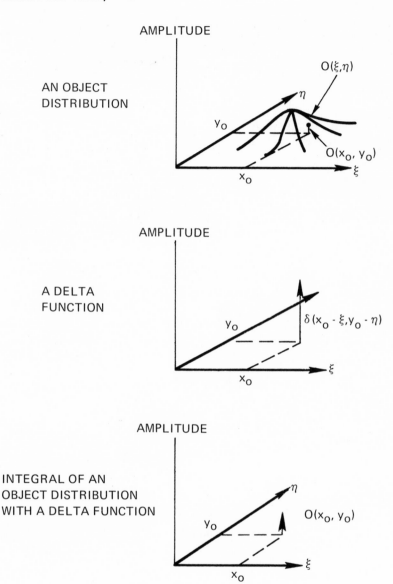

AMPLITUDE

AN OBJECT
DISTRIBUTION

$O(\xi,\eta)$

η

y_0

$O(x_0, y_0)$

x_0

ξ

AMPLITUDE

A DELTA
FUNCTION

y_0

$\delta(x_0 - \xi, y_0 - \eta)$

x_0

ξ

AMPLITUDE

INTEGRAL OF AN
OBJECT DISTRIBUTION
WITH A DELTA FUNCTION

η

y_0

$O(x_0, y_0)$

x_0

ξ

Figure 3.2 The sifting property of the delta function.

Assuming linearity, this becomes

$$I(x_i,y_i) = \int\limits_{-\infty}^{\infty}\int O(\xi,\eta)\, S\left\{\delta(x_o\text{-}\xi)\,\delta\,(y_o\text{-}\eta)\right\}\,d\xi d\eta. \tag{3.4}$$

Thus the image function is a weighted sum of the system responses to the component delta functions of the object.

The above equation is called a superposition integral, and $S\left\{\delta(x_o\text{-}\xi)\,\delta\,(y_o\text{-}\eta)\right\}$ is called the point spread function or impulse response. Equation 3.4 is called the superposition integral because it expresses a general function $I(x_i,y_i)$ as an integral of superimposed responses to infinitesimally spaced three-dimensional delta functions. In electromagnetic field theory, the quantity analogous to the point spread function is Green's function for a point charge or current source, and in electric circuit theory, it is the response to a current or voltage impulse.

We generalize to the three-dimensional case of interest by the expression

$$I(x_i,y_i,t) = S\left\{O(x_o,y_o,t)\right\}$$

$$\tag{3.5}$$

$$= \int\limits_{-\infty}^{\infty}\int\int O(\xi,\eta,t')\, S\left\{\delta(x_o\text{-}\xi)\,\delta\,(y_o\text{-}\eta)\,\delta\,(t\text{-}t')\right\}d\xi d\eta dt',$$

where it is assumed that any time delay between input and output is negligible.

The superposition integral is further simplified by requiring that the system be spatially invariant, that is by requiring independence of the impulse response of the system on the time and position of the impulse.* If we assume unity magnification M for simplicity in all that follows, then the system response to an impulse becomes

$$S\left\{\delta(x_o\text{-}\xi)\,\delta\,(y_o\text{-}\eta)\,\delta\,(t\text{-}t')\right\} = S\left\{\delta(x_i\text{-}\xi)\,\delta\,(y_i\text{-}\eta)\,\delta\,(t\text{-}t')\right\}$$

$$\tag{3.6}$$

$$\triangleq r(x_i,y_i,t;\xi,\eta,t').$$

---

*Actually, framing thermal imaging systems are not time-invariant, because a given object point produces a response only a small fraction of the time. However, we are interested in the effect of the system when viewed by a time-integrating device such as the human visual system, so that when the object is static and there is no motion between the object and the system, we may consider the system to be time-invariant.

The following graphical argument demonstrates that the superposition integral simplifies to a form called the convolution integral when the system is spatially invariant so that the form of the impulse response r is constant with coordinate translations. Consider the object distribution and spread function shown in Figure 3.3. The image distribution is found by decomposing the object into weighted delta functions, replacing each of these deltas with weighted spread functions, and summing, as represented in Figure 3.4.

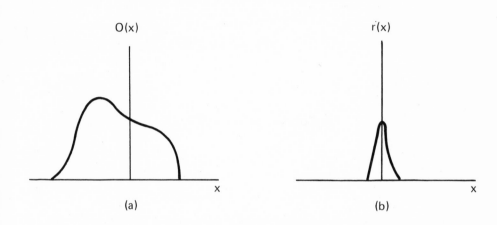

**Figure 3.3 (a) An object distribution and
(b) a spread function.**

Thus the value of the image distribution at a point x_i is found by summing the contributions at x_i of impulse responses centered at $x_i - \xi$ and weighted by the values of O at all points ξ. Mathematically this is given by

$$I(x_i) = \int_{-\infty}^{\infty} O(\xi)\, r\,(x_i - \xi)\, d\xi, \tag{3.7}$$

which is a convolution integral and is symbolized by

$$I(x) = O(x) * r(x). \tag{3.8}$$

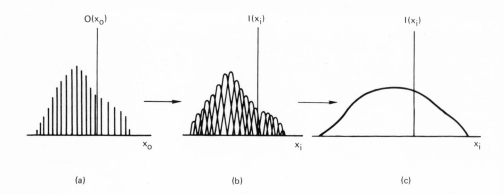

Figure 3.4 Construction of the image function by decomposition of the object function (a), replacement with weighted spread functions (b), and summation (c).

An alternative way to look at this which may not be intuitively obvious is to recognize that for a particular x_i, the variable in the above equation is ξ, and that $r(x_i-\xi) = r[-(\xi-x_i)]$ is the reflection of $r(\xi)$ centered on x_i. Thus this integral may be interpreted as shown in Figure 3.5 as the integral over the product of the object and the reflected and displaced spread function. Thus it is clear that

$$r(x_i,y_i,t; \xi,\eta,t') = r(x_i - \xi, y_i - \eta, t - t'), \tag{3.9}$$

both from the above argument and because r must be invariant with coordinate translations. Then for the three-dimensional case, the convolution integral is

$$I(x_i,y_i,t) = \int\limits_{-\infty}^{\infty}\!\!\int\!\!\int O(\xi,\eta,t')\, r(x_i\text{-}\xi,y_i\text{-}\eta,t\text{-}t')\, d\xi d\eta dt'$$

$$\tag{3.10}$$

$$= O(x_i,y_i,t) * r(x_i,y_i,t).$$

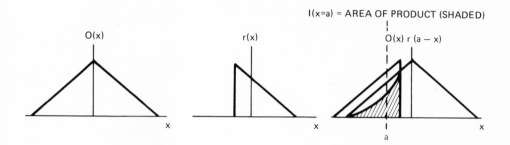

Figure 3.5 Convolution.

3.3 The Fourier Transform

The Fourier transform representation we will use here is that the forward transform is defined in one dimension by

$$F\left\{O\left(x\right)\right\} = \int_{-\infty}^{\infty} O\left(x\right)e^{-2\pi i x f_x} \, dx \triangleq \tilde{O}\left(f_x\right). \tag{3.11}$$

The quantity f_x is a Cartesian spatial variable called spatial frequency, and is analogous to the familiar temporal or electrical frequency f_t. The units of f_t are Hertz, and spatial frequencies have units of cycles per milliradian. The meaning of spatial frequency should be clear from Figure 3.6 which represents a one-dimensional sinusoidally varying light source having a spatial period θ_x. The source is viewed at a distance R such that the small angle approximation

$$2 \sin\left(\frac{\theta_x}{2R}\right) \simeq \left(\frac{\theta_x}{R}\right) \tag{3.12}$$

can be made. Figure 3.7 shows how object and image spatial frequencies and angular magnification are related for the case of an image viewed on a screen.

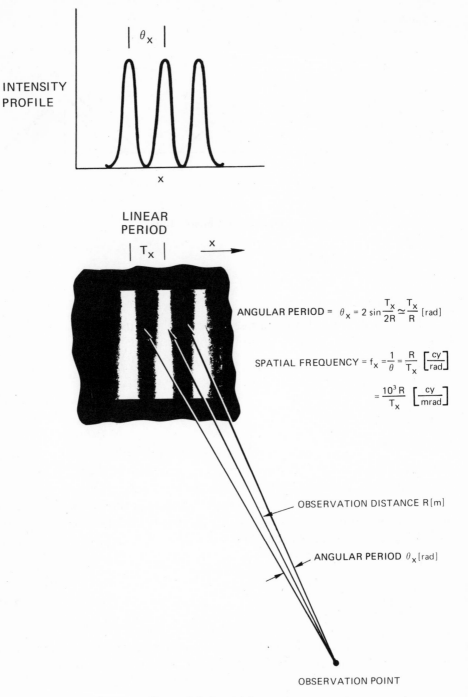

INTENSITY
PROFILE

x

LINEAR
PERIOD

$|\ T_x\ |$ x ⟶

θ_x

ANGULAR PERIOD $= \theta_x = 2 \sin \dfrac{T_x}{2R} \simeq \dfrac{T_x}{R}$ [rad]

SPATIAL FREQUENCY $= f_x = \dfrac{1}{\theta} = \dfrac{R}{T_x}$ $\left[\dfrac{cy}{rad}\right]$

$\qquad\qquad\qquad = \dfrac{10^3 R}{T_x}$ $\left[\dfrac{cy}{mrad}\right]$

OBSERVATION DISTANCE R[m]

ANGULAR PERIOD θ_x [rad]

OBSERVATION POINT

Figure 3.6 Spatial frequency.

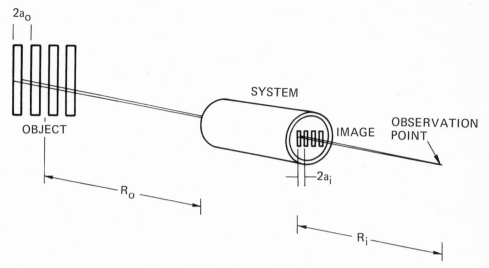

Figure 3.7 Angular magnification.

The periods of the object and image patterns are $2a_o/R_o$ and $2a_i/R_i$ respectively. The corresponding spatial frequencies are $R_o/2a_o$, and $R_i/2a_i$, and the angular magnification is

$$M = \frac{R_i a_o}{R_o a_i} . \tag{3.13}$$

In three dimensions,

$$F\left\{O\ (x,y,t)\right\} = \int\int\int_{-\infty}^{\infty}\int O\ (x,y,t) e^{-2\pi(xf_x + yf_y + tf_t)}\ dx\ dy\ dt \tag{3.14}$$

$$\stackrel{\Delta}{=} \tilde{O}\ (f_x,\ f_y,\ f_t).$$

The inverse transform* is

$$F^{-1}\left\{\tilde{O}(f_x)\right\} = \frac{1}{2\pi}\int_{-\infty}^{\infty} \tilde{O}\ (f_x) e^{2\pi i x f_x}\ df_x = O(x) \tag{3.15}$$

---

*Normalization factors of $1/2\pi$ will be ignored in subsequent sections for simplicity.

in one dimension, and

$$F^{-1}\left\{O(f_x,f_y,f_t)\right\} = \frac{1}{(2\pi)^3}\int\int\limits_{-\infty}^{\infty}\int \tilde{O}(f_x,f_y,f_t)e^{2\pi(xf_x+yf_y+tf_t)}df_x\,df_y\,df_t$$

(3.16)

$$\overset{\Delta}{=} O(x,y,t)$$

in three dimensions.

Successive forward and inverse transformations yield the original function, that is

$$F^{-1}\left\{F\left\{O(x)\right\}\right\} = O(x).$$

A detailed discussion of the Fourier transform and its applications is given by Bracewell[3].

3.4 The Convolution Theorem and the Optical Transfer Function

When a system is defined by a series of successive convolutions involving a complicated object and different spread functions, the process is difficult to visualize and to handle analytically. The convolution theorem of Fourier analysis allows this process to be transformed to the frequency domain where it is easier to manipulate. This theorem states that the Fourier transform of the convolution of two functions equals the product of the transforms of the two functions. For the image equation (3.10) this is expressed as

$$F\left\{O*r\right\} = F\left\{O\right\}\cdot F\left\{r\right\}.$$

(3.17)

An abbreviated notation for this is

$$\tilde{I} = \tilde{O}\cdot\tilde{r}.$$

(3.18)

The Fourier transform $\tilde{r}$ of an impulse response r is called the optical transfer function (OTF). If each component impulse response of a system is linear and is independent of the other component impulse responses, the system impulse response r_s is produced by a series of convolutions,

$$r_s = r_1 * r_2 * \ldots * r_n.$$

(3.19)

Then by the convolution theorem, the cascaded transfer functions can be multiplied together to produce the system OTF

$$\tilde{r}_s = \prod_{m=1}^{n} \tilde{r}_m \tag{3.20}$$

and the image spectrum is

$$\tilde{I} = \tilde{O} \prod_{m=1}^{n} \tilde{r}_m . \tag{3.21}$$

Thus the complicated operation of multiple convolutions can be replaced by the simpler process of finding the inverse transform of the product of the transforms of the convolved functions. This also allows us to think of the imaging process as a selective weighting in the frequency domain of the object spatial frequency spectrum by the system, as shown in Figure 3.8. Thus the OTF is a measure of an imaging system's ability to recreate the spatial frequency content of a scene.

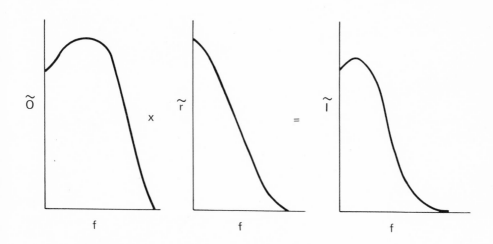

Figure 3.8 Attenuation of object frequencies by the MTF, as a function of spatial frequency f.

The OTF is a complex quantity whose modulus or absolute value is a sine wave amplitude response function called the modulation transfer function (MTF). The value of this function is normalized to unity at or near zero spatial frequency by convention. The argument in radians of the OTF is called the phase transfer function (PTF) and indicates the spatial frequency phase shift introduced by the system. The complex representation of the OTF is:

$$OTF(f) = MTF(f) \, e^{i\, PTF(f)}. \tag{3.22}$$

Much of the OTF theory in the literature of optical and television systems is equally applicable to thermal systems if four conditions are satisfied. These conditions are that: 1) the radiation detection is incoherent, 2) the signal processing is linear, 3) the imaging is spatially invariant, and 4) the system mapping is single valued (specifically, non-noisy). The last three conditions are regularly violated in thermal imagers. The infrared optical system will be spatially variant if its impulse response varies from the center to the edge of the field due to aberrations, and if scan position is nonlinear with time, causing the electrical filter to have different OTF's for the same frequency in different parts of the field. The detector array samples periodically in the direction perpendicular to the scan, thereby producing non-convolutionary imagery, and each detector is electrically noisy, violating the one-to-one mapping requirement. The analog electronics may be both nonlinear and noisy. Nonlinear signal compression may be used intentionally in the video processing to improve the system dynamic range, and the electro-optical signal conversion by the display may be nonlinear.

In addition to these characteristics, thermal imaging systems may differ from other optical and electro-optical devices in four nontrivial ways. First and most important, a thermal imager usually subtracts the average scene value from the video, displaying only variations around the average value, and usually adds an arbitrary constant signal to the displayed image. Second, electronic signal shaping or special display techniques may be used to generate spread functions with negative amplitudes, producing MTF's with greater than unity amplitude and causing an OTF definition and normalization problem. Third, many different quasi-linear system operating points may be identified whose corresponding OTF's may not be identical. Fourth, the image may be formed by an insufficiently-filtered sampling process so that the OTF exists only in the direction normal to the direction in which the sampler operates.

Consequently when we speak of the OTF of a thermal imaging system, we very likely speak of a shaky construct. This construct is nevertheless irreplaceable as a design, analysis, and specification tool because it is one of the best measures of system image quality. Therefore, extreme care must be used in the analysis, specification, and testing of OTF to ensure that all of the above-mentioned deviations from the OTF existence conditions are avoided or accounted for.

Since many thermal imaging systems do not satisfy the OTF-existence requirements in all directions, it is often necessary to resolve a point spread function in a direction where the conditions are satisfied. This results in the determination of a line spread function (LSF) which is the response to a long infinitesimally narrow line source composed of a string of delta functions. Such a geometrical line source contains only spatial frequencies in a direction perpendicular to the long dimension and contains all frequencies from zero to infinity with unity normalized amplitude as shown in Figure 3.9. The Fourier transform of the LSF is therefore unidirectional.

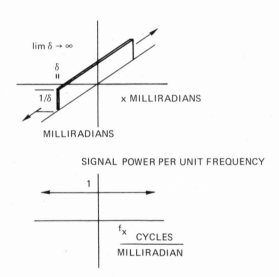

Figure 3.9 A line source and its spectrum.

As the analog signal corresponding to this line passes through each component of the system, its frequency content is attenuated selectively. The image that appears at the output of the system consequently is not precisely reproduced and the effect observed is a spreading of the line image. An example LSF, MTF, and PTF for a thermal imaging system in the scan direction are shown in Figure 3.10. Note that the asymmetrical LSF indicates the presence of phase shifts. In thermal imaging systems, the LSF and MTF are usually dominated by the detector angular subtense α, so the x axis may be normalized to α and the frequency axes normalized to $1/\alpha$.

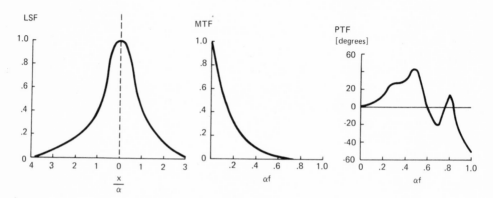

Figure 3.10 An LSF, MTF, and PTF.

The Fourier theory of optics was introduced independently by Duffieux[4] in 1946, Schade[5] in 1948, and Selwyn[6] in 1948. The subject is treated generally in such texts as Goodman[1], O'Neill[2], Françon[7] and Linfoot[5]. Central problems in optics have been analyzed in references 9 through 17, and several excellent tutorial papers on the OTF of optical and electro-optical systems are provided by references 20 through 33.

Fourier analysis in optics is similar to Laplace analysis in electrical circuit theory, with a few exceptions. Time filters differ from spatial filters in two crucial ways, as indicated by O'Neill[2]. First, time filters are single-sided in time and must satisfy the casuality requirement that no change in the output of a time filter may occur before the application of an input, whereas optical filters are double-sided in space. As is shown in many texts on operational mathematics or on electrical filter analysis, this difference requires the use of the single-sided Laplace transform for time filters rather than the Fourier transform. Second, electrical signals may be either positive or negative, whereas optical intensities are always positive.

3.5 Convolutionary and Non-convolutionary Image Formation

Since linear filter theory is such a useful image analysis tool, it is often worthwhile to approximate a non-convolutionary process as convolutionary wherever the deviations from ideal behavior are minor. For example, an imaging process which exhibits moderate spatial variance may be approximated as convolutionary over small object areas called isoplanatic patches. An image formed by samples of a convolution will sometimes be sufficiently filtered following the sampling process to be nearly convolutionary. Similarly, the effects of slight noise or slight nonlinearity may usually be safely ignored.

One of the most important convolutionary processes in thermal imaging is the dissection of an image by a scanning detector. Consider Figure 3.11, where a square detector of dimensions "a" by "a" scans an image plane (x,y) on a scan line center $y = y'$. If the detector has a uniform responsivity over its surface and instantaneously integrates all incident radiation to provide a single electrical signal, the detector signal at a point (x',y') in the scan is given by $I(x',y')$ and depends on the image plane energy distribution $O(x,y)$ as follows,

$$I(x',y') = \int_{x'-\frac{a}{2}}^{x'+\frac{a}{2}} \int_{y'-\frac{a}{2}}^{y'+\frac{a}{2}} O(x,y) \, dx \, dy. \tag{3.23}$$

Now define the detector impulse response factor as $r(x)$ given by:

$$r(x) = 1, \text{ for } -\frac{a}{2} \leqslant x \leqslant \frac{a}{2}$$

$$= 0, \text{ otherwise.} \tag{3.24}$$

As the detector scans, x has a continuum of values x', but y has only the value y'. Thus, we may rewrite the image equation as

$$I(x',y') = \int_{y'-\frac{a}{2}}^{y'+\frac{a}{2}} \left[\int_{-\infty}^{\infty} O(x,y) \, r(x'-x) \, dx \right] dy. \tag{3.25}$$

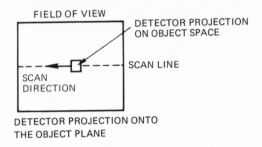

Figure 3.11 A scanning detector.

The inner integral is a convolution, so the process is convolutionary in the scan direction. The outer integral indicates spatial variance, so the process is nonconvolutionary perpendicular to the scan direction. Broadcast television is a familiar example of this type of scanning image formation. Other examples of nonconvolutionary imaging abound. The eye is non-convolutionary in the strictest sense because finite detectors are used and because resolution degrades from the center to the edge of the visual field. Halftone photography is non-convolutionary because it is binary in intensity at a given point and is therefore both nonlinear and spatially variant. Ordinary photography is processed nonlinearly and exhibits an adjacency effect in development which makes it spatially variant. However, such processes may be treated as convolutionary if an appropriate magnification scale is chosen or if attention is confined to limited areas.

3.6 Shorthand Fourier Analysis

Before considering specific impulses responses and their transforms it is useful to make a brief digression into Fourier analysis notation. In many problems, explicit Fourier series and integrals are used to describe image formation in the frequency domain. Unfortunately, these equations are often difficult to visualize and to manipulate in complicated problems, as for example when many functions are convolved or multiplied together. Fourier analysis of such cases is considerably simplified by using the shorthand notation introduced by Bracewell[3] and Goodman[1]. This notation has two elements: mathematical operators and basic function transform pairs.

From the previous sections we have that the two-dimensional* Fourier transform operator is defined by:

$$F\{g(x,y)\} = \int\limits_{-\infty}^{\infty}\!\!\int g(x,y) \exp\,[-2\pi i(xf_x + yf_y)]\ dx\ dy$$

$$= \widetilde{g}\,(f_x,f_y).$$

(3.26)

Convolution of two functions $f(x,y)$ and $g(x,y)$ is denoted by $f*g$ where

$$f(x,y) * g(x,y) = \int\limits_{-\infty}^{\infty}\!\!\int f(x-\xi,\ y-\eta)\ g(\xi,\eta)\ d\xi\ d\eta. \qquad (3.27)$$

Convolution is also defined for frequency domain functions, and is associative, commutative, and distributive. Scalar multiplication is denoted by the symbol $\cdot$, as in $f\cdot g$. The operations of convolution and multiplication are transform pairs, thus

$$F\{f\cdot g\} = F\{f\} * F\{g\}. \qquad (3.28)$$

The basic functions which are useful in thermal imaging are given in Figure 3.12 as mnemonic codes with their definitions, graphs, and transform pairs. To keep the notation simple, the sifting property of the Dirac delta function which is defined by

$$f(a) = \int\limits_{-\infty}^{\infty} f(x)\ \delta\,(x-a)\ dx \qquad (3.29)$$

is here represented by

$$f(a) = \delta(a)\ f(x). \qquad (3.30)$$

---

*Whenever an imaging process occurs in a temporal filter, we will consider stationary scenes only, and implicitly perform a coordinate transformation to the spatial domain so that only a two-dimensional transform is needed.

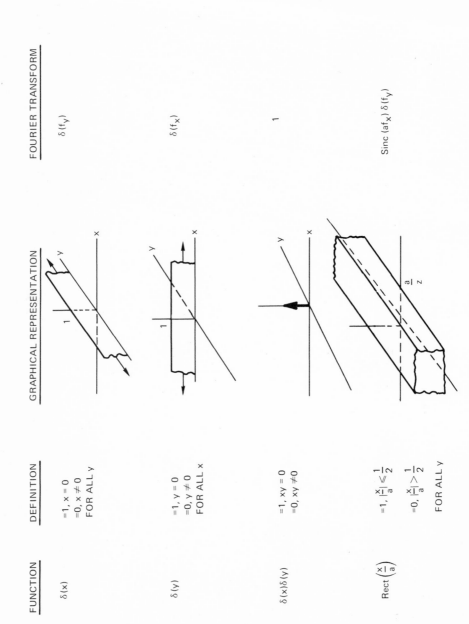

Figure 3.12 Shorthand functional notation. (Sheet 1 of 3)

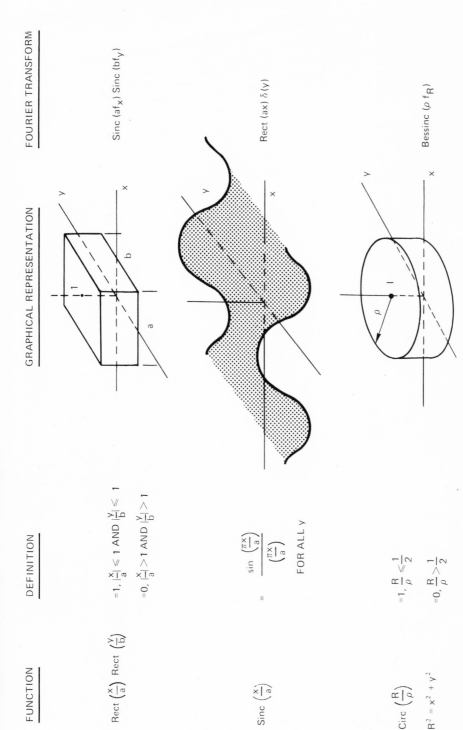

Figure 3.12 Shorthand functional notation. (Sheet 2 of 3)

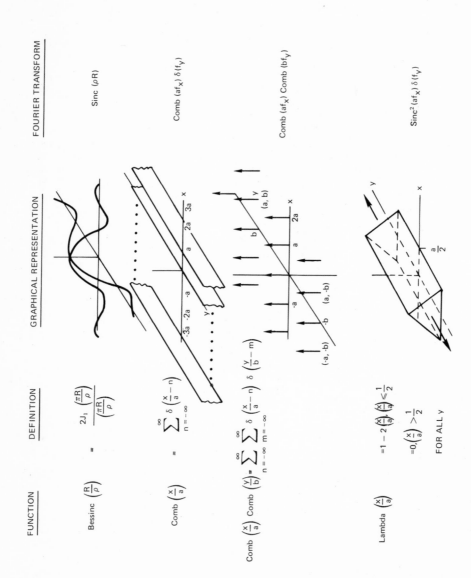

FUNCTION	DEFINITION	GRAPHICAL REPRESENTATION	FOURIER TRANSFORM
$\text{Bessinc}\left(\dfrac{R}{\rho}\right)$	$= \dfrac{2J_1\left(\dfrac{\pi R}{\rho}\right)}{\left(\dfrac{\pi R}{\rho}\right)}$		$\text{Sinc}\,(\rho R)$
$\text{Comb}\left(\dfrac{x}{a}\right)$	$= \displaystyle\sum_{n=-\infty}^{\infty} \delta\left(\dfrac{x}{a}-n\right)$		$\text{Comb}\,(af_x)\,\delta(f_y)$
$\text{Comb}\left(\dfrac{x}{a}\right)\text{Comb}\left(\dfrac{y}{b}\right)$	$= \displaystyle\sum_{n=-\infty}^{\infty}\sum_{m=-\infty}^{\infty} \delta\left(\dfrac{x}{a}-n\right)\delta\left(\dfrac{y}{b}-m\right)$		$\text{Comb}\,(af_x)\,\text{Comb}\,(bf_y)$
$\text{Lambda}\left(\dfrac{x}{a}\right)$	$= 1 - 2\left|\dfrac{x}{a}\right|, \left|\dfrac{x}{a}\right| \leq \dfrac{1}{2}$ $= 0, \left|\dfrac{x}{a}\right| > \dfrac{1}{2}$ FOR ALL y		$\text{Sinc}^2(af_x)\,\delta(f_y)$

Figure 3.12 Shorthand functional notation. (Sheet 3 of 3)

As a simple example of how the notation is used, consider the stylized detector array shown in Figure 3.13. The individual detectors are described by Rect $(x/\alpha) \cdot$ Rect (y/β). An infinitely long detector array is described by [Rect $(x/a) \cdot$ Rect (y/β)] $*$ [Comb $(y/\gamma) \cdot \delta(x)$]. A finite detector array is limited by Rect (y/B). Combining these, Figure 3.13 is completely described by

$$\{[\text{Rect}\,(x/\alpha) \cdot \text{Rect}\,(y/\beta)] * [\text{Comb}\,(y/\gamma) \cdot \delta(x)]\} \cdot \text{Rect}\,(y/B).$$

We transform this expression by taking transforms of the outermost bracketed terms in succession, yielding:

$$F\left\{\{[\text{Rect}\,(x/\alpha) \cdot \text{Rect}\,(y/\beta)] * [\text{Comb}\,(y/\gamma) \cdot \delta(x)]\} \cdot \text{Rect}\,(y/B)\right\}$$

$$= F\{[\text{Rect}\,(x/\alpha) \cdot \text{Rect}\,(y/\beta)] * [\text{Comb}\,(y/\gamma) \cdot \delta(x)]\} * F\{\text{Rect}\,(y/B)\}$$

$$= \left[F\{\text{Rect}\,(x/\alpha) \cdot \text{Rect}\,(y/\beta)\} \cdot F\{\text{Comb}\,(y/\gamma) \cdot \delta(x)\}\right] * F\{\text{Rect}\,(y/B)\}.$$

Using the transform pairs given earlier, this reduces to:

$$\{\text{Sinc}\,(\alpha f_x) \cdot \text{Sinc}\,(\beta f_y) \cdot \text{Comb}\,(\gamma f_y)\} * \text{Sinc}\,(B f_y).$$

This demonstrates some commonly performed manipulations.

3.7   Typical Component OTF's

There are three spread functions which are useful approximations to the spread functions commonly found in imaging system components. The first is the two-dimensional rectangular function [Rect (x/α) Rect (y/β)] shown in x-profile in Figure 3.14. As a first approximation, rectangularly-shaped detector elements are assumed to have such a spatial impulse response. In fact, the detector responsivity contours may be anything but rectangular, but manufacturers may not be able to control this precisely or the precise contour may not be known, so one usually makes the rectangular assumptions.

If the scan is in the x-direction only, the detector OTF is the transform of Rect (x/α), is real, and is given by the function

$$\widetilde{\tau}_d = \sin(\pi\alpha f_x)/(\pi\alpha f_x) = \text{sinc}\,(\alpha f_x). \tag{3.31}$$

Values of this function and its square are tabulated in Table 3.1 and graphed in Figure 3.15. The rectangular function may also describe rectangular display elements such as electroluminescent diodes or glow modulator cavities.

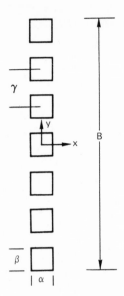

Figure 3.13 A detector array in angular coordinates.

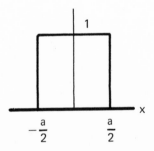

Figure 3.14 A rectangular detector profile.

Table 3.1

Values of $\dfrac{\sin(\pi\alpha f)}{\pi\alpha f}$ and $\left[\dfrac{\sin(\pi\alpha f)}{\pi\alpha f}\right]^2$

αf	$\dfrac{\sin(\pi\alpha f)}{\pi\alpha f}$	$\left[\dfrac{\sin(\pi\alpha f)}{(\pi\alpha f)}\right]^2$	αf	$\dfrac{\sin(\pi\alpha f)}{\pi\alpha f}$	$\left[\dfrac{\sin(\pi\alpha f)}{(\pi\alpha f)}\right]^2$
0	1	1	1	0	0
.05	.996	.992	1.05	-.047	.002
.1	.984	.968	1.1	-.089	.008
.15	.963	.928	1.15	-.126	.016
.2	.935	.875	1.2	-.156	.024
.25	.900	.810	1.25	-.180	.032
.3	.858	.737	1.3	-.198	.039
.35	.810	.657	1.35	-.210	.044
.4	.757	.573	1.4	-.216	.047
.45	.699	.488	1.45	-.217	.047
.5	.637	.405	1.5	-.212	.045
.55	.672	.327	1.55	-.202	.041
.6	.505	.255	1.6	-.189	.036
.65	.436	.190	1.65	-.172	.030
.7	.368	.135	1.7	-.151	.023
.75	.300	.090	1.75	-.129	.017
.8	.234	.055	1.8	-.078	.006
.85	.170	.029	1.9	-.052	.003
.9	.109	.012	1.95	-.025	.001
.95	.052	.003	2.0	0	0

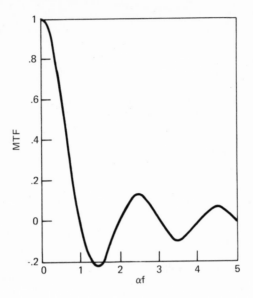

Figure 3.15 Sinc function MTF.

The second useful spread function is the two-dimensional gaussian distribution, the simplest of which is the circularly-symmetric gaussian shown in x-profile in Figure 3.16. The spread function is

$$r(x,y) = \frac{1}{2\pi\sigma^2} \ \exp \left(-\frac{x^2 + y^2}{2\sigma^2} \right),$$ (3.32)

and the associated OTF's are

$$\tilde{r}_x = e^{-2\pi^2 \sigma^2 f_x^2} \ \text{ and } \ \tilde{r}_y = e^{-2\pi^2 \sigma^2 f_y^2} \ .$$

The elliptically symmetrical gaussian is

$$r(x,y) = \frac{1}{2\pi\sigma_x\sigma_y} \ \exp \left(-\frac{x^2}{2\sigma_x^2} \right) \exp \left(-\frac{y^2}{2\sigma_y^2} \right).$$ (3.33)

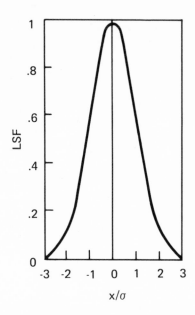

Figure 3.16 The generalized gaussian LSF.

The x and y dependences are separable and the associated OTF's are

$$\tilde{r}(f_x) = e^{-2\pi^2 \sigma_x^2 f_x^2} \text{ and } \tilde{r}(f_y) = e^{-2\pi^2 \sigma_y^2 f_y^2} . \tag{3.34}$$

A normalized gaussian MTF is shown in Figure 3-17. Three gaussians are tabulated in Table 3.2 where the sigmas are normalized to a detector subtense α and the frequencies are normalized to $f_o = 1/\alpha$. These are useful for quickly estimating the effects, relative to the detector MTF, of video monitors or other components having gaussian MTF's.

Although impulse responses are seldom precisely gaussian in shape, it is often helpful to approximate them by gaussians. For example, an optical spot size may be given as the diameter 2ρ of a circle in which a certain percentage of the total spot energy is found. If the spot is suspected to be gaussian, we may approximate it by finding the gaussian which has the same percentage of energy P in a circle of radius ρ.

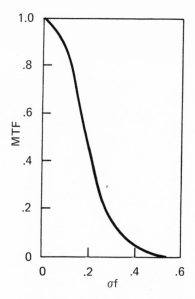

Figure 3.17 The generalized gaussian MTF.

Thus for the gaussian form

$$r(x,y) = \exp[-(x+y)^2/2\sigma^2] = \exp(-R^2/2\sigma^2)$$

shown in Figure 3.18 we must find the σ for which the following condition is satisfied:

$$P = \frac{\int_0^{2\pi}\int_0^{\rho} r(R) R \, dR \, d\phi}{\int_0^{2\pi}\int_0^{\infty} r(R) R \, dR \, d\phi} \, . \tag{3.35}$$

Integration over ϕ and then over R yields:

$$P = 1 - \exp(-\rho^2/2R^2),$$

and solving for σ yields

$$\sigma = \frac{\rho}{\sqrt{2\ln\left(\frac{1}{1-P}\right)}} \, . \tag{3.36}$$

Thus one can easily translate a percentage of energy P within a given diameter for a gaussian spot into a standard deviation. For example, if 85 percent falls within a circle of radius ρ, $\sigma = 0.51\rho$.

Table 3.2

MTF's of Components Having Gaussian Spread Functions. Sigma's are Normalized to Detector Angular Subtense α.

$$MTF = \exp\left[-2\pi^2 \left(\frac{\sigma_m}{\alpha}\right)^2 \left(\frac{f}{f_o}\right)^2\right]$$

$\dfrac{f}{f_o}$	MTF for $(\sigma/\alpha) = .125$	MTF for $(\sigma/\alpha) = .25$	MTF for $(\sigma/\alpha) = .5$
0	1	1	1
.05	.999	.997	.987
.1	.997	.987	.952
.15	.993	.973	.894
.2	.987	.952	.821
.25	.981	.925	.735
.3	.973	.894	.641
.35	.963	.859	.546
.4	.952	.821	.454
.45	.939	.779	.368
.5	.925	.735	.291
.55	.911	.689	.225
.6	.894	.641	.169
.65	.878	.594	.124
.7	.859	.546	.089
.75	.841	.500	.062
.8	.821	.454	.042
.85	.800	.410	.028
.9	.779	.368	.018
.95	.757	.328	.012
1.0	.735	.291	.007

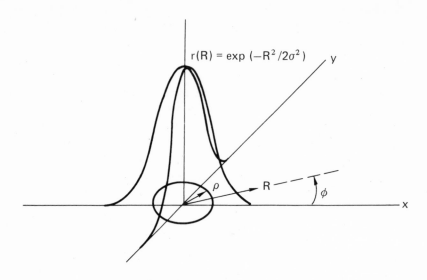

Figure 3.18 A circular gaussian distribution.

Some impulse responses which are often appropriately represented as gaussian are symmetrically aberrated optical impulse responses, random scan encoder position errors, sensor vibrations, and cathode ray tube spots. Schade[34] presents numerous practical gaussian-type cathode ray tube spread functions and their transforms. Detailed knowledge of this sort usually is not available from the CRT manufacturer, who instead may specify spot size in terms of the semi-subjective shrinking raster test in which the raster is contracted electronically until modulation reaches the threshold of perceptibility. If the spot is a symmetrical gaussian, the sigma is related[35,36] to the shrinking raster center line to center line separation s by

$$\sigma = 0.54 \ s. \tag{3.37}$$

If the manufacturer specifies the resolution in terms of the highest spatial frequency square wave input which is resolvable on the monitor the sigma is related to the bar pattern period p by

$$\sigma = 0.42 \ p. \tag{3.38}$$

Jenness, et al.[35] calculated the ripple which results from television raster by gaussian shaped scan lines. For line spacing h and spot standard deviation σ, they found for h/σ ratios of 0 to 4 that the ripple is sinusoidal. The percentage ripple versus h/σ is shown in Figure 3.19.

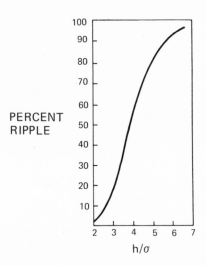

Figure 3.19 Ripple for a raster constructed of gaussian spots versus the ratio of line spacing to spot sigma, adapted from reference 35.

The third useful spread function is the circularly-symmetric pillbox of Figure 3.20 given by

$$r(R) = \operatorname{Circ}\left(\frac{R}{\rho}\right). \tag{3.39}$$

The OTF is real and is given by

$$\tilde{r} = 2\,\operatorname{Bessinc}(\rho\,f_R) = 2J_1\,(\pi\rho f_R)/(\pi\rho f_R), \tag{3.40}$$

where f_R is spatial frequency in any radial direction. This is shown in x-profile in Figure 3.21. The circular pillbox may describe a severely defocussed optic, a circular detector, or a circular display element. The Bessinc function is accurately approximated[30] by

$$2\,\operatorname{Bessinc}(\rho\,f_R) \simeq \operatorname{Sinc}(0.867\,\rho\,f_R). \tag{3.41}$$

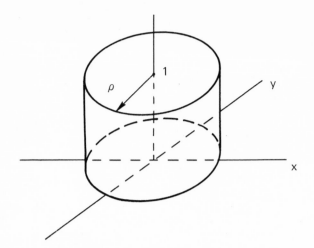

Figure 3.20 A circular function.

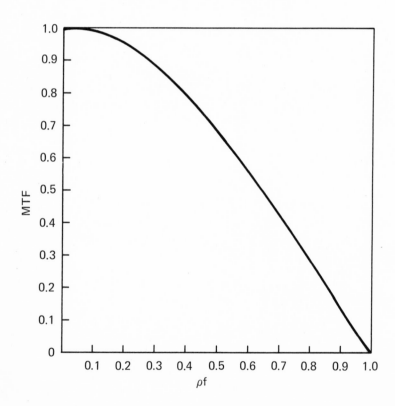

Figure 3.21 Truncated Bessinc function.

3.8 Optical Component OTF's

The majority of thermal imaging systems operate with broad spectral band-passes and detect radiation noncoherently, so partial coherence effects[37] are not observed. Therefore classical diffraction theory is adequate for analyzing the optics of noncoherent thermal images. The OTF of a diffraction-limited optic depends on the wavelength of the radiation and on the shape of the aperture. Specifically, the OTF is the autocorrelation[1] of the entrance pupil* function p(x,y) with entrance pupil coordinates x and y replaced by spatial frequency coordinates x/λ and y/λ.

$$\text{OTF} \left(f_x = \frac{x}{\lambda}, f_y = \frac{y}{\lambda}\right) =$$

$$\frac{\displaystyle\int\limits_{-\infty}^{\infty}\int p\left(\xi - \frac{x}{\lambda}, \eta - \frac{y}{\lambda}\right) p\left(\xi + \frac{x}{\lambda}, \eta + \frac{y}{\lambda}\right) d\xi d\eta}{\displaystyle\int\limits_{-\infty}^{\infty}\int p(\xi,\eta)\, d\xi d\eta}. \tag{3.42}$$

For the simple case of plane wave monochromatic illumination on a clear rectangular aperture, the OTF is real and the autocorrelation simplifies[1] to an MTF,

$$\tilde{r} = 1 - \frac{f}{f_c}, \text{ for } \frac{f}{f_c} \leqslant 1, \tag{3.43}$$

where

$$f_c = \frac{D_0}{\lambda}, \tag{3.44}$$

D_0 is the aperture width, and λ is the wavelength.

---

*If spatial frequencies are expressed in object plane coordinates.

For a clear circular aperture,

$$\tilde{r} = \frac{2}{\pi} \left\{ \arccos\left(\frac{f}{f_c}\right) - \left(\frac{f}{f_c}\right) \left[1 - \left(\frac{f}{f_c}\right)^2 \right]^{1/2} \right\}, \text{ for } \frac{f}{f_c} \leqslant 1, \qquad (3.45)$$

where $f_c = D_0/\lambda$ and D_0 is the entrance pupil diameter. The PTF for a diffraction limited optical system is always zero. This MTF is tabulated in Table 3.3 and shown in Figure 3.22. An approximate expression accurate to ±0.01 when both the exact and the approximate expressions are rounded to two decimal places is

$$\tilde{r} \simeq 1 - 1.218 \left(\frac{f}{f_c}\right) \text{ over } 0 \leqslant \frac{f}{f_c} \leqslant 0.6. \qquad (3.46)$$

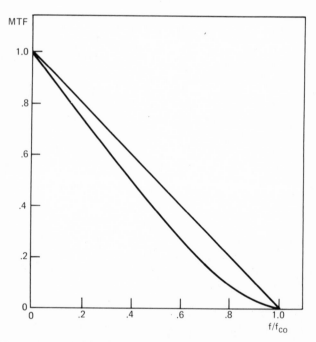

Figure 3.22 MTF's of diffraction-limited circular and rectangular lens apertures with $f_c = D_0/\lambda$.

O'Neill[1] derived the MTF for the case of a circular diffraction-limited lens with a circular central obscuration. The resulting formulas are too complex to reproduce here, but several cases calculated by Viswanathan[38] are shown in Figure 3.23. The lens diameter is D_0, the obscuration diameter is ηD_0, and the fractional obscuration is $1-\eta^2$.

Table 3.3
MTF of an Unobscured Circular Aperture

f/f_c	MTF	f/f_c	MTF
0.00	1.0000	.5	0.3910
0.025	0.9682	.525	.3638
.05	.9364	.55	.3368
.075	.9046	.575	.3105
.10	.8729	.6	.2848
.125	.8413	.625	.2696
.15	.8097	.65	.2351
.175	.7783	.675	.2112
.2	.7471	.7	.1881
.225	.7160	.725	.1658
.25	.6850	.75	.1443
.275	.6543	.775	.1237
.3	.6238	.8	.1041
.325	.5936	.825	.0855
.35	.5636	.85	.0681
.375	.5340	.875	.0520
.4	.5046	.9	.0374
.425	.4756	.925	.0244
.45	.4470	.95	.0133
.475	.4188	.975	.0047
		1.00	.0000

Equation 3.45 is the diffraction-limited MTF for plane wave monochromatic illumination, whereas the cases of real interest involve broadband illumination. If incoherence may be assumed, the polychromatic impulse response is the weighted sum of the component monochromatic impulse responses. For the special case of a uniformly weighted spectral distribution of plane wave illumination incident on a rectangular or circular aperture, the impulse response can be obtained analytically. The broadband incoherent impulse response is

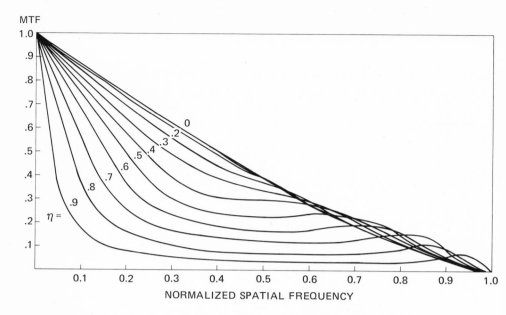

Figure 3.23 MTF's of centrally-obscured circular apertures versus obscuration diameter ratio, adapted from reference 38.

$$r(x) = \int_{\lambda_1}^{\lambda_2} r(x,\lambda)\, d\lambda, \tag{3.47}$$

and the transform is

$$\tilde{r}(f_x) = F \left\{ \int_{\lambda_1}^{\lambda_2} r(x,\lambda)\, d\lambda \right\} = \int_{\lambda_1}^{\lambda_2} \tilde{r}(f_x,\lambda)\, d\lambda. \tag{3.48}$$

For the 8 to 14 μm and the 3 to 5 μm bands the difference between the exact expressions resulting from the integration and the monochromatic expressions using the mean wavelength $(\lambda_1 + \lambda_2)/2$ to replace λ are insignificant.

The impulse response of a non-diffraction-limited optical system is determined by wavefront errors such as the primary aberrations: defocussing, spherical aberration, chromatic aberration, coma, distortion, astigmatism, and field curvature; and by higher order aberrations. The OTF of an aberrated optical system is calculated by replacing[1] the real pupil function p(x,y) for the diffraction-limited case with a generalized complex pupil function

$$p'(x,y) = p(x,y) \exp \left[-\frac{2\pi i}{\lambda} W(x,y), \right] \qquad (3.49)$$

where $W(x,y)$ is called the aberration function. $W(x,y)$ is the wavefront deviation in the exit pupil at coordinates (x,y) from an ideal unaberrated spherical converging wavefront. The OTF is found by substituting the complex pupil function for the unaberrated pupil function in the autocorrelation equation.

Aberration functions $W(x,y)$ for the various aberrations are derived as demonstrated by Klein[39]. Discussion of these wavefront aberrations and the complicated OTF formulas which result are beyond the scope of this book. The reader may consult a text such as Smith[40], Klein[39], Barakat[41], or O'Neill[2] for summaries of the aberrations, and references 1, 9, 10, 11, 13, 14, and 15 for details. Slight aberrations can produce serious degradations in MTF accompanied by severe phase shifts.

The aperture size in an 8-14 μm system is usually determined jointly by resolution and thermal sensitivity requirements, and such a system typically is designed to be diffraction-limited. Systems in the 3 to 5 band, however, tend to have apertures sized for only thermal sensitivity, because the diffraction limit is higher at shorter wavelengths. These systems tend to be simpler but non-diffraction limited.

3.9 Electrical Processing OTF's

The electrical filters used in detector signal processing differ from other thermal imaging system components in that their spread functions are single-sided due to the time causality condition in stable electrical circuits. Consequently, their time domain convolution integrals are single-sided and their impulse responses are Laplace transformed rather than Fourier transformed to yield the frequency response. Electrical filtering occurs in detector signal preamplifiers, in video processing amplifiers, and in MTF-boosting or aperture correction circuits. Also, detectors are in effect electrical filters because their response times are limited by such mechanisms as charge diffusion lengths, recombination rates, and thermal time constants.

Electrical or temporal frequencies in Hertz are conveniently related to scan direction spatial frequencies in cycles/milliradian by multiplying by the ratio of the detector subtense α to the detector dwelltime τ_d,

$$f[\text{Hertz}] = \frac{\alpha}{\tau_d} f \,[\text{cy/mrad}]. \qquad (3.50)$$

Subsequently, we will let f refer either to electrical frequency or to the equivalent spatial frequency, and let the appropriate units be implied by the context.

Any single electrical filter or ensemble of cascaded filters may usually be approximated by cascaded combinations of three types of filters. These are the RC filters, so named because they are represented by circuits using ideal resistors and capacitors. These three types of RC filters are the high-pass, the all-pass lead, and the low-pass filter, and are shown in Figure 3.24 together with their characteristic frequencies f_c and their transfer functions. Note that these OTF's are complex, indicating nonzero phase transfer functions and the presence of phase shift. The MTF of the single RC low-pass filter is tabulated in Table 3.4 and graphed in Figure 3.25.

Electrical phase shifts from these components are linearly converted into displayed spatial phase shifts, and the spatial direction of the phase shift is determined by the scan direction. Thus if there is an electrical phase lag and the scan is from left to right, the shift on the display will be to the right. With unidirectional scan, moderate phase shift does not seriously degrade image quality. With bi-directional scan, however, phase shifts on successive frames are in opposite directions, and the full frame-to-frame phase shift in a stationary target is doubled when seen by a sensor (such as the eye) whose integration time is greater than the frame time.

The consequence of this is a serious image blurring which we will now describe mathematically. Let OTF(f) be the complex transfer function of the electrical filter

$$OTF = Re(OTF) + j\ Im(OTF), \tag{3.51}$$

where Re and Im denote the real and the imaginary parts of the OTF. The bi-directional scan produces a multiframe-averaged real unnormalized transfer function

$$\tilde{r}\ (f) = OTF\ (f) + OTF^*\ (f) = 2\ Re\ (OTF), \tag{3.52}$$

where the asterisk denotes the complex conjugate.

Expressed in polar form this is

$$\tilde{r} = OTF\ exp\ (j\ arg\ OTF) + OTF\ exp\ (-j\ arg\ OTF)$$
$$= 2\ OTF\ cos\ [arg\ (OTF)]. \tag{3.53}$$

Normalized to a value of one, this is:

$$\tilde{r} = OTF\ cos\ [arg\ (OTF)]. \tag{3.54}$$

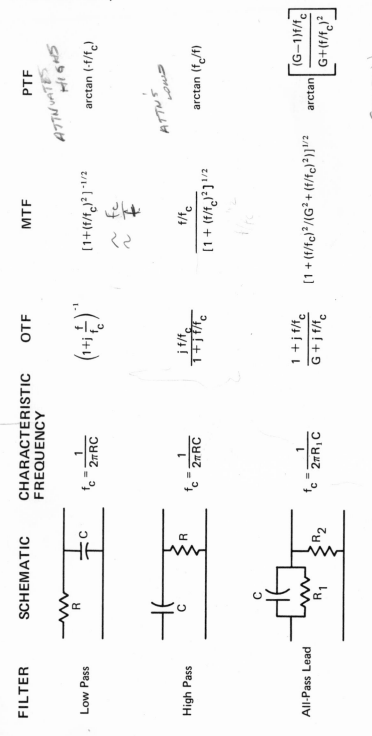

$$OTF(f) = MTF(f)\, e^{i(PTF(f))}$$

Figure 3.24 Simple RC circuits.

Table 3.4
MTF of a Single-RC Low-pass Filter.

f/f_c	$[1+(f/f_c)^2]^{-1/2}$	f/f_c	$[1+(f/f_c)^2]^{-1/2}$
0	1	2.2	.414
.1	.995	2.3	.399
.2	.981	2.4	.385
.3	.958	2.5	.371
.4	.928	2.6	.359
.5	.894	2.7	.347
.6	.857	2.8	.336
.7	.819	2.9	.326
.8	.781	3.0	.316
.9	.743	3.2	.298
1.0	.707	3.4	.282
1.1	.673	3.6	.268
1.2	.640	3.8	.254
1.3	.610	4.0	.243
1.4	.581	4.5	.216
1.5	.555	5.0	.196
1.6	.530	5.5	.178
1.7	.507	6.0	.164
1.8	.486	7.0	.141
1.9	.466	8.0	.124
2.0	.447	9.0	.110
2.1	.430	10.0	.100

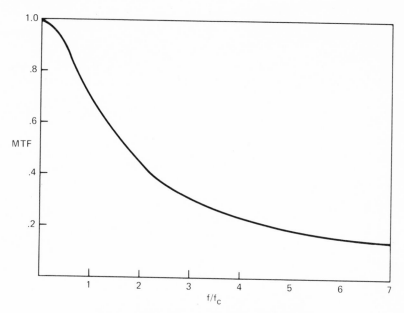

Figure 3.25 Single RC low-pass filter MTF.

Thus the time-averaged bi-directional phase shift results in a real OTF. Since the argument of a complicated filter OTF is the sum of the component OTF arguments, the MTF loss due to phase shift alone in bi-directional scanning with phase shift is

$$\widetilde{r} = \cos \left[\sum_i \arg (OTF_i) \right]. \tag{3.55}$$

For the single-RC low pass filter, the complex OTF is

$$OTF = (1 + jf2\pi RC)^{-1} \tag{3.56}$$

and the normalized unidirectional MTF $\widetilde{r}$ is

$$\widetilde{r} = [1 + (2\pi f\, RC)^2]^{-1/2}. \tag{3.57}$$

With bi-directional scanning, the MTF is degraded to

$$\widetilde{r} = [1 + (2\pi\, fRC)^2]^{-1}. \tag{3.58}$$

This bi-directional scan-induced MTF loss is more serious than it may appear at first glance. Consider again the single-RC low-pass filter. This filter has a PTF of arctan $(-f/f_c)$, which at low frequencies is approximately $-f/f_c$, that is, which is linear with frequency. Now this linear frequency-dependent phase shift is simply a time lag and is displayed as a translation in the scan direction, so combined with a bi-directional scan the imagery has a slightly "doubled" appearance. For example, a slit input produces the image shown in Figure 3.26. Thus, even a slight phase shift may cause objectionable image degradation. This effect, and the edge flicker effect discussed in Section 4.6.3, strongly militate against bi-directional scanning.

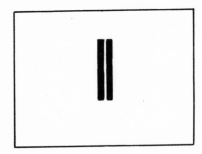

Figure 3.26 Image doubling due to bi-directional scan.

The three filters of Figure 3.24 can be used individually or in combination to describe most realizable filters. Many detectors, for example, behave like single-RC low-pass filters with responsive time constants of

$$\tau_R = 1/(2\pi RC). \tag{3.59}$$

The simplest ac coupling device is the single-RC high pass filter, and many preamplifiers behave as single- or double-RC low pass filters. Many boosting circuits may be approximated by an all-pass lead filter followed by double-RC low-pass filter. If these simple circuits are inadequate, equivalent RLC networks can often be devised.

3.10 Image Motion and Image Position Uncertainty MTF Losses

In addition to the sources of resolution loss discussed in Section 3.5 which are invariably present in thermal systems, there are four other mechanisms which must be dealt with. These are sensor vibration relative to the scene, display vibration relative to the observer, uncertainly in the scan (read in) position at a point in time, and uncertainly in the display (write out) position in time. Motion of the sensor relative to the scene occurs when the sensor is inadequately stabilized to compensate for platform vibrations and wind buffeting. Display vibration relative to the observer occurs when the platform shakes at rates and amplitudes such that the observer's body and head move more than his visual system can compensate for. Uncertainty in the read-in position occurs when there is uncertainty in the scan mirror angle from frame to frame at a given nominal angular position. Write-out position errors occur when the scan position encoder introduces uncertainty in the scan position signal.

Random motions which are periodic with frequencies which do not exceed about 2 cycles per second do not cause serious image blurring because the eye's integration time is too short to fuse such motions. Since the eye's ability to extract information in the presence of image motion is not well understood, it is not possible to associate a given motion with an effective OTF loss unequivocally. However, it seems likely that purely random motions or position errors may be described by the Fourier transform of the probability density function of the image position when the density function is continuous. For example, a scan position pickoff or synchronization error which has a gaussian probability density centered on the correct value of scan position should have an effective an MTF which is gaussian with a sigma the same as that of the density function.

3.11 Equivalent Bandwidth, Equivalent Resolution, and the Central Limit Theorem

Schade[5] discovered that the apparent image sharpness of a television picture tends to be describable by the integral over frequency of the square of the television system's MTF. Schade called this integral the equivalent line number N_e [cycles/picture width],

$$N_e \triangleq \int_0^\infty [\tilde{\tau}(f)]^2 \, df. \tag{3.60}$$

This quantity is in effect an equivalent square bandwidth, as shown in Figure 3.27. N_e is one of the best measures of image sharpness and resolution in non-noisy imagery, and its significance is discussed in detail in Section 4.10.

If N_e is a good measure of resolution in the frequency domain, then there should exist a related parameter in the spatial domain which we could call equivalent resolution. Sendall[42] examined all of the reasonable potential definitions of an equivalent resolution $\bar{\tau}$, and concluded that the best one is $\bar{\tau} = 1/2N_e$, since this gives $\bar{\tau} = \alpha$ for a rectangular spread function. Figure 3.28

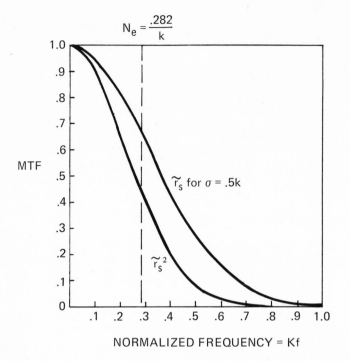

Figure 3.27 An example of equivalent bandwidth
for a Gaussian MTF.

LSF	MTF	N_e	$\bar{r}$
Rect (x/α)	Sinc (αf)	$1/2\alpha$	α
$e^{-x^2/2\sigma^2}$	$e^{-2\pi^2\sigma^2 f^2}$	$1/(4\sigma\sqrt{\pi}) = .142/\sigma$	$\sigma/.284$
$a^2/(x^2 + a^2)$	$e^{-2a\lvert f\rvert}$	$1/2a$	a

Figure 3.28 Examples of equivalent bandwidths and resolutions.

gives N_e and $\bar{r}$ for some common OTF's. This definition for $\bar{r}$ has the disadvantage that $\bar{r}$ is not necessarily the angular dimension of the smallest detail resolvable by the system.

The well known central limit theorem of probability and statistics has an analog[3] in linear filter theory, which is that the product of N bandlimited continuous MTF's tends to a gaussian form as N becomes large*. Many thermal imaging systems have at least four component MTF's, and often as many as seven, so the system line spread function can often be adequately approximated by $r(x) = \exp(-x^2/2\sigma^2)$, where σ is the standard deviation. The corresponding MTF is:

$$\tilde{r}(f_x) = \exp(-2\pi^2 \sigma^2 f_x^2). \qquad (3.61)$$

As an example, Table 3.5 and Figure 3.29 show component MTF's which might be used with the one milliradian example system and whose product is a system MTF which is approximately gaussian. Thus it is worthwhile for analytical and specification purposes to note some properties of this function. The standard deviation of a spread function may be estimated by finding the points $r(\sigma) = .61$ or $r(1.175\,\sigma) = .5$. The scale of an MTF may be estimated by finding the frequency $f = .159/\sigma$ at which $\tilde{r} = .61$. Various estimates of spread function size in terms of the percentage of total spread function energy P between $x = -x_0$ and $x = x_0$ may be converted to a gaussian spread function from the following values:

P	x_0
.383	$\sigma/2$
.683	σ
.842	$\sqrt{2\sigma}$
.955	2σ
.997	$3\sigma.$

Another way to approximate an MTF by a gaussian is to find the σ which makes the gaussian's N_e of $0.141/\sigma$ equal to the N_e of the functionally unknown MTF. Three universal gaussian curves are shown in Figure 3.30 as functions of the dimensionless parameter Kf for σ's of 0.4 K, 0.5 K, and 0.6 K. These curves may be used to see the effect of σ on the MTF of a

---

*A rigorous statement of the central limit theorem is given in reference 43.

system by choosing a value for K. A K of 0.25 mrad gives a nominal 0.25 mrad system with a cutoff near 4 cy/mrad, and the figure then shows MTF's for σ's of 0.1, 0.125, and 0.15 mrad.

Table 3.5

MTF's for the Example System.

$\alpha = 1.0$ mrad, $f_o = 1$ cy/mrad

$D_o = 2$ cm, $f_c = 2$ cy/mrad at $\overline{\lambda} = 10$ μm

electronics characteristic frequency = 0.5 cy/mrad

$$\frac{\sigma_m}{\alpha} = .25$$

f cy/mrad	optics $\tilde{r}_o$	detector $\tilde{r}_d$	electronics $\tilde{r}_{elec}$	monitor $\tilde{r}_m$	system $\tilde{r}_s$	gaussian MTF with $\sigma = .55\alpha$
0	1	1	1	1	1	1
.1	.9634	.984	.981	.987	.918	.942
.2	.8729	.935	.928	.952	.721	.788
.3	.8097	.858	.857	.894	.532	.584
.4	.7471	.757	.781	.821	.363	.385
.5	.6850	.637	.707	.735	.227	.224
.6	.6238	.505	.640	.641	.129	.117
.7	.5636	.368	.583	.546	.066	.054
.8	.5046	.234	.530	.454	.028	.022
.9	.4470	.109	.486	.368	.009	.008
1.0	.3910	0	.447	.291	0	.003

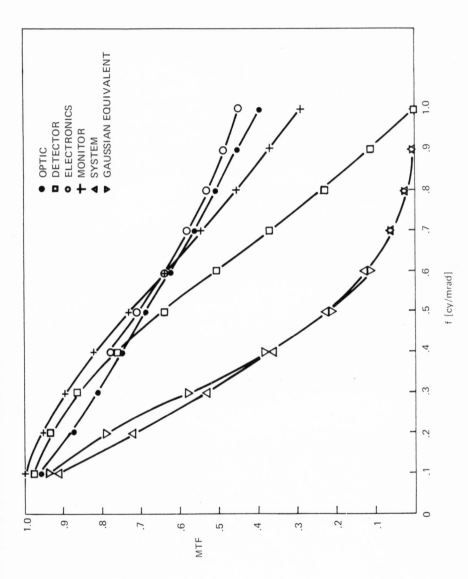

Figure 3.29 MTF's for the example system.

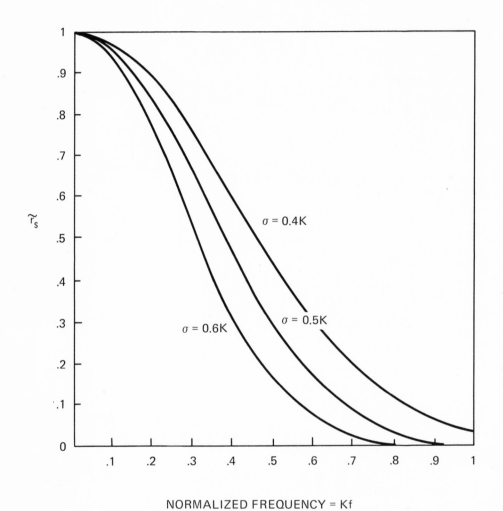

NORMALIZED FREQUENCY = Kf

Figure 3-30 Example gaussian MTF's.

REFERENCES

[1] J.W. Goodman, *Introduction to Fourier Optics*, McGraw-Hill, 1968.

[2] E.L. O'Neill, *Introduction to Statistical Optics*, Addison-Wesley, 1963.

[3] R.N. Bracewell, *The Fourier Transform and its Applications*, McGraw-Hill, 1965.

[4] P.M. Duffieux, *L'Integrale de Fourier et Ses Applications a L'Optique*, Societe Anonyme des Imprimeries Oberthur, 1946.

[5] O.H. Schade, "Electro-Optical Characteristics of Television Systems", RCA Review, *9*, in four parts: pp 5-37, March 1948; pp 245-286, June 1948; pp 490-530, September 1948; pp 653-686, December 1948.

[6] E.W.H. Selwyn, "The Photographic and Visual Resolving Power of Lenses", Photographic Journal, Section B, *88*, in two parts: pp 6-12, and pp 46-57.

[7] M. Françon, *Modern Applications of Physical Optics*, Wiley, 1963.

[8] E.H. Linfoot, *Fourier Methods in Optical Image Evaluation*, Focal Press, 1964.

[9] H.H. Hopkins, "On the Diffraction Theory of Optical Images", Proc. Roy. Soc., *217A*, pp 408-432, 1953; "The Frequency Response of a Defocussed Optical System", Proc. Roy. Soc., *231*, pp 91-103, July 1955.

[10] H.H. Hopkins, "The Frequency Response of Optical Systems", Proc. Phys. Soc., Section B, *69*, pp 562-576, May 1956; "Application of Frequency Response Techniques in Optics", *79*, pp 889-919, May 1962.

[11] M. De, "The Influence of Astigmatism on the Response Function of an Optical System", Proc. Roy. Soc., *233A*, pp 91-104, 1955.

[12] E.L. O'Neill, "Transfer Function for an Annular Aperture", JOSA, *46*, pp 285-288, April 1956.

[13] G. Black and E.H. Linfoot, "Spherical Aberration and the Information Content of Optical Images", Proc. Roy. Soc., *239A*, pp 522-540, 1957.

[14] A.M. Goodbody, "The Influence of Spherical Aberration on the Response Function of an Optical System", Proc. Phys. Soc., *72*, pp 411-422, September 1958.

[15] R. Barakat and M.V. Morello, "Computation of the Transfer Function of an Optical System from the Design Data for Rotationally Symmetric Aberrations", JOSA *52*, in two parts, pp 985-991 and pp 992-997, September 1962.

[16] B.P.Hildebrand, "Bounds on the Modulation Transfer Function of Optical Systems in Incoherent Illumination", JOSA, *56*, pp 12-13, January 1966.

[17] J.B. DeVelis and G. B. Parrent, Jr., "Transfer Function for Cascaded Optical Systems", JOSA, *57*, pp 1486-1490, December 1967.

[18] J.W. Coltman, "The Specification of Imaging Properties in Response to a Sine Wave Input", JOSA, *44*, pp 468-471, June 1954.

[19] P. Elias, D.S. Grey, and D.Z. Robinson, "Fourier Treatment of Optical Processes", JOSA, *42*, pp 127-134, February 1952.

[20] R.V. Shack, "Characteristics of an Image-Forming System", Journal of Research of the National Bureau of Standards, *56*, pp 245-260, May 1956.

[21] R.L. Lamberts, "Relationship between the Sine-Wave Response and the Distribution of Energy in the Optical Image of a Line", JOSA, *48*, pp 490-495, July 1958.

[22] R.L. Lamberts, "Application of Sine-Wave Techniques to Image-Forming Systems" JSMPTE, *71*, pp 635-640, September, 1962.

[23] R.L. Lamberts, G.C. Higgens, and R.N. Wolfe, "Measurement and Analysis of the Distribution of Energy in Optical Images", JOSA, *48*, pp 487-490, July 1958.

[24] F.H. Perrin, "Methods of Appraising Photographic Systems" JSMPTE, *69*, in two parts: pp 151-156, March 1960, and pp 239-249, April 1960.

[25] F.D. Smith, "Optical Image Evaluation and the Transfer Function", Appl. Opt., 2, pp 335-350, April 1963.

[26] O.H. Schade, "Modern Image Evaluation and Television (the Influence of Electronic Television on the Methods of Image Evaluation)", Appl. Opt., *3*. pp 17-21, January 1964.

[27] G.C. Higgins, "Methods for Engineering Photographic Systems", Appl. Opt. *3*, pp 1-10, January 1964.

[28] D.H. Kelly, "Spatial Frequency, Bandwidth, and Resolution", Appl. Opt., *4*, pp 435-437, April 1965.

[29] R.M. Scott, et al., a series of papers entitled "The Practical Applications of Modulation Transfer Functions", Phot. Sci. and Eng., *9*, pp 235-263, 1965.

[30] J.R. Jenness, Jr., and J.A. Ake, "The Averaging-Aperture Model of an Electro-Optical Scanning System", JSMPTE, pp 717-720, July 1968.

[31] L.M. Biberman and S. Nudelman, editors, *Photoelectronic Imaging Devices*, 2 Volumes, Plenum, 1971.

[32] H.E. Brown, F.A. Collins, and J.A. Hawkins, "Analysis of Optical and Electro-Optical Imaging Systems Using Modulation Transfer Functions", Memorandum DRL-TR-68-13, Defense Research Laboratory, the University of Texas at Austin, March 1968.

[33] N. Jensen, *Optical and Photographic Reconnaissance Systems*, Wiley, 1968.

[34] O.H. Schade, "A Method of Measuring the Optical Sine-Wave Spatial Spectrum of Television Image Display Devices", JSMPTE, *67*, pp 561-566, 1958.

[35] J.R. Jenness, Jr., W.A. Eliot, and J.A. Ake, "Intensity Ripple in a Raster Generated by a Gaussian Scanning Spot", JSMPTE, *76*, pp 549-550, June 1967.

[36] L.E. White, "Comparison of Various Types of Resolution Measurement", Westinghouse Product Engineering Memo ETD-6402, Westinghouse Electric Corporation, Electronic Tube Division, Elmira, New York, September 1964.

[37] M. Beran and G.B. Parrent, Jr., *The Theory of Partial Coherence*, Prentice Hall, 1964.

[38] V.K. Viswanathan, personal communication, Honeywell Radiation Center, Lexington, Massachusetts.

[39] M.V. Klein, *Optics*, Wiley, 1970.

[40] W.J. Smith, *Modern Optical Engineering*, McGraw-Hill, 1966.

[41] R. Barakat, Chapter 15 of *Handbook of Military Infrared Technology*, W.L.Wolfe, editor, USGPO, 1965.

[42] R.L. Sendall, personal communications, Xerox Electro-Optical Systems, Pasadena, Cal.

[43] A. Papoulis, *Systems and Transforms with Applications in Optics*, McGraw-Hill, 1968.

CHAPTER FOUR — VISUAL PSYCHOPHYSICS

4.1 Introduction

The problem to be addressed in this chapter is the following: how much can the electro-optical image deviate from "reality" before the human visual system becomes disoriented and cannot efficiently extract information from it? This question cannot be answered unambiguously today, but to approach the answer we will consider the following visual processes and effects:

- resolving power and spatial frequency response;
- perception of noise;
- image magnification;
- discrimination of signals from noise and/or background;
- disagreeability of effects such as too large or small field of view, too fast scene motion, raster, and flicker.

Some of these effects are well understood, but a frequent problem is that the understanding is usually in the terminology of the field of visual psychophysics, so that it must be translated into the imaging systems view-point. On the other hand, some of the significant and original research on vision has been and continues to be conducted by engineers seeking solutions to practical imaging problems. Before considering the specifics of visual psychophysics, a brief summary of general visual system features is in order.

4.2 General Characteristics of the Human Visual System

We will consider here only those characteristics of the eye which are important for thermal imaging system design. The reader interested in a broader treatment should consult Gregory[1,2], Cornsweet[3], Davson[4], Fry[5], or Schade[6]. The eye has a moderate quality field of view of approximately 30 degrees in elevation and 40 degrees in azimuth, a high quality circular field called the foveal field of about 9 degrees, and a best vision field of about 1 to 2 degrees in diameter. The 30 by 40 degree field is used in most visual tasks, and it is for this reason that sensors with 4:3 aspect ratio fields (such as commercial television) are adjudged more aesthetically pleasing than similar sensors with other aspect ratios[7]. Color vision extends over about 90 degrees of the central field, with color sensitivity decreasing as the retinal edge is approached. The very rim of the retina does not produce a conscious sensation of vision, and is used only for motion detection. A normal youthful eye can focus on objects as close as about 25 centimeters, but this minimum adaptation distance increases with age.

The luminance range within which the eye can usefully operate extends over about nine decades, and it is sensitive over about ten decades. The eye's sensitive range is divided into three somewhat indistinct luminance regions called the scotopic, the mesopic, and the photopic ranges. The scotopic range extends from the absolute visual luminance threshold of about 10^{-6} foot-Lambert (fL) up to about 10^{-4} fL; the mesopic range from around 10^{-4} fL to 1 fL; and the photopic from about 1 fL up to 10^4 fL. These three regions correspond roughly to the shift from rod or extrafoveal vision at extremely low levels, to a combination of rod and cone vision at twilight levels, to cone or foveal vision at the highest levels.

The resolution capabilities of the eye are describable several different ways. Perhaps the simplest measure of visual resolution is the visual acuity, the reciprocal of the smallest angular detail in arcminutes resolvable by the eye. The maximum value of visual acuity obtained depends on the definition of what constitutes the smallest resolvable angular detail, but nominally it has a value of one reciprocal arcminute. Visual acuity decreases as the position of the target moves away from the line of sight.

The resolution capabilities of the eye are controlled to some extent by the pupil diameter. When light levels are high and the eye does not need all of the available energy to function properly, the pupil is constricted. This blocks off-axis rays which produce aberrations, thereby improving resolution in bright light. At normal levels the pupil diameter is about 3 mm, and the range of variation with light level is about 2 mm to 8 mm.

It is important to understand how the visual system works so that subjective impressions of image quality may be translated into measurable objective correlates. There are four distinct subjective impressions of quality. These are the perceptions of sharpness, noisiness, contrast, and spurious image structure. Sharpness is the term used to describe the overall detail rendition of a system and generally describes the ability of an observer to extract information from quiet, high contrast, interference-free imagery. The terms resolution and resolving power are often used synonymously with sharpness, although they denote slightly different qualities. Noisiness refers to the degree of masking of signals by fixed-pattern or time-varying noise. Other common terms for image noise are granularity and "snow". Contrast describes the brightness of interesting details relative to the brightness of background details. Spurious image structure includes such artifacts as flicker due to perception of framing action, aliasing due to too low a spatial or temporal sampling rate, inter-line flicker in interlaced systems, line crawl, and optical distortion.

4.3 Spatial Frequency Response

One of the best understood and most investigated aspects of visual psychophysics is the sine wave response of the eye. For analytical convenience, it is desirable to treat vision as a linear process, thereby utilizing the power of linear filter theory and the concept of the optical transfer function. However, visual processes are not always linear, as in the case of the Mach phenomenon.* It is nevertheless useful to approximate the behavior as linear for low image contrasts, but to emphasize this approximation we will refer to the eye's sine wave response (SWR) rather than to an eye OTF. Sine wave response is defined whether an operation is linear or not, and in the non-linear case it is multi-valued depending on light level and other parameters.

The sine wave response is determined by at least eight processes:

- Diffraction at the pupil;
- Aberrations in the lens;
- Finite retinal receptor size;
- Defocussing;
- Ocular tremor;
- Retinal receptor interconnections;
- Optic nerve transmission;
- Interpretation in the brain.

The overall SWR of the eye may be measured by the technique of liminal (threshold) recognition of sine wave patterns presented to the eye, or by a suprathreshold technique of comparing the peaks and troughs of the perceived sine wave to adjustable calibrated sources. With the liminal method, the contrast at which a sine wave is resolvable with a specified level of confidence is plotted versus spatial frequency, as shown in Figure 4.1. The assumption is made that the variation with spatial frequency is due to spatial filtering and that the threshold contrast equals a minimum value C_{MIN} divided by an eye MTF. This MTF is obtained by dividing the curve by C_{MIN} and inverting the values, as shown in Figure 4.2

---

*The overshoot and undershoot perceived at the transition between light and dark at a sharp high contrast edge target.

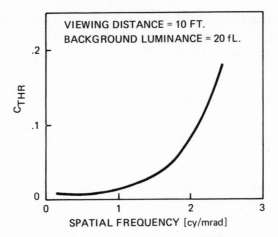

Figure 4.1 A measured sine wave threshold contrast curve,
adapted from reference 19.

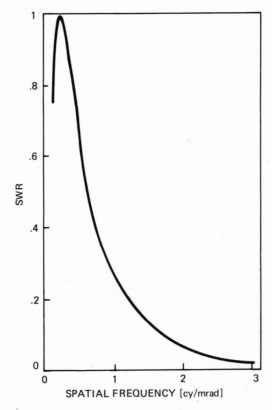

Figure 4.2 The calculated eye SWR for the contrast thresholds of Figure 4.1.

The problems with these kinds of measurements are fairly obvious. A threshold measurement is necessarily made at a low signal-to-noise ratio, so the visibility of sine waves is related not only to the signals in the eye-brain system, but also to the sources of noise in the imaging system and in the brain's decision-making processes. Also, the threshold results are not necessarily valid for high contrast imagery, whereas the suprathreshold measurements necessarily reflect the influence of the nonlinear Mach phenomenon. Even though the visual system is demonstrably nonlinear, nonstationary, and nondeterministic, the concept of an eye SWR is still very useful whenever we are able to use a small signal analysis.

The eye's sine wave response has been measured under a variety of conditions, and the following parameters have been found to have a significant effect on the SWR:

- Adaptation (mean background) luminance;
- Viewing distance at constant magnification;
- Target angular orientation;
- Target color;
- Exposure time;
- Temporal frequency of presentation.

The following six basic conclusions may be drawn about the eye SWR:

1. The observed spatial frequency response of the eye is not the result of a single spatial filter, but is rather the envelope of multiple narrowband tuned filters. Each of these filters is tuned to a different center frequency and is stochastically independent of the others. Each is independently selectable and the brain constructs an approximate optimum filter for a given detection task by appropriate selection of filters from the available filter ensemble.

2. This filter ensemble has an envelope of the general shape shown in Figure 4.3 for a single set of conditions. In general, the response is poor near dc, rises sharply to a maximum at low frequencies, and falls off in approximately gaussian fashion at high frequencies. Depending on the individual, the double peak shown in Figure 4.3 may be pronounced, barely noticeable, or absent altogether.

3. The spatial frequency at which the peak occurs shifts higher as the average adaptation luminance increases, and the overall SWR covers a broader frequency range, as shown in Figure 4.4. For the luminance range ordinarily encountered with an electro-optical display, the peak occurs in the neighborhood of 0.1 to 0.4 cy/mrad.

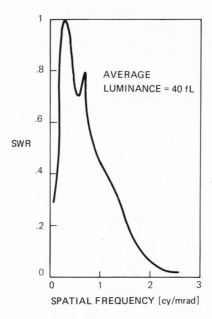

Figure 4.3 An eye sine wave response function showing double peaking behavior, measured using a CRT viewed at 39 inches, adapted from reference 22.

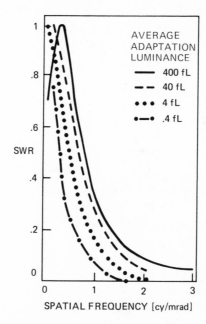

Figure 4.4 An eye sine wave response demonstrating the effect of average target luminance, measured using a CRT, adapted from reference 6.

4. At constant magnification, the peak frequency shifts higher in frequency as the accommodation distance increases. Also, the SWR broadens with increasing distance due to the lesser eye lens curvature and the rejection of off-axis rays.

5. The threshold contrast at which sine waves are detectable decreases as adaptation luminance and adaptation distance increase.

6. The contrast threshold increases as the angle that the sine wave chart makes with either the horizontal or the vertical approaches 45 degrees.

The first direct measurements of eye SWR were made by Schade[6] in 1956 using the threshold technique. Many other investigators have varied Schade's basic experiment and most of them verified Schade's general findings. The most significant deviation occurred when Campbell and Robson[8] compared the contrast thresholds for sine wave, square wave, rectangular, and sawtooth spatial frequency patterns, and concluded from a harmonic analysis that multiple narrowband tuned filters are present. Then Sachs, et al.[9], demonstrated more directly that the spatial frequency response of the eye represents not a single filter, but is rather the envelope or sum of several parallel channels, each of which is a narrowband tuned filter. Furthermore, they showed that the noise statistics of these channels are independent of each other. The measurements of Stromeyer and Julesz[10] also support this theory.

Most eye SWR experiments have been performed using the threshold method, but useful visual tasks are not always performed at threshold. Thus the research on suprathreshold SWR by Watanabe, et al.[11], is particularly interesting. They made suprathreshold (as well as threshold) measurements by the comparison method and observed that the higher above threshold the pattern becomes, the wider the frequency response is. They suggest that their SWR data are well fitted by the following formulas:

low frequency liminal region:

between $SWR = 1-[1 + (f/f_o)^2]^{-1/2}$ (4.1)

and $SWR = 1-[1 + (f/f_o)^2]^{-1}$; (4.2)

high frequency liminal region:

$SWR = [1 + (f/f_1)^2]^{-2}$; (4.3)

high frequency suprathreshold region:

$SWR = e^{-(f/f_2)^2}$; (4.4)

where f_0, f_1, and f_2 are characteristic frequencies.

Some good summaries of the eye's sine wave response characteristics are given by Fry[12], Fiorentini[13], Levi[14], Westheimer[15], and Ostrovskaya[16]. The latter reference includes results of some studies not available in English. Figure 4.5 shows the range of variation in the results of those experiments using natural pupils from the experiments reported in references 6 and 8 through 28.

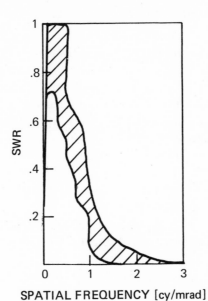

SPATIAL FREQUENCY [cy/mrad]

Figure 4.5 Range of variation of measured sine wave responses, adapted from references 6 and 8 through 28.

In general, the SWR as a function of light level seems to be explainable as follows: at very low levels, the pupil is wide and aberrations dominate; at moderate levels, such as 20 fL, the retina and brain dominate; and at very high levels with narrow pupils, diffraction dominates. The results of the following investigators are reasonably consistent with each other and are applicable to electro-optical imaging problems: Schade[6], Lowry and DePalma[19], Campbell[25], Gilbert and Fender[26], Pollehn and Roehrig[27], Kulikowski[28], Robson[12], and Watanabe, et al.[11]. Peaking behavior is not observed below about 5 fL and the highest resolved frequency reported is 3.44 cy/mrad.

Understanding the filtering properties of the visual system is important for two reasons. First, the eye itself approximates an optimum filter, so trying to improve the eye's performance by prefiltering the image spectrum is a questionable procedure. Second, the efficient coupling of the final image to the eye depends in large measure on the overall system magnification. Obviously, one should not present information at 4 cy/mrad to an eye limited to 2 cy/mrad or less. This coupling problem and the important choice of an efficient magnification are the subjects of the following discussion.

Schade[6] observed that the eye sine wave response is approximately gaussian above the peak frequency, so that

$$SWR \simeq \exp(-2\pi^2 \sigma_e^2 f^2) \tag{4.5}$$

where σ_e is the standard deviation of the equivalent line spread function as defined in Section 3.14. Consider the gaussian MTF's shown in Figure 4.6 for sigmas of 0.15, 0.2 and 0.25 mrad. If we compare these gaussian MTF's with the measured eye responses, we see that most of the results of interest are bounded by the $\sigma = 0.2$ and the $\sigma = 0.3$ mrad curves. Alternatively, one could use the form

$$SWR \simeq [1 + (f/f_1)^2]^{-2} \tag{4.6}$$

suggested by Watanabe, et al.[11] Some equations of this form are graphed in Figure 4.7.

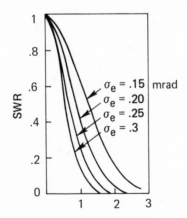

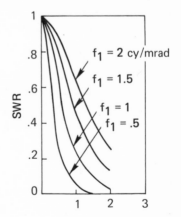

Figure 4.6 Sine wave responses given by

$$\exp[-2\pi^2 \sigma_e^2 f^2].$$

Figure 4.7 Sine wave responses given by

$$[1 + (f/f_1)^2]^{-2}$$

For analytical purposes, we shall assume following Schade that:

$$SWR \cong e^{-2\pi^2 \sigma_e^2 \; f^2} \quad \text{for } f \geqslant 0.2 \text{ cy/mrad}. \tag{4.7}$$

4.4 Magnification Effects

Schade suggested that the angular magnification M of a device be chosen so the frequency of greatest interest coincides with the peak frequency of SWR. This is an approximate way to match the eye and the system MTF's together to achieve the best overall MTF. We may carry this one step further by asking what the value of M is which optimizes the equivalent bandwidth N' of the eye-system combination. A useful system MTF, and one particularly amenable to analysis, is the gaussian form. The effect of increasing magnification is to widen the eye's effective SWR with respect to the system's MTF. If we let the system sigma be σ_s we may write

$$
\begin{aligned}
N' &= \int_{.2M}^{\infty} SWR^2(f) \, \tilde{r}_s^{\,2}(f) \, df \\
&= \int_{.2M}^{\infty} \exp \left\{ -4\pi^2 f^2 \left[\left(\frac{\sigma_e}{M} \right)^2 + \sigma_s^{\,2} \right] \right\} df.
\end{aligned} \tag{4.8}
$$

Here we are referencing all frequencies to object plane spatial frequencies, so that N' may be compared with the original system bandwidth N_s before the eye is included. By a variable of integration change, equation 4.8 becomes

$$N' = \frac{M}{2\pi[\sigma_e^{\,2} + M^2 \, \sigma_s^{\,2}]^{1/2}} \int_{0.4M[\sigma_e^{\,2} + M^2 \sigma_s^{\,2}]^{1/2}}^{\infty} e^{-x^2} \, dx \quad . \tag{4.9}$$

Normalizing to $N_s = \dfrac{1}{4\sigma_s \sqrt{\pi}}$,

$$\frac{N'}{N_s} = \frac{2M\sigma_s}{\sqrt{\pi} \, [\sigma_e^{\,2} + M^2 \, \sigma_s^{\,2}]^{1/2}} \int_{0.4M[\sigma_e^{\,2} + M^2 \sigma_s^{\,2}]^{1/2}}^{\infty} e^{-x^2} \, dx \quad . \tag{4.10}$$

The parameter N/N_s' is plotted as a function of $M\sigma_s$ on Figure 4.8 for $\sigma_e = 0.25$ mrad.

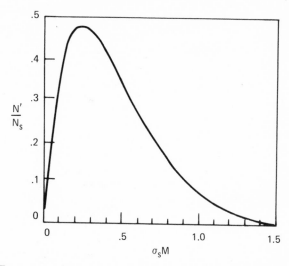

Figure 4.8 Eye-system equivalent bandwidth normalized to the system bandwidth and plotted as a function of the product of magnification and system sigma.

There are several conclusions which may be drawn from this function. First, the optimum magnification is $1/(4\sigma_s)$, there is broad optimal range, and the 3 dB-down points are separated by a magnification change of approximately 5X. Second, the N′ at $1/(4\sigma_s)$ is slightly less than half of the N_s of the system, $0.141/\sigma_s$. Third, it is clear that over- and under-magnification can result in severe loss of overall resolving power. As an example, we may consider a system with $\sigma_s = 0.125$ mrad, nominally a "0.25 mrad" system. The optimum magnification in this case is 2.0, and a range of 0.8 to 4.0 would probably give good results.

Bennett, et al.[29], in the course of photointerpreter performance studies, found that magnifying one resolution element beyond 3 arcminutes (0.873 mrad) at the eye resulted in degraded performance. If we interpret this loosely, we may say that the system cutoff frequency divided by M should not be less than about 1.1 cy/mrad. Thus, a 4 cy/mrad system would need at most a magnification of 3.6, which is consistent with the previous discussion.

4.5 Visual Sensitivity to the Spatial Frequency of Random Noise

Random noise interferes with the proper functioning of a thermal imaging system in two ways. First, random noise interferes with an observer's ability to detect, recognize, and identify targets on a display. Second, certain kinds of noise are so objectionable by themselves that an observer may not look at the display at all. Therefore it is important to know how noise is perceived.

The first investigators to report that the eye perceives different spatial frequency distributions of noise differently were Mertz[30] and Baldwin[31]. Baldwin subjectively measured the visibility of various spectral types of television noise, and Mertz deduced from these measurements the eye's noise weighting function, or sensitivity to the spatial frequency content of random noise. Baldwin observed the peculiar effect that perception of white noise with a sharp cutoff frequency f_c is independent of f_c, apparently because the masking effect of higher frequency noise offsets the increased noise amplitude due to the widened bandwidth. Coltman and Anderson[32] reverified this observation by measuring the detectability of noisy television patterns. They summarized their findings this way: "the masking effect of white noise depends only on the noise power per unit bandwidth, and is independent of the upper frequency limit of the noise spectrum, provided that this exceeds the frequency limit set by the eye." That frequency limit is approximately 0.3 cy/mrad.

Barstow and Christopher[33],[34] investigated noise visibility using a 6 by 8 inch television display viewed at 24 inches against a 0.0025 fL background with noise present in grey areas having luminances of from 0.3 to 0.6 fL. They found that noise power within any narrow spectrum is more objectionable than the same amount of power in a broader spectral band with the same center frequency. Also, low frequency narrow band noise is more objectionable than the same noise power in a higher frequency band.

Barstow and Christopher gave vivid descriptions of the appearances of various types of noise. They observed that very low-frequency noise appears streaky and annoying. When the frequency of the noise is increased the noise has the appearance of fine-grained photographic noise in random motion with the apparent grain size decreasing as the frequency increases. They also observed that narrow band noise resembles a herringbone pattern. By comparing the interference of narrow bands of noise, they found the eye's sensitivity to television noise for 1962 vintage television. These results are shown in Figure 4.9. Brainard, et al.[35] conducted similar experiments with 225 line, 30 fps imagery.

Brainard[36] conducted a noise weighting function measurement using a 60 fps 160-line TV with a 4.5 inch square picture viewed at 32 inches. Brainard detected a fine structure superimposed on the behavior observed by Mertz, by Barstow and Christopher, and by Brainard, et al. He found that the eye is more sensitive to narrowband noise centered on line frequency multiples than to noises between line frequency multiples. This effect will be observed in television-type scanning systems only. An interesting possible application of this to FLIR systems is the following. Noise at line-frequency multiples appears as a varying level from line to line which is approximately constant across one line. The same effect will result if the dc level or the ac

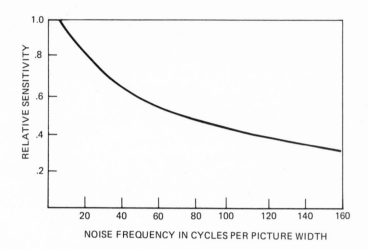

Figure 4.9 The noise perceptibility data of Barstow and Christopher, adapted from reference 34.

gain from line to line varies, as it tends to do in systems where each line is created by a different detector. The objectionability of such line-to-line noise is evident from Brainard's data showing increased eye sensitivity to these noises.

One of the most interesting and comprehensive studies of the effects of still noise was performed by Huang[37]. Huang generated three still monochrome pictures (a face, a cameraman, and a crowd), and added independent additive rectangular low-pass gaussian noise in both the horizontal and the vertical directions. By controlling the bandwidths in each direction independently, he was able to generate "isopreference curves" which indicate the various combinations of vertical and horizontal noise which give equivalent image degradation. Using 3.5 by 3.5 inch photographs with signal-to-noise ratios of 26, 22, and 18 dB (19.95, 12.59, 7.94 linear signal-to-noise ratios) and with horizontal and vertical noise bandwidths N_x and N_y ranging from 0.076 to 1.908 cy/mrad, observers rank-ordered the imagery under 30 foot-candle illumination. Figures 4.10 and 4.11 show the differences between the isopreference curves for a moderately detailed picture (a cameraman) and a highly detailed picture (a crowd). Huang found from data reduction that the noise weighting function has the same type of behavior as the sine wave response.

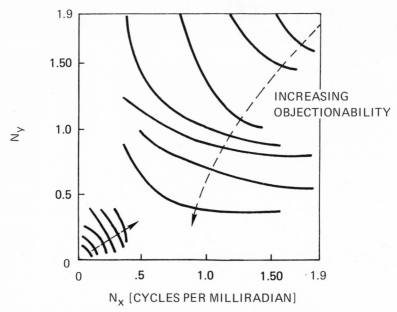

Figure 4.10 Noise isopreference curves for a moderately detailed
picture (a camerman), adapted from reference 37.

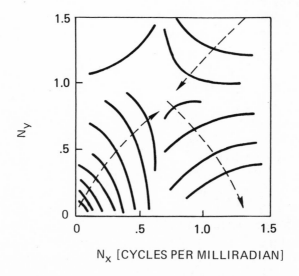

Figure 4.11 Noise isopreference curves for a highly detailed
picture (a crowd), adapted from reference 37.

Huang reached three basic conclusions:

1. Fixed power, square bandwidth noise is most objectionable when the bandwidth is approximately 0.70 cy/mrad, but is decreasingly objectionable as the bandwidth decreases or increases. Thus if the noise power is kept constant and bandwidth is increased equally in both directions, the objectionability increases, reaches a maximum at 0.7 cy/mrad, and then decreases.

2. Noise with horizontal streaks is less objectionable than noise with vertical streaks.*

3. Noises whose frequency content is similar to the signal frequency content of the picture are less objectionable than other noises.

4.6 Spatial and Temporal Integration by the Eye

It is well known that a large target can be perceived at a threshold luminance lower than the threshold for a smaller target, and that the threshold luminance of a simple target is inversely proportional to the square root of the target area within certain limits. This phenomenon may be interpreted as a spatial integration of signal and noise luminances within the target area, and is called spatial integration or areal summation. Using simple rectangular or square targets, Coltman and Anderson[32], Schade[38], and Rosell[39] have proven that for television imagery, the perceived signal-to-noise ratio is proportional to the square root of the ratio of the target area to the noise coherence area in the image. By perceived signal to noise ratio we mean the following. Since the point SNR does not change significantly as the target size is increased, the fact that the eye detects broader signals more easily suggests that the eye operates on a perceived SNR for the total target rather than the point SNR. In the case of a FLIR, noise can usually be assumed to be coherent over a dimension on the display equivalent to the detector angular subtense.

The limits of spatial integration were determined by Bagrash, et al.[40]. They used squares, circles, and compound shapes, and found that areal summation ceases for areas of approximately 11.6 mrad in diameter or greater. More interestingly, their results with compound stimuli strongly support the parallel tuned-filter approach to the eye SWR discussed earlier, suggesting that areal summation is actually a selective spatial frequency filtering effect.

---

*Since a raster is in effect a fixed-pattern streaky noise, this may be one of the reasons that the horizontal raster orientation of commercial TV is preferable to the alternative vertical raster.

Visual persistence, or the capacity of the eye to retain an image for a brief time, is a common experience. We are all familiar with the blurring of a rapidly moving object and with the persistence of a flash of light. Many experiments to determine the threshold of visibility of pulsed optical signals have indicated that the eye retains or stores a signal for a finite time, and sums signals slightly separated in time. Experiments with live noisy electro-optical imagery have verified the existence of this temporal integration property and further indicated that the eye does not perceive the instantaneous value of random noise but rather its root-mean-square value taken over a finite period. This behavior is critical to the satisfactory performance of a framing sensor, and we will devote this section to the available experimental results. Most of these experiments demonstrate that it is reasonable to associate an effective eye integration time T_e with the phenomenon of temporal integration.

An early expression of signal integration in time by the eye is the Blondel-Rey law. Glasford[7] expressed this by:

$$L = L_\infty \left(1 + \frac{0.21}{t_0}\right),$$

(4.11)

where L is the threshold luminance for a source exposed for t_0 seconds, and L_∞ is the threshold luminance for a source exposed for an infinite time. The value 0.21 may be interpreted as the summation time T_e.

In 1943, DeVries[41] reviewed existing data and concluded that a suitable eye summation or integration time T_e is 0.2 second. Rose[42] compared the visual impression of noise on a kinescope with photographs of kinescope noise taken with varying exposures. He concluded that a good match occurred for an exposure of 0.25 second. Coltman and Anderson[32] recommended a T_e of 0.2 second, and the data of Rosell and Willson[39] are consistent with this value.

Using the data of others, Schade[6] deduced that T_e is 0.2 at low luminances, and decreases at high luminances to 0.1 second for rod vision and 0.05 second for cone vision. Luxenberg and Kuehn[43] recommend 0.1 second for display applications. Blackwell[44] investigated visual thresholds for two noiseless pulses of 0.0025 second duration separated by 0.004 to 0.475 second. He found perfect summation of the pulses for separations of less than 0.02 second, with decreasing summation up to 0.5 second. Budrikis[45] analyzed the data of Graham and Margaria[46] who measured thresholds over a wide exposure range. Budrikis found that their data were fitted well by assuming an exponentially decaying impulse response e^{-t/t_0}, with $t_0 = 0.135$ second.

Tittarelli and Marriott[47] double-flashed a one arcminute disc with a flash duration of 1.4 millisecond and a flash separation of from 20 to 140 milliseconds. They measured the threshold of resolvability of the two flashes and concluded that for cone vision an optical disturbance decays with a half life of approximately 20 milliseconds. For pairs of flashes separated by more than 80 msec, no significant improvement in the threshold was noted, indicating that temporal integration is not effective on stimuli separated by more than 80 msec. One might be tempted to conclude from this that systems with frame rates of less than 12.5 fps (i.e., 1/(80 msec)) do not benefit from temporal integration. These results are somewhat at odds with other studies which demonstrated effective integration times of 0.1 to 0.2 second.

The conclusion which can be drawn from these studies is that in a fast-framing system, if the noise is uncorrelated in space from frame to frame while the signal remains constant, the perceived image resulting from the eye's combining of several frames is less noisy than a single-frame image. The author's experience is that measurements of observer responses to thermal imagery are consistent with a T_e of 0.2 second. That value is used in the rest of this text.

The concept of temporal integration is important when we consider the choice of a suitable information update rate, or frame rate, for a system. Since the eye appears to integrate signal and to rms uncorrelated noise within its integration time, if the noise power is constant from frame to frame and the image is stationary, an image signal-to-noise ratio improvement by the square root of the number of independent samples (frames) occurs. This effect allows us to consider trading off the observable flicker, the per-frame signal to noise ratio, and the frame rate in order to achieve a desired perceived signal to noise ratio with low flicker.

For example, if we contrive to make the rms displayed noise amplitude per frame at a point remain constant while we increase the frame rate $\dot{F}$, the perceived SNR improves as $\sqrt{\dot{F}T_e}$*. On the other hand, if the noise amplitude per frame is somehow proportional to $\sqrt{\dot{F}}$, as in a system where noise power per root cycle/milliradian is proportional to frame rate, increasing $\dot{F}$ does not change the perceived SNR and $\dot{F}$ should be increased until it exceeds the critical flicker frequency.

---

*Realistically, as the frame rate and the number of integrated independent samples approach infinity while the per frame noise remains constant, the contrast limited case is achieved where sensor noise is insignificant.

4.7 Perception of Framing Action

Some of the most objectionable image defects are those associated with image dissection by a framing raster scanner. Ideally, a framing system should take advantage of the temporal and areal summation of the eye to produce the subjective impression that the image formed is continuous in space and time. The worst defect occurs when the eye can sense that the image is discontinuous in time. This particular perception of the framing action is called the flicker phenomenon, or simply flicker.

4.7.1 Flicker

Flicker is a complex effect whose properties vary among observers and are strong functions of the display characteristics. Understanding flicker is important to visual psychophysicists because it is indicative of the nature of signal processing mechanisms of the brain. Flicker has therefore been widely investigated in a nonelectro-optical context solely for its psychophysical interest. The perception of flicker is apparently a probabilistic function with a very narrow transition from visibility to invisibility. As the frame rate of an image increases, the fluctuations become indiscriminable at a well-defined threshold frequency called the critical flicker-fusion frequency CFF.

Flicker must be suppressed in an imaging system because at best it is distracting and annoying, and at worst it may cause headaches, eye fatigue, and nausea. It is particularly important to avoid flicker rates in the alpha rhythm range of approximately 3 to 10 Hz because they may trigger convulsions, or cause photic driving (stimulation of sympathetic brain rhythms) with unknown consequences. Obviously, flicker is a serious defect which must not be allowed to compromise system and operator performance.

Flicker in an imaging display was first investigated by Engstrom[48], a television pioneer who varied the duty cycle, frame rate, and screen illumination of a rectangularly-pulsed blank display. This white light optical display was 12 × 16 inches and was viewed at 60 inches in an ambient of about 1 foot-candle. The display was blank. Engstrom[48] found that the CFF depends on the angular size of the flickering display, the adaptation luminance of the display, and the luminance variation waveform. The data shown in Figure 4.12 demonstrate that over a limited range, CFF is proportional to the logarithm of the peak luminance. Engstrom repeated the experiment on a kinescope with significant phosphor storage to demonstrate the effect of integration, and Schade[6] extrapolated Engstrom's data for other values of phosphor exponential decay.

The apparent luminance of a flickering field above the CFF is described by the Talbot-Plateau law discussed by Luxenberg and Kuehn[43] and by

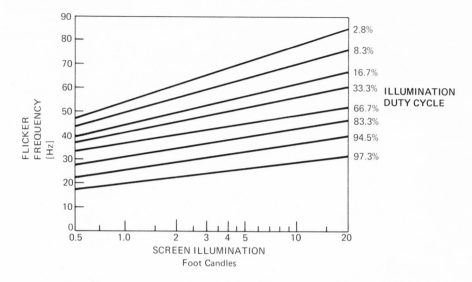

Figure 4.12 Flicker frequency as a function of screen illumination
and duty cycle, adapted from reference 48.

Davson[4]. If the equivalent perceived steady-state luminance is L', the displayed luminance as a function of time is $L(t)$, and the frame period is T_f, then the Talbot-Plateau law predicts:

$$L' = \frac{1}{T_f} \int_{0}^{T_f} L(t) \, dt. \tag{4.12}$$

That is, the apparent luminance equals the time-averaged luminance when the stimulus rate exceeds the CFF. According to Luxenberg and Kuehn, this law has never been found to be in error by more than 0.3 percent.

An important effect often overlooked in system design is the dependence of the CFF on the location in the visual field of the flickering stimulus. It has been observed that the CFF increases as the extrafoveal (off-axis) angle increases. Thus peripheral vision is more sensitive to flicker than foveal vision. It is not uncommon to see a system where the magnification has been chosen for eye-system resolution matching, independent of flicker considerations, in which flicker at the field edge is extremely annoying. The moral here is to run a large display at a high frame rate, or alternatively, to keep a flickering display small. An effect related to flicker perception is the phi phenomenon, wherein scene motion when viewed by a framing sensor appears to be smooth so long as flicker is not perceptible.

4.7.2 Interlace

Early in television research, Engstrom[48] showed that flicker at a given frame rate may be eliminated by a technique called two-to-one interlacing. In this scheme, a frame is divided into two sequential image components called fields, each of which is presented at a field rate of twice the frame rate. The first field contains only alternate raster lines, and the second field contains the remaining lines, as shown in Figure 4.13.

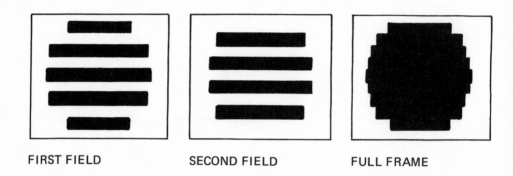

FIRST FIELD   SECOND FIELD   FULL FRAME

Figure 4.13 Interlace.

If a uniform area of the display is viewed at a distance such that the raster lines are unresolvable, the eye fuses the two frames to give a visual impression of spatial uniformity in that area. This technique doubles the effective frame rate so that it exceeds the CFF. However, if the lines are resolvable, that uniform area will appear to "shimmer" or "scintillate" or "vibrate vertically", depending on the viewer's perception. This effect is due to the perception of the half-frame-period phase difference between adjacent lines, and is called interline flicker. Engstrom demonstrated experimentally this correlation of interline flicker with line-resolvability which is the cause of a flicker and resolution trade-off problem. Engstrom also reported that blinking, rapid eye motion, or rapid head motion may cause the observer to perceive only one field during the change. This is a very annoying effect which becomes more pronounced as the interlace factor is increased. The subjective impression is that the picture is "banded".

If for some reason such as bandwidth conservation or TV-compatibility it is necessary to eliminate broad area flicker by interlacing rather than by increasing frame rate, a decision must be made. One can either magnify the image so that the full resolution capability of the system is used, and tolerate the interline flicker, or one can demagnify the image until interline flicker disappears, and sacrifice resolution. Another effect was noted by Schade[6], who observed that a finely-spaced, but pronounced, raster significantly reduces both noise visibility and scene detail visibility. When Schade eliminated the raster structure either by defocussing or by increasing the viewing distance, he found that the apparent noise level increased markedly but that details were more visible.

An excellent summary of raster defects in imagery presented at 30 frames and 60 fields per second with 2:1 interlace is given by Brown[49]. He noted that if all lines but one are optically masked, the unmasked line appears stationary and non-flickering and that if all but two adjacent lines are masked, the two lines appear to oscillate. When a complete picture is shown this interline flicker is perceived only in small areas of uniform luminance, and interline flicker is most noticeable at the boundaries of these areas.

Brown also described the phenomenon of line crawling which accompanies interlace, wherein the lines appear to crawl either up or down depending on which direction the individual's eye tends to track. Line crawling is observed when interline flicker is perceptible, and becomes more noticeable as either the picture luminance or the line center-to-center angular spacing increases. Other investigators have found that line crawling becomes pronounced to the point of extreme objectionability as the interlace ratio increases.

Brown also investigated the subjective effects of reducing video bandwidth in a TV system by introducing a 2:1 interlace, while keeping the frame rate constant. He compared noninterlaced pictures with from 189 to 135 lines per picture height in increments of $(2)^{1/9}$ with a 225 line interlaced picture to discover illuminance effects. He also compared noninterlaced 225 to 135 line pictures in $(2)^{1/4}$ increments with a 225 line interlaced picture for effects of noise and related parameters.

His most interesting conclusion was that the subjective effect of interline flicker to the observer is that of noise. This observation can easily be verified with a TV by looking closely at high signal-to-noise ratio images and noticing the apparent presence of vertical noise. This applies to highlight luminance values of from 40 to 100 fL.

It has often proved desirable from cost considerations to interlace by more than two to one, but the degrading effects are not well known. Inderhees[50] investigated flicker thresholds using an oscilloscope with a P1 phosphor to generate a rectangular 12 line raster with interlaces of 2:1, 3:1, and 4:1 (with scan patterns of 1-2-3-4 and 1-3-2-4). He used frame rates of 20, 25, 30, and 40 frames per second, and determined the interline flicker threshold as a function of display luminance and interlace pattern, as shown in Figure 4.14. Inderhees also observed the "line crawl" or "waterfall" effect noted by television researchers which occurs when the interlace is sequential in one direction. The eye then tends to perceive each line separately, resulting in an apparent rolling motion of the raster.

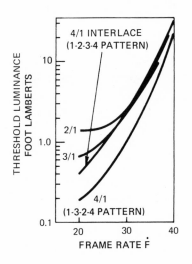

Figure 4.14 **Threshold of interline flicker as a function of interlace and frame rate, adapted from reference 50.**

4.7.3 Edge Flicker

Many thermal imaging system users have observed another framing effect related to simple flicker which occurs near the scan edges of a high duty-cycle bidirectional scanner. Points near either edge of the display are illuminated in rapid succession by the forward and reverse scans, and if the separation in time is small enough, those two events may be interpreted as a single event by the brain.

Consider the simple case of a rectilinear bidirectional scanner with scan dead time δ seconds per frame, frame period T_f, and display width W. Assume the dead time is equally divided on the right and left hand sides of the scan, and consider the time history of the luminance at the arbitrary display point ξ shown in Figure 4.15, where points (1) and (2) represent the display (but not the scan) extremes.

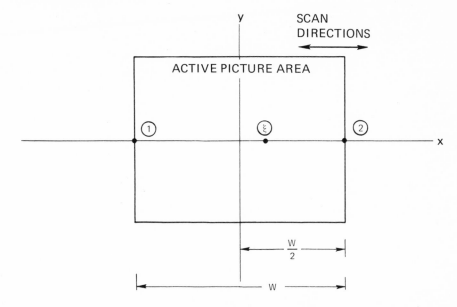

Figure 4.15 Notation for edge flicker.

Figure 4.16 shows the points in time at which (1), (2), and ξ are illuminated. If the duty cycle is high and the dead time low, (2) will be illuminated twice in rapid succession, and the eye may interpret these two closely-spaced events as a single event. The same effect occurs at (1) at the other display edge, and may also occur at the arbitrary point ξ if it is close enough to the display edge.

If two successive illuminations at ξ are fused, the reader may easily verify that the apparent frame rate $\dot{F}'$ will be lower than the time frame rate and is given at a point ξ by:

$$\dot{F}' = \frac{\dot{F}(W - \xi)}{2\left(\frac{W}{2}\right)} = \frac{\dot{F}\,(W - \xi)}{W}. \tag{4.13}$$

Every point ξ will not necessarily appear to be presented at a rate below the actual rate, but only those points near the edge. For example, for $\xi = W/2$ (i.e., at either edge), $\dot{F}' = \dot{F}/2$. This edge flicker effect is not well enough understood to enable prediction of its occurrence, but it is another factor which militates for higher frame rates and unidirectional scans.

4.8 Lines per Picture Height and Preferred Viewing Distance

Engstrom[48] simulated 24 frame-per-second noninterlaced television with 60, 120, 180, 240, and infinite lines per picture height. Using display

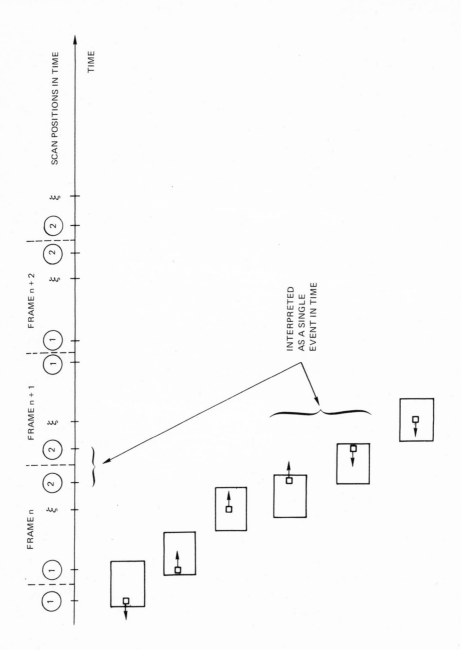

Figure 4.16 Sequence of events in edge flicker.

heights of 6, 12, and 24 inches and display illuminations of 5 to 6 foot candles, Engstrom determined the viewing distance preferred by viewers. He found that viewers tended to select the distance at which the raster is just noticeable, and that at closer distances viewers objected to the raster. The results are shown in Figures 4.17 and 4.18.

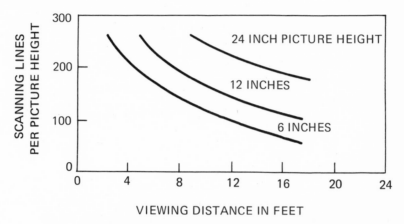

Figure 4.17 Visibility of scan lines as a function of picture height and viewing distance, adapted from reference 48.

The curve of Figure 4.18 shows that viewers spontaneously elect to view a display from a distance such that the scan lines are on 2 arcminute (0.58 mrad) centers. Engstrom also found that if rastered and nonrastered images were viewed at a distance such that both appeared to be identical, that distance exceeded the preferred viewing distance for the rastered image by approximately 50 percent.

In a later experiment, Thompson[51] ascertained preferred viewing distances for rastered television imagery and for imagery with the raster suppressed by spot wobble. He found that for 480 active scan lines displayed at 60 fields and 30 frames per second, the average preferred distance was such that the rastered picture subtended 7.6 degrees vertically and the non-rastered picture subtended 13.1 degrees vertically. This corresponds to approximately 0.28 mrad per line with raster and to 0.48 mrad per line without. The former number conflicts with the results of Engstrom, but the disagreement may be due to the larger number of experimental subjects used by Thompson. In this regard, the distribution with subjects of the Thompson data shown in Figure 4.19 is most revealing. One conclusion which may be drawn is that a system having a fixed magnification (or viewing distance with a monitor) will please only a small proportion of its viewers.

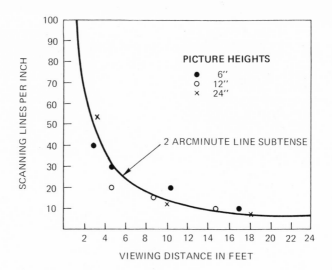

Figure 4.18 Resolvable number of scan lines per inch as a function of viewing distance for three picture heights, adapted from reference 48.

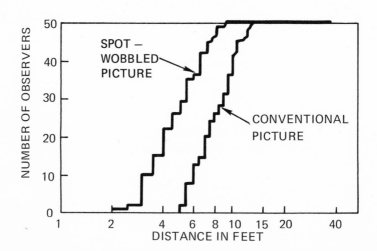

Figure 4.19 Number of observers who choose to sit closer than a particular distance for spot-wobbled and conventional pictures, adapted from reference 51.

For a particular class of scenes, it is obvious that reducing the lines per picture height indefinitely will eventually result in useless imagery. For the purposes of broadcast television, Engstrom concluded from his studies that story line and action could be followed satisfactorily when 240 lines are available. A line rate of 180 was marginal, 120 lines were barely acceptable, and 60 lines were totally inadequate. In most FLIR applications, there is no story line as such, but problems such as navigation and battlefield surveillance would seem to come close enough to the conditions considered by Engstrom so that his findings are significant to FLIR.

4.9 Surround Factor

A common experience among users of electro-optical imaging devices is that theoretically predicted performance is not achieved under field use conditions. A frequent cause of this performance loss is the strong influence by the luminance of the environment of the observer on information extraction from the display. The eye adapts to the average luminance to which it is exposed, so an attempt to get good eye performance by using a bright display will fail if the display surroundings are dark, because the eye will adapt to the average (lower) level where it doesn't perform as well.

Luxenberg and Kuehn[43] have summarized this problem by defining a surround factor F_s as the ratio of the luminance of the surround to the luminance of the working surface. For example, the surround might be the interior of a cockpit while the working surface is the face of an electro-optical display. They assert that the surround factor influences visual performance in the following ways:

1. A surround factor of less than 1.0 but greater than approximately 0.1 gives optimum vision.
2. An F_s of less than 0.1 gives poorer vision than condition (1).
3. An F_s of greater than 1.0 gives the worst vision.

The importance of surround factor has been demonstrated conclusively by Kiya[52] in a thermal imaging application. In this particular instance it was found that the target detection and recognition performance of observers using a FLIR was considerably improved when the FLIR display was surrounded by a uniform surface having approximately the same luminance as the display. This observation is important because it is typical in FLIR practice to have an observer view the imagery through a light-tight eyepiece or against a darkened background. Thus a critical consideration in specifying the luminance range of a thermal imaging system is knowledge of the ambient light level of the display environment so that an effective surround factor may be obtained.

4.10 Detection of Targets in Random Noise

The detection of simple targets embedded in random noise is a well understood visual process, and the target-masking effects of two basic types of noise have been investigated. These are the signal-independent additive gaussian noise typical of high background photon detectors, and the signal-dependent multiplicative poisson noise inherent in optical signals and typical of low background detectors. The discipline called fluctuation theory attempts to understand how the visual system detects optical signals embedded in gaussian or poisson noise, and to predict the results of detection experiments. Legault[53] gives a historical account of the development of fluctuation theory and summarizes the major contributions. The following sections treat the two types of noise separately, summarize the results of the critical experiments, and show how the results are interrelated.

4.10.1 Detection in Signal-Independent Noise

The probability of detection of various simple targets masked by additive white gaussian noise has been investigated by Coltman[34], Coltman and Anderson[32], Schade[38], and Rosell and Willson[39]. Coltman and Anderson generated square-waves which filled the screen of a television monitor, masked them by white gaussian noise, and determined detection thresholds as functions of spatial frequency. Schade determined the detectability of tribar targets in noisy still photographs. Rosell and Willson found the detection thresholds of rectangles embedded in noise on live television screens. Some of their experimental results are shown in Figure 4.20 for a particular eye time constant T_e; time information update rate $\dot{F}$; target displayed area A_t; display noise correlation area A_c (nominally one resolution element); noise voltage spectrum $g(f)$; and monitor MTF $\tilde{r}_m$. All of these experiments demonstrated conclusively that the probability of detection is a strong function of the signal-to-noise ratio when all other image quality parameters are held constant.

Coltman and Anderson, Schade, and Rosell and Willson generalized this relation to a broad class of conditions by applying the concepts of spatial and temporal integration. The effect of spatial integration is accounted for by assuming that the eye improves the image SNR by a factor $(A_t/A_c)^{1/2}$ and the improvement due to temporal integration is accounted for by a factor $(T_e\dot{F})^{1/2}$. Then we may define* a perceived signal-to-noise ratio SNR_p by:

$$SNR_p = SNR_i \, (T_e\dot{F} \, A_t/A_c)^{1/2} \qquad (4.14)$$

---

*Rosell's symbol for this quantity is "displayed" signal-to-noise ratio SNR_D. That symbol is not used here because the author feels that that notation suggests SNR_D is a quantity existing on the display, whereas it is a hypothetical signal-to-noise ratio assumed to be calculated by the visual system.

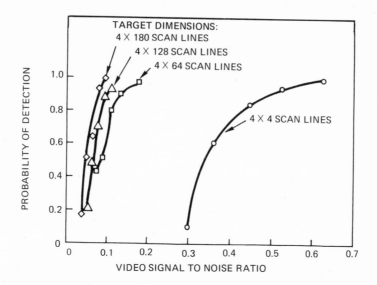

Figure 4.20 Probability of detection versus video signal-to-noise ratio for rectangular images, adapted from reference 39.

where SNR_i is the image point signal-to-noise ratio and

$$A_c = \beta[2 \int_0^\infty g^2 \, (f) \, \tilde{r}_m^2(f) \, df.]^{-1}. \tag{4.15}$$

Experiments of this type are now described by the universal curve shown in Figure 4.21 which relates the probability of detection to R_p. The normalized data of Figure 4.20 are shown superimposed on the theoretical curve in Figure 4.21.

This curve is the cumulative probability function of a gaussian probability density. Denoting the probability of detection as a function of SNR_p as P_{SNR_p} [Det],

$$P_{SNR_p} [Det] = \frac{1}{\sigma\sqrt{2\pi}} \int_{-\infty}^{SNR_p - \mu} e^{-\xi^2/2\sigma^2} d\xi. \tag{4.16}$$

The Rosell and Willson data yield a standard deviation σ of 1 and a mean μ of 3.2. A simple verbal statement of equation 4.16 is that P[Det] is the probability that the signal plus the instantaneous value of noise exceeds 1.6σ. Another interpretation of this cumulative probability function is that the probability of correctly interpreting a display disturbance as a signal

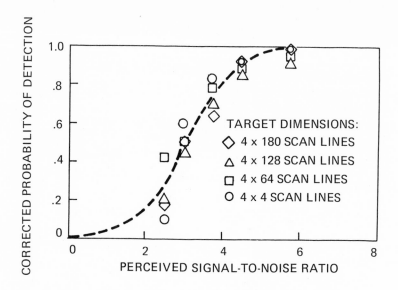

**Figure 4.21 Probability of detection versus SNR$_p$ required
for rectangular images, adapted from reference 39.**

equals the probability that the signal-to-noise ratio is greater than or equal to
a decision value of 3.2. Various equivalent expressions for this are:

$$P_{SNR_p}[Det] = \frac{1}{\sqrt{2\pi}} \int_{-\infty}^{SNR_p-3.2} e^{-\xi^2/2} \, d\xi \tag{4.17}$$

$$= \frac{1}{\sqrt{2\pi}} \int_{3.2-SNR_p}^{\infty} e^{-\xi^2/2} \, d\xi \tag{4.18}$$

$$= \frac{1}{\sqrt{2\pi}} \int_{-\infty}^{SNR_p} e^{-(\xi-3.2)^2/2} \, d\xi. \tag{4.19}$$

$$= \frac{1}{\sqrt{2\pi}} \int_{SNR_p-3.2}^{\infty} e^{-(\xi-SNR_p)^2/2} \, d\xi. \tag{4.20}$$

For $SNR_p > 1$, this function can be approximated by a simpler function,

$$P_{SNR_p}[Det] \cong 1 - \exp[-B(SNR_p-A)^2], \tag{4.21}$$

where many pairs of parameters A and B are acceptable, but most simply $A = 1$ and $B = .15$. For analytical purposes there is probably little loss in accuracy in using this approximate function, given the deviations of the experimental data from the idealized form.

The hypothesis suggested by this is that the visual system establishes a signal-to-noise ratio threshold as a reference to test for the significance of neural impulses. The effect is that low SNR optical signals are not detectable, but also that low level noise events are not mistaken for signals. Thus, we are not constantly mentally disturbed by fleeting impressions (false alarms) of objects which do not exist.

It must be noted, however, that the Rosell and Willson data represent the composite behavior of different observers. The change in the SNR between a low probability of detection condition and a high probability of detection condition for a particular individual may be smaller than the composite data would suggest, and the gaussian group behavior could be due to differences in individual thresholds and to changes in observer condition with time. The reader with access to a thermal imaging system may test his own thresholds by using a controllable thermal source to produce various SNR's.

4.10.2 Detection in Signal-Dependent Noise

A source of light which emits N photons/second exhibits a poisson distribution of emission rate with a mean value N and a standard deviation $\sqrt{N}$. Therefore no optical signal is noise free; there is always noise present due to the random emission rate of the source. Because the mean signal to rms noise ratio is $N/\sqrt{N} = \sqrt{N}$, such quantum noise is termed signal-dependent. Any optical detection system, including the eye, is ultimately limited by this inherent noisiness of optical signals[41,55,56].

Blackwell has reported numerous experiments which determined the eye's capability to extract simple signals from uniform backgrounds where the sources are free from other than quantum noise. These classic experiments[44,57-59] determined the limits of the eye's ability to detect simple targets embedded in uniform backgrounds. The targets were circular discs of variable size and luminance, and the backgrounds were large (10 to 60 degrees) uniformly illuminated fields of variable luminance. The experiments consisted of varying background (adaptation) luminance L_B, target presentation (exposure) time t, and target angular diameter θ. The observers' collective probability of detecting the target 50 percent of the time was determined. All of the data are reported in terms of threshold target contrast defined by $C = (L_T - L_B)/L_B$.

In the first of the 1946 experiments, observers were asked to choose among eight possible target positions for a target appearing at a known time. The background luminance varied between 10^{-6} and 120 fL, the target angle varied between 1.05 and 35.1 mrad, and the exposure time was six seconds. The results are graphed in Figure 4.22. Similar results were obtained for negative contrasts. In the second of the 1946 experiments, both the position and time of target presentation were known, and exposure was extended to not more than 15 seconds to permit the maximum possible temporal integration. Target angles were from 0.173 to 105 mrad, and luminance varied from 10^{-6} to 120 fL, with the results shown in Figure 4.23.

In the 1950 to 1952 experiments, observers were asked to distinguish the time of target presentation from among 4 intervals for targets of known size and position. The luminance varied from 0 to 100 fL, the targets from 0.233 to 14.9 mrad, and exposures from 10^{-3} to 1 second. In the 1956 to 1957 experiments, background luminances of 0.1 to 300 fL were used with target sizes of 0.29 to 16.98 mrad and exposure times of 1, 0.1, and 0.01 second. The results of these later tests are similar to the 1946 data. It is instructive to plot the 1950-52 data as shown in Figures 4.24 and 4.25. In Figure 4.24, the functional dependences of threshold target luminance difference on target angle for constant background clearly show that an areal summation mechanism is operative up to a size limit of about 10 milliradians. Figure 4.25 indicates that temporal integration still has a significant effect for exposures as long as 0.5 second.

Blackwell drew the important conclusion from his 1946 data that the probability of detection as a function of contrast has the form of the normal ogive given by

$$P_C[\text{Det}] = \frac{1}{\sqrt{2\pi}} \int_{-\infty}^{\left(\frac{C-C_{THR}}{\sigma}\right)} e^{-\xi^2/2} \, d\xi, \qquad (4.22)$$

where the standard deviation σ has the value 0.5 and C_{THR} is the measured threshold contrast.

Thus detection of signals embedded in signal dependent noise is describable as a random gaussian process with a mechanism similar to that observed for additive noise. Blackwell suggested that the average neural effect or excitation E of a stimulus L is a linear function of L. For the visual task of detecting a ΔL in a background L_B, Blackwell surmised that the visual

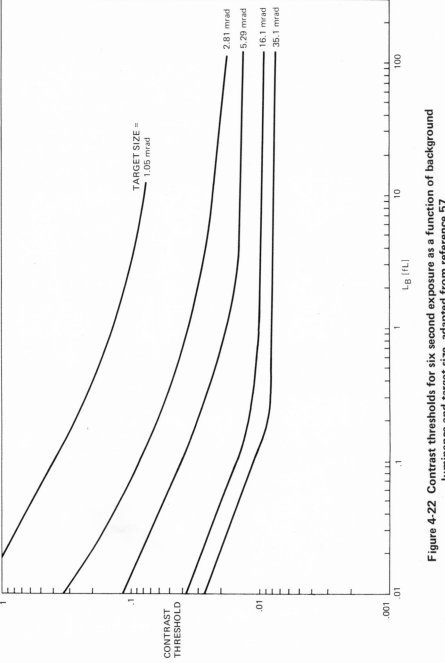

Figure 4-22 Contrast thresholds for six second exposure as a function of background luminance and target size, adapted from reference 57.

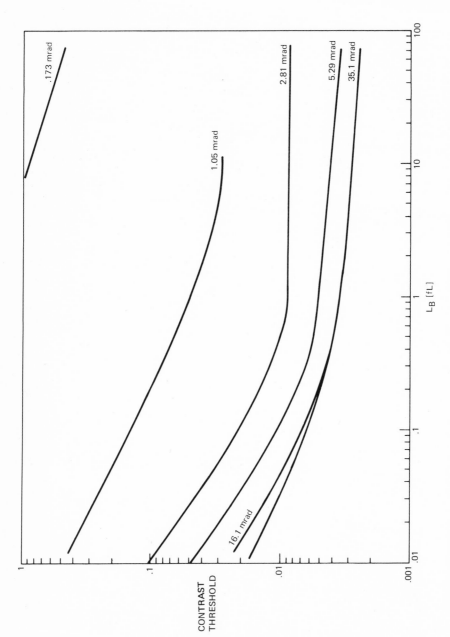

Figure 4-23 Contrast thresholds for maximum exposure (≤15 seconds) as a function of background luminance and target size, adapted from reference 57.

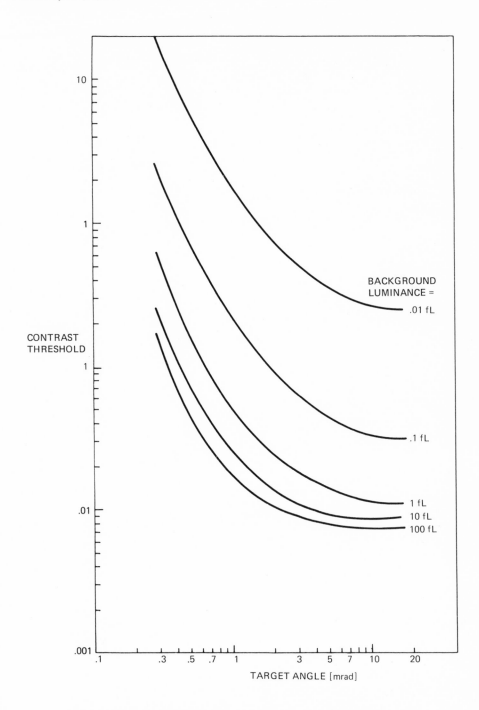

Figure 4.24 Contrast thresholds for one second exposure as a function of
target size and background luminance, adapted from reference 44.

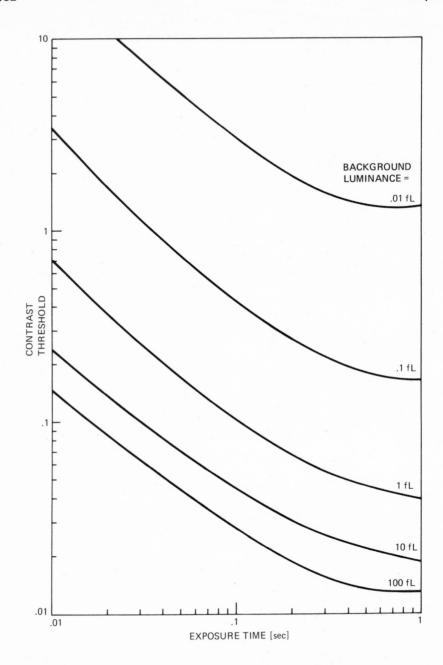

Figure 4.25 Contrast thresholds for a 1.16 mrad. target size as a function
of exposure time and background luminance, adapted from reference 44.

system examines the distribution of the E around the mean E_B corresponding to a background L_B, and establishes a comparison criterion E_c which is proportional to E_B. The criterion E_c is then used to test for instantaneous events E to see if they contain significant deviations from E_B. The assumption is that every $E = E_B + \Delta E$ is compared to E_c, and that the percentage of time E exceeds E_c is the probability that a decision will be made that a signal ΔL corresponding to ΔE is present. Another way of saying this is that ΔE must exceed $E_c - E_B$ before the detection probability exceeds 50 percent. This is the same mechanism postulated by Schade and by Rosell to explain their own experiments.

In the later paper, Blackwell[59] observed that curves of threshold ΔL as functions of L_B, target angle, and exposure time are fitted well by equations of the form

$$\frac{\Delta L}{L_B} = a_1 \left(1 + \frac{a_2}{L_B} \right), \tag{4.23}$$

where a_1 and a_2 are complicated functions of exposure and target size.

The following discussion is a condensation of a derivation by Ory[60] which offers more insight into the problem of visual detection. We discussed above how the experiments of Schade, of Rosell and Willson and of Blackwell support the hypothesis that the visual system discriminates between the presence or absence of an optical signal masked by gaussian noise in accordance with decision theory. Ory hypothesized that the probability of detection P[Det] (i.e., of correctly deciding that an optical signal is or is not present) is determined in a neural excitation space such that

$$P[Det] = \frac{1}{\sigma_T \sqrt{2\pi}} \int_K^\infty e^{-(E-E_T)^2/2\sigma_T^2} dE$$

$$= \frac{1}{\sqrt{2\pi}} \int_{-\infty}^{(E_T-K)/\sigma_T} e^{-\xi^2/2} d\xi , \tag{4.24}$$

where the E's are neural excitations. E or ξ is the variable of integration, E_T is the instantaneous total number of neural excitations consisting of E_B excitations corresponding to the background, and ΔE corresponding to the signal, $E_T = E_B + \Delta E$. The standard deviation of the variations (noise) around E_T is σ_T. This is the probability that the instantaneous value of total neural excitations E_T exceeds a threshold value K. The meaning of K is that the eye sets a threshold which if exceeded indicates the significance or signal nature of the excitations. This is a straightforward extension of the Schade and Blackwell hypotheses, with the exception that Ory postulated that the significance criterion depends on the illumination level in a more complex way which permits the eye to function over six orders of magnitude in luminance. A gaussian noise is used for all luminance levels by assuming that the photon flux is large enough to approximate the poisson-distributed photon noise with a gaussian distribution. Thus, both the quantum noise inherent in light and system-induced electronic noise may be treated as gaussian. For the purely quantum noise case,

$$\sigma_T = \sqrt{E_T} = \sqrt{E_B + \Delta E} . \qquad (4.25)$$

Ory assumes the threshold excitation to have the form $K = E_B + C$, where C depends on the noise and is given by

$$C = x + y\sigma_T + z\sigma_T^2 = x + y\sigma_T + z (E_B + \Delta E). \qquad (4.26)$$

Written this way, the upper limit of the second integral in equation 4.24 becomes

$$\frac{E_T - K}{\sigma_T} = \frac{\Delta E - C}{\sigma_T} . \qquad (4.27)$$

Now the first term in C in equation 4.26 indicates an absolute lower limit on the detectable signal. The second term indicates a threshold in the detectable signal component which proportional to the noise, so that detection is dependent on achievement of a specific signal to noise ratio. The third term indicates a contrast dependent threshold. Thus Ory's model predicts the three cases familiar from the literature:

1. An absolute lower threshold limit on signal detection;
2. Signal detection at moderate light levels which is a function of signal-to-noise ratio;
3. Contrast limited detection at high light levels.

Other analyses related to Blackwell's research and data are presented by Bailey[61], Overington and Lavin[62], and Kornfeld and Lawson[63]. They derive expressions which fit the Blackwell data and which interrelate threshold target luminance, the target solid angle, the viewing time, and the background luminance.

Blackwell's original tests[57] were conducted under idealized laboratory conditions, so the results are not necessarily applicable to practical contrast-limited target detection tasks. Blackwell reported results for practical tasks in a later paper[44], and DeVos, et al.[64] conducted experiments similar to Blackwell's original tests, but under less artificial conditions. They reported upper and lower limiting thresholds as compared to Blackwell's 50 percent threshold to indicate the contrasts required to perform practical tasks.

The breakpoints in fluctuation theory were loosely delineated by Rose[56], who observed that quantum noise variations tend to become un-noticeable in the neighborhood of 10^{-2} fL. At that luminance and above, a white surface appears smooth and noiseless. Rose asserted that statistical fluctuations are significant from 10^{-6} to 10^2 fL, but are relatively unimportant above 10^2 fL in the contrast limited regime.

4.11 The Subjective Impression of Image Sharpness and its Objective Correlates

The earliest measure of image sharpness was the two-point Rayleigh resolution criterion[65], the angular separation between two point sources at which the two points are just distinguishable under a specified set of conditions. Depending on the spread function shape, the two-point resolution separation can be the same as another early measure of resolution, the angular separation of the 50 percent intensity points of the impulse response. A somewhat more sophisticated definition of resolution which is related to the OTF is the "resolving power" or "limiting resolution". Limiting resolution is the maximum spatial frequency of an extended sine wave or square wave chart which can be resolved visually using a system under a specified set of conditions.

A widely used objective correlate of sharpness for photographic systems is "acutance", which under some circumstances correlates well with observer estimates of image sharpness. Acutance is defined[43] in terms of the response in photographic density units D to an object consisting of a step function. If the edge response of the photographic system is D(x), acutance A is defined as shown in Figure 4.26 by

$$A = \bar{G}_x{}^2 / D',$$
(4.28)

where

$$\bar{G}_x{}^2 = 1/N \sum_{i=1}^{\dot{N}} \left(\frac{\Delta D_i}{\Delta x_i}\right)^2.$$
(4.29)

O SADS   A M RL M00L
BOB KINSLEY/
            OSAR

PHOTOGRAPHIC
DENSITY

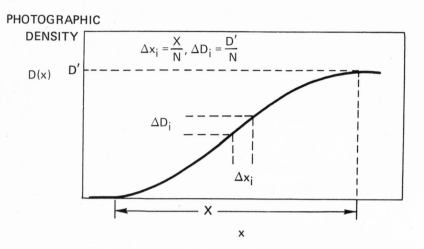

Figure 4.26 An edge response showing the notation
for the definition of acutance.

Acutance is therefore a measure of the steepness of the edge response. Roetling, et al.[66], have shown that acutance is proportional to N_e, so that two seemingly different descriptions of resolution are actually equivalent. Higgens[67] has demonstrated the differences between acutance and resolving power by generating imagery of bar charts and squares with high resolving power and low acutance in one case, and low resolving power and high acutance in another.

We introduced the concept of the optical transfer function (OTF) in Chapter Three. The correlation of image quality with the MTF has been investigated quite extensively, but the effects of the phase transfer function have been relatively neglected. Since the OTF is the Fourier transform of the system impulse response, it is evident that in a very general way, the narrower the LSF is, the better the MTF. Thus, resolution is in a gross sense indicative of the MTF. The limiting resolution under ideal conditions is indicative of the frequency at which the MTF drops to a low value, commonly taken to be between 2 percent and 5 percent modulation.

The MTF can exhibit virtually any kind of behavior within realizability constraints, but there are no experimental data which suggest that image quality is related to the shape of the MTF curve in a precise manner. However, experience yields some general guidelines concerning the types of MTF's shown in Figure 4.27. Schade[68] asserts that the approximately gaussian MTF of curve A is the most desirable for television-type imagery because it closely follows the form of the eye MTF and is adjudged to look natural. Further, Schade considers major departures from the gaussian form to give unnatural looking imagery. The MTF of curve B will cause "ringing" at sharp edges which may be objectionable. The MTF of curve C suppresses broad background features, emphasizes mid-frequency detail, sharpens the edge response, and in some cases may cause undershoot in the wake of small high intensity targets. Unusual behavior like that of curve D should be avoided because the subjective effects are impossible to assess in advance. Hopper[69] has observed that whenever a system's MTF falls below the curve of Figure 4.28, the imagery will be objectionable. The verbal response to the imagery will be that it looks "washed out", it "needs more gain", or "it needs more gray shades." The lesson is that N_e is useful chiefly for comparing systems whose MTF's have the same functional form.

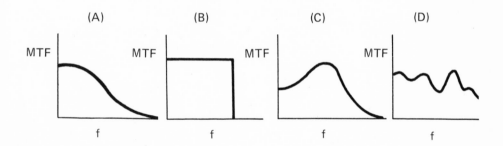

Figure 4.27 Some possible MTF shapes.

The most universally accepted objective correlate of sharpness is the quantity introduced in Section 3.14 called equivalent line number (N_e), equivalent signal bandwidth, and other names. Schade[70] first introduced N_e and recognized its significance. In his words, ". . .the integral of the squared sinewave response function of a system. . .is a measure indicating correctly the effective width of an edge transition, correlating well with the general subjective impression of image sharpness. . ." Further, Schade asserted[6] that "A barely detectable difference. . .of sharpness. . .corresponds to a change of 3.5 percent in equivalent bandwidth." This conclusion was based on the data of Baldwin[71], and of Schade. In equation form,

$$N_e = \int_0^\infty \tilde{r}_s^2 (f) \, df, \qquad (4.30)$$

where $\tilde{r}_s$ is the overall system MTF. Note that phase shift effects are not included.

The significance of N_e has been verified by the experiments of Scott[72]. Scott used four aerial photographs showing a ground area of 530 by 530 feet at a photographic scale of 1 to 3000, and degraded them with gaussian spread functions and photographic grain, both separately and simultaneously. Photo-interpreters then used unaided vision or magnifiers, as they desired, to rank the blurred images, the noisy images, and to rank and equate noisy, blurred images. Scott parametrized the MTF of each by the characteristic frequency f_c at which the MTF equals 0.61. This corresponds to a standard deviation for the spread function of $0.155/f_c$, and to an N_e of $0.21f_c$. His f_c's ranged from 107.7 cycles per picture width to 11.31 cycles per picture width.

Scott concluded that in grainless photographs the liminal increment in f_c for a single scene is approximately 5 percent, and that the threshold of f_c discrimination for grainless photographs of different scenes is about 10 percent. A second experiment found good correlation between photo-interpreter performance with still aerial imagery and the photo-interpreters' subjective rankings of the imagery. Elworth, et al.[73], have verified that N_e correlates better with interpreter performance than any other resolution measure. The N_e concept should not be applied without regard for the scene frequency content, however. Consider the two MTF's shown in Figure 4.29 where A and B have the same N_e. B will provide slightly better imagery than A for very low frequency objects, but A will be better than B for very high frequency objects.

We noted earlier that Schade used the data of Baldwin. Baldwin's experiments[71] are worth considering in more detail, because he determined the discriminability of spread function changes in continuously degraded motion picture films. Baldwin produced square spread functions with width "a" (corresponding to an N_e of $1/2a$) in a 14-degree by 19-degree field viewed at 30 inches. The N_e's ranged from 0.175 to 0.75 cy/mrad at the eye, corresponding to first zero frequencies of from 0.35 to 1.5 cycles/mrad. Baldwin's results are expressed in units of liminal discriminability of the resolution change. One liminal unit indicates a 50 percent probability of discriminability, and approximately 3 to 4 units indicates nearly 100 percent discriminability. Baldwin observed the dependence of discriminability on N_e shown in Figure 4.30.

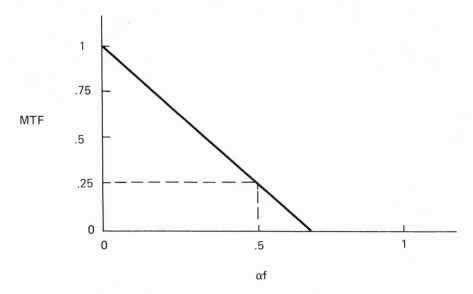

Figure 4.28 Minimum acceptable MTF, adapted from reference 73.

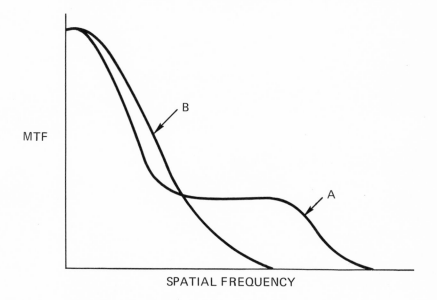

Figure 4.29 Two MTF's with the same equivalent bandwidths.

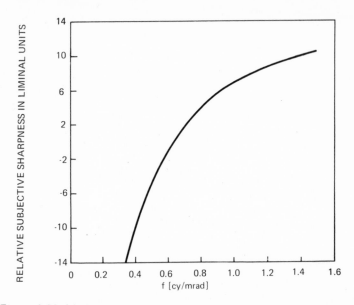

Figure 4.30 Liminal sharpness versus cutoff frequency for a square spread function, adapted from reference 71.

Interpreting his data, we find a range of delta-N_e/N_e for one liminal increment of 0.03 (at $1/a = .35$ cy/mrad) to 0.075 (at $1/a = 1.33$ cy/mrad). Thus, for a fixed magnification system, it becomes more difficult to notice resolution improvement as the baseline resolution increases. That is, at 1.33 cy/mrad (3/4 mrad system) it would take a 7.5 percent improvement to have a 50 percent probability of discriminating the difference, and around 20 percent improvement to have a positively noticeable difference. Baldwin also varied vertical and horizontal resolution using a rectangular spread function, and found a slight preference for better resolution horizontally than vertically, as shown in Figure 4.31.

4.12 Application of Visual Psychophysics to Thermal Imaging System Design

Five facts of critical importance were introduced in this chapter. These are that the eye:

- behaves like an ensemble of stochastically independent displaced and overlapped tuned filters which appear to approximate an optimum filter for particular visual tasks;
- integrates a signal extended in space and repeated in time;
- detects and recognizes objects in a probabilistic manner;
- has contrast-limited and noise-limited detection modes; and
- is sensitive to framing and interlacing.

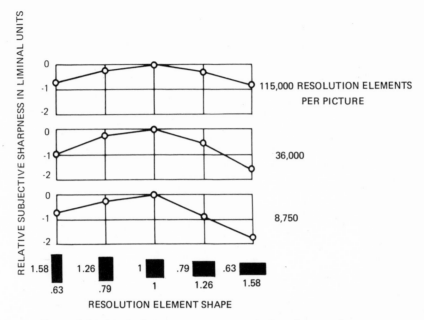

Figure 4.31 Liminal sharpness versus resolution element shape,
adapted from reference 71.

Each of these five facts has a corresponding action which must be taken in system design. These are:

- Assure that the system magnification and displayed MTF are adequate to bring the target spatial frequency spectrum well within the eye's bandpass.
- Allow for the beneficial SNR-enhancing effects of temporal and spatial integration in system design and performance specifications.
- Provide SNR sufficient to achieve desired performance on a probabilistic basis.
- Provide an appropriate surround factor and displayed luminance, dynamic range, and contrast to permit system-noise-limited performance.
- Eliminate perception of framing and interlacing.

When these five actions have been taken, system performance will be very close to optimal. It is not necessary to eliminate defects completely because the eye is tolerant of minor defects in a television-type picture. It is readily verifiable that the perception of noise and shading is reduced when picture detail increases, whereas perception of contour sharpness is improved. Thus, minor artifacts which would be objectionable in a featureless TV picture tend to be unnoticed in a detailed picture.

REFERENCES

[1] R.L. Gregory, *Eye and Brain: The Psychology of Seeing*, World University Library, Third Impression, 1967.

[2] R.L. Gregory, *The Intelligent Eye*, McGraw-Hill, 1970.

[3] T.N. Cornsweet, *Visual Perception*, Academic Press, 1970.

[4] H. Davson, editor, *The Eye*, four volumes, Academic, 1969.

[5] G.A. Fry, "The Eye and Vision", Chapter 1 of Volume 2 of *Applied Optics and Optical Engineering*, R. Kingslake, editor, Academic, 1965.

[6] O.H. Schade, Sr., "Optical and Photoelectronic Analog of the Eye", JOSA, *46*, pp 721-739, September 1956.

[7] G.M. Glasford, *Fundamentals of Television Engineering*, McGraw-Hill, 1955.

[8] F.W. Campbell and J.G. Robson, "Application of Fourier Analysis to the Visibility of Gratings", Journal of Physiology (Great Britain), *197*, pp 551-566, 1968.

[9] M.B. Sachs, J. Nachmias, and J.G. Robson, "Spatial-Frequency Channels in Human Vision", JOSA, *61*, pp 1176-1186, September 1971.

[10] C.F. Stromeyer III and B. Julesz, "Spatial-Frequency Masking in Vision: Critical Bands and the Spread of Masking", JOSA, *62*, pp 1221-1232, October 1972.

[11] A. Watanabe, T. Mori, S. Nagata, and K. Hiwatashi, "Spatial Sine-Wave Responses of the Human Visual System", Vision Research, *8*, pp 1245-1263, 1968.

[12] G.A. Fry, "The Optical Performance of the Human Eye", *Progress In Optics*, Volume 8, E. Wolf, editor, American Elsevier Publishing Company, 1970.

[13] A. Fiorentini, "Dynamic Characteristics of Visual Processes" in *Progress in Optics*, Volume 1, E. Wolf, editor, North Holland Publishing Company, 1965.

[14] L. Levi, "Vision in Communication" in *Progress in Optics,* Volume 8, E. Wolf, editor, American Elsevier Publishing Company, 1970.

[15] G. Westheimer, "Image Quality in the Human Eye", Optica Acta, *17*, pp 641-658, September 1970.

[16] M.A. Ostrovskaya, "The Modulation Transfer Function (MTF) of the Eye", Soviet Journal of Optical Technology, *36*, pp 132-142 (English Translation), January 1969.

[17] J.G. Robson, "Spatial and Temporal Contrast Sensitivity Functions of the Visual System", JOSA, *56*, pp 1141-1142, August 1966.

[18] E.M. Lowry and J.J. DePalma, "Sine-Wave Response of the Visual System. I. The Mach Phenomenon", JOSA, *51*, pp 740-746, July 1961.

[19] E.M. Lowry and J.J. DePalma, "Sine-Wave Response of the Visual System. II. Sine-Wave and Square-Wave Contrast Sensitivity", JOSA, *52*, pp 328-335, March 1962.

[20] O. Bryngdahl, "Characteristics of the Visual System: Psychophysical Measurements of the Response to Spatial Sine-Wave Stimuli in the Mesopic Region", JOSA, *54*, pp 1152-1160, September 1964.

[21] H.A. W. Schober and R. Hilz, "Contrast Sensitivity of the Human Eye for Square-Wave Gratings", JOSA, *55*, pp 1086-1091, September 1965.

[22] A.S. Patel, "Spatial Resolution by the Human Visual System. The Effect of Mean Retinal Illuminance", JOSA, *56*, pp 689-694, May 1966.

[23] F.L. Van Nes and M.A. Bouman, "Spatial Modulation Transfer in The Human Eye", JOSA, *57*, pp 401-406, March 1967.

[24] F.L. Van Nes, J.J. Koenderink, H. Nas, and M.A. Bouman, JOSA, *57*, pp 1082-1088, September 1967.

[25] F.W. Campbell, "The Human Eye as an Optical Filter", Proc. IEEE, *56*, pp 1009-1014, June 1968.

[26] D.S. Gilbert and D.H. Fender, "Contrast Thresholds Measured with Stabilized and Non-Stabilized Sine-Wave Gratings", Optica Acta, *16*, pp 191-204, March 1969.

[27] H. Pollehn and H. Roehrig, "Effect of Noise on the Modulation Transfer Function of the Visual Channel", JOSA, *60*, pp 842-848, June 1970.

[28] J.J. Kulikowski, "Some Stimulus Parameters Affecting Spatial And Temporal Resolution of Human Vision", Vision Research, *11*, pp 83-93, January 1971.

[29] C.A. Bennett, S.H. Winterstein, and R.E. Kent, "Image Quality and Target Recognition", Human Factors, *9*, pp 5-32, February 1967.

[30] P. Mertz, "Perception of Television Random Noise", JSMPTE, *54*, pp 9-34, January 1950.

[31] M.W. Baldwin, unpublished, reported in Reference 30.

[32] J.N. Coltman and A.E. Anderson, "Noise Limitations to Resolving Power in Electronic Imaging", Proc. IRE, *48*, pp 858-865, May 1960.

[33] J.M. Barstow and H.N. Christopher, "The Measurement of Random Monochrome Video Interference", Trans. of the AIEE, *72*, Part I – Communication and Electronics, pp 735-741, January 1954.

[34] J.M. Barstow and H.N. Christopher, "The Measurement of Random Video Interference to Monochrome and Color Television Pictures", AIEE Transactions on Communication and Electronics, *63*, pp 313-320, November 1962.

[35] R.C. Brainard, F.W. Kammerer, and E.G. Kimme, "Estimation of the Subjective Effects of Noise in Low-Resolution Television Systems", IRE Trans. on Info. Thry., *IT-8*, pp 99-106, February 1962.

[36] R.C. Brainard, "Low-Resolution TV: Subjective Effects of Noise added to a Signal", BSTJ, *46*, pp 223-260, January 1967.

[37] T.S. Huang, "The Subjective Effect of Two-Dimensional Pictorial Noise", IEEE Trans. Info. Thry., *IT-11*, pp 43-53, January 1964.

[38] O.H. Schade, Sr., "An Evaluation of Photographic Image Quality and Resolving Power", JSMPTE, *73*, pp 81-120, February 1964.

[39] F.A. Rosell and R.H. Willson, "Recent Psychophysical Experiments and the Display Signal-to-Noise Ratio Concept", Chapter 5 of *Perception of Displayed Information*, L.M. Biberman, editor, Plenum, 1973.

[40] F.M. Bagrash, L.G. Kerr, and J.P. Thomas, "Patterns of Spatial Integration in the Detection of Compound Visual Stimuli", Vision Research, *11*, pp 625-634, July 1971.

[41] H.L. DeVries, "The Quantum Character of Light and its Bearing Upon the Threshold of Vision, the Differential Sensitivity, and Visual Acuity of the Eye", Physica, *10*, pp 553-564, July 1943.

[42] A. Rose, "The Sensitivity Performance of the Human Eye on an Absolute Scale", JOSA, *38*, pp 196-208, February 1948.

[43] H.R. Luxenberg and R.A. Kuehn, *Display Systems Engineering*, McGraw-Hill, 1968.

[44] H.R. Blackwell, "Development and Use of a Quantitive Method for Specification of Interior Illumination Levels", Illuminating Engineering, *54*, pp 317-353, June 1959.

[45] Z.L. Budrikis, "Visual Thresholds and the Visibility of Random Noise in TV", Proc. IRE (Australia), pp 751-759, December 1961.

[46] C.H. Graham and R. Margaria, "Area and the Intensity – Time Relation in the Peripheral Retina", Am. J. Physiology, *113*, pp 299-305, 1935.

[47] R. Tittarelli and F.H.C. Marriott, "Temporal Summation in Foveal Vision", Vision Research, *10*, pp 1477-1480, December 1970.

[48] E.W. Engstrom, "A Study of Television Image Characteristics", Proc. IRE, Part I, *21*, pp 1631-1651, December 1933; Part II, *23*, pp 295-310, April, 1935.

[49] E.F. Brown, "Low Resolution TV: Subjective Comparision of Interlaced and Non-Interlaced Pictures", BSTJ, *66*, pp 119-132, January 1967.

[50] J.A. Inderhees, personal communication, Cincinnati Electronics Corporation, Cincinnati, Ohio.

[51] F.T. Thompson, "Television Line Structure Suppression", JSMPTE, *66*, pp 602-606, October 1952.

[52] M. Kiya, personal communication, United States Air Force Space and Missile Systems Organization, Los Angeles, California.

[53] R.R. Legault, "Man – The Final Stage of an Electro-Optical Imaging System", IEEE EASCON 1969 Convention Record, pp-16-29.

[54] J.W. Coltman, "Scintillation Limitations to Resolving Power in Imaging Devices", JOSA, *44*, pp 234-237, March 1954.

[55] A. Rose, "The Relative Sensitivities of Television Pickup Tubes, Photographic Film, and the Human Eye", Proc. IRE, *30*, pp 293-300, 1942.

[56] A. Rose, "A Unified Approach to the Performance of Photographic Film, Television Pickup Tubes, and the Human Eye", JSMPTE, *47*, pp 273-295, October 1946.

[57] H.R. Blackwell, "Contrast Thresholds of the Human Eye", JOSA, *36*, pp 624-643, November 1946.

[58] H.R. Blackwell, "Studies of the Form of Visual Threshold Data", JOSA, *43*, pp 456-463, June 1953.

[59] H.R. Blackwell, "Neural Theories of Simple Visual Discrimination", JOSA, *53*, pp 129-160, January 1963.

[60] H.A. Ory, "Statistical Detection Theory of Threshold Visual Performance", Rand Corporation Memorandum RM-5992-PR, Santa Monica, Ca., September 1969.

[61] H.H. Bailey, "Target Detection Through Visual Recognition: A Quantitative Model", Rand Corporation Memorandum RM-6158-PR, Santa Monica, Ca., February 1970.

[62] I. Overington and E.P. Lavin, "A Model of Threshold Detection Performance for the Central Fovea", Optica Acta, *18*, pp 341-357, May 1971.

[63] G.H. Kornfeld and W.R. Lawson, "Visual Perception Models", JOSA, *61*, pp 811-820, June 1971.

[64] J.J. DeVos, A. Lazet, and M.A. Bouman, "Visual Contrast Thresholds in Practical Problems", JOSA, *46*, pp 1065-1068, December 1956.

[65] For a discussion of the Rayleigh resolution criterion, see pp 13-14 of *Perception of Displayed Information*, L.M. Biberman, editor, Plenum, 1973.

[66] P.G. Roetling, E.A. Trabka, and R.E. Kinzly, "Theoretical Prediction of Image Quality", JOSA, *58*, pp 342-346, March 1968.

[67] G.C. Higgens, "Methods for Engineering Photographic Systems", Appl. Opt., *3*, pp 1-10, January 1964.

[68] O.H. Schade, Sr., "Modern Image Evaluation and Television, (The Influence of Electronic Television on the Methods of Image Evaluation)", Appl. Opt., *3*, pp 17-21, January 1964.

[69] G.S. Hopper, Personal Communication, Texas Instruments, Inc., Dallas, Texas.

[70] O.H. Schade, Sr., "Electro-Optical Characteristics of Television Systems", RCA Rev., *9*, pp 5-37, March 1948.

[71] M.W. Baldwin, Jr., "The Subjective Sharpness of Simulated Television Images", BSTJ, *19*, pp 563-587, October 1940.

[72] Frank Scott, "The Search for a Summary Measure of Image Quality — A Progress Report", Phot. Sci. and Eng., *13*, pp 154-164, May-June 1968.

[73] From a Boeing Corporation report summarized in Chapter Three of *Perception of Displayed Information*, L.M. Biberman, editor, Plenum, 1973.

CHAPTER FIVE – PERFORMANCE SUMMARY MEASURES

5.1 Introduction

In previous chapters, we discussed such fundamental system properties as the optical transfer function, the image signal-to-noise ratio, the frame rate, and the magnification. Each of these properties describes only a small part of a system's ability to perform practical tasks. They are useful ways to characterize a design, but other measures which combine them in ways appropriate to particular tasks are needed in order to conduct performance tradeoffs. The ideal such summary performance measure must include the observer's limitations, must be predictable from fundamental system parameters, must be easily measurable, and must relate to the performance of the system as it is ultimately intended to be used. In this chapter we will consider some commonly used summary measures and attempt to ascertain how they are useful and how they may be misleading.

5.2 Noise Equivalent Temperature Difference

The oldest and most widely used (and misused) measure of the ability of a system to discriminate small signals in noise is the "noise-equivalent temperature difference" or "noise-equivalent differential temperature". Several different symbols for this quantity are in current use, and we shall use the most common of these, NETD. Similarly, several different operational definitions for NETD exist. The simplest and most commonly used definition follows: the NETD is the blackbody target-to-background temperature difference in a standard test pattern which produces a peak-signal to rms-noise ratio (SNR) of one at the output of a reference electronic filter when the system views the test pattern.

The intent thus is to find the ΔT which produces a unity SNR in a specified filter under normal conditions when the scanner is operating. The

NETD test pattern is shown in Figure 5.1, where the dimension W is several times the detector angular subtense to assure good signal response. For standardization and comparison purposes, it is desirable to use an external measurement filter such that the product of all electronics MTF's including the external filter gives a reference MTF of:

$$\tilde{r}_R = [1 + (f/f_R)^2]^{-1/2},$$ (5.1)

where

$$f_R = \frac{1}{2\tau_d}$$ (5.2)

and

τ_d = detector dwelltime.

For analytical purposes all electronics MTF's are then attributed to this filter.

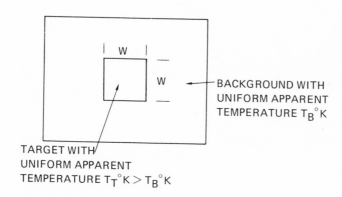

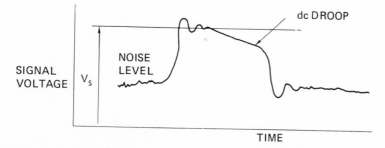

Figure 5.1 The NETD test pattern and the resulting voltage waveform.

The rms noise voltage V_n is measured using a true rms voltmeter which operates during the active scan time. The signal voltage V_s typically is determined from an oscilloscope trace of the voltage waveform corresponding to the target, as shown in Figure 5.1. To achieve good results, the target $\Delta T = T_T - T_B$ should be at least several times the expected NETD to assure that $V_s \gg V_n$. The NETD is then calculated by:

$$NETD = \frac{\Delta T}{V_s/V_n} \cdot \qquad\qquad (5.3)$$

5.3 NETD Derivation

Weihe[1], Hawkins[2], Hudson[3], and Soboleva[4] present NETD derivations using parameters each of those authors finds useful. The NETD derivation which follows is one which is simple and which the author finds useful in practice. The NETD is most conveniently derived for the case where the electronic processing of the detector signals has a flat frequency response within the system bandwidth, and where all electrical bandwidth limitations are due to the external measurement filter. This allows system video output signal and noise to be related directly to detector parameters. The resultant equation is easily modified for more complicated cases by adjusting the equivalent noise bandwidth term which appears.

This derivation makes the following assumptions:

- The detector responsivity is uniform over the detector's rectangular sensitive area.
- The detector D* is independent of other parameters in the NETD equation.[†]
- Atmospheric transmission losses between the target and the sensor are negligible.
- The target and background are blackbodies.
- The detector angular subtense, the target angular subtense, and the inverse of the collecting optic focal ratio can be approximated as small angles.
- The electronic processing introduces no noise.

[†]This is assumed so that the NETD equation will be simple to permit simple tradeoffs. If the detector is background limited, this assumption may be violated and the equations of Section 5.4 must be used.

The following quantities appear in the NETD equation:

a,b   = detector dimensions [cm]

α,β = detector angular subtenses [radian]

T_B = background temperature [°K]

$D^*(\lambda)$ = specific detectivity as a function of wavelength, evaluated at the electrical frequency at which the noise voltage spectrum of the detector is normalized to be unity, as shown in Figure 5.2 [cm Hz$^{1/2}$/watt]

g(f)   = normalized electrical noise voltage spectrum of the detector [dimensionless]

A_o = effective collection area of the infrared optics, including obscuration [cm^2]

$\tau_o(\lambda)$ = infrared optical transmission as a function of λ [dimensionless]

f   = infrared optics effective focal length [cm]

Δf_R = equivalent noise bandwidth [Hz] of the NETD test reference filter with the spectrum g(f) as a source:

$$\Delta f_R = \int_0^\infty g^2(f)\tilde{\tau}_R^2(f)\,df. \tag{5.4}$$

Consider the stylized view of a FLIR scanning the NETD target shown in Figure 5.3. This target is assumed to be a diffuse blackbody radiator obeying Lambert's cosine law. The target has a spectral radiant emittance W_λ [watts/cm^2 μm], and a spectral radiance

$$N_\lambda = W_\lambda/\pi \left[\frac{watts}{cm^2\,\mu m\,sr}\right]. \tag{5.5}$$

The optical aperture subtends a solid angle of A_o/R^2 measured from the target, so that the spectral irradiance on the aperture is $W_\lambda A_o/\pi R^2$ [watts/cm^2 μm]. At any instant in time while the detector is scanning the target, the detector receives radiation from a target area of $\alpha\beta R^2$, so the spectral irradiant power received by the detector as modified by the lens transmission $\tau_o(\lambda)$ is:

$$P_\lambda = \frac{W_\lambda}{\pi}\frac{A_o}{R^2}\alpha\beta R^2\,\tau_o(\lambda)$$

$$= \frac{W_\lambda}{\pi} A_o\,\alpha\beta\,\tau_o(\lambda)\left[\frac{watt}{\mu m}\right]. \tag{5.6}$$

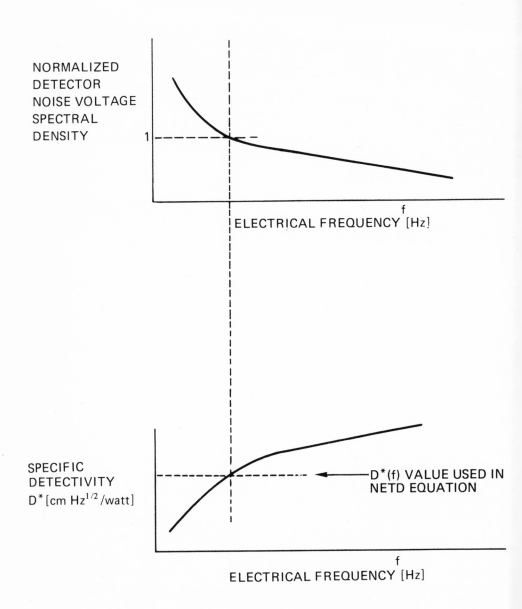

Figure 5.2 Normalization of the noise voltage spectrum.

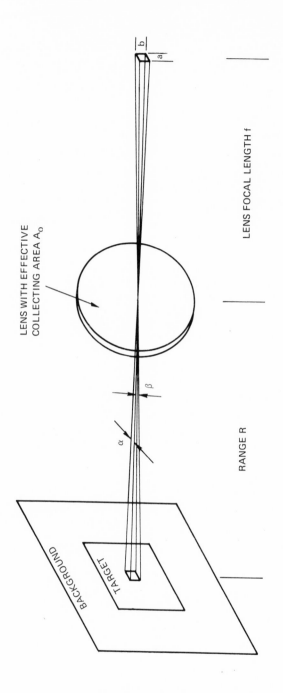

Figure 5.3 Geometry of the NETD derivation.

We are interested in differential changes with respect to target temperature, so we take

$$\frac{\partial P_\lambda}{\partial T} = \frac{\alpha\beta}{\pi} A_o \tau_0(\lambda) \frac{\partial W_\lambda}{\partial T} \left[\frac{watt}{\mu m^\circ K}\right].$$

(5.7)

The differential signal voltage produced in the detector by a target-to-background temperature differential is found by multiplying the above expression by the detector responsivity $R(\lambda)$:

$$\frac{\partial V_s}{\partial T} = \frac{1}{\pi}\alpha\beta A_o \tau_0(\lambda) R(\lambda) \frac{\partial W_\lambda}{\partial T} \left[\frac{Volts}{\mu m^\circ K}\right].$$

(5.8)

Since

$$R(\lambda) = \frac{V_n D*(\lambda)}{\sqrt{ab \, \Delta f_R}} \, ,$$

(5.9)

where V_n is the detector rms noise voltage produced by Δf_R, we get

$$\frac{\partial V_s(\lambda)}{\partial T} = \frac{\alpha\beta A_o \tau_0(\lambda) V_n D*(\lambda)}{\pi} \frac{\partial W_\lambda}{\sqrt{ab \, \Delta f_R}} \frac{\partial W_\lambda}{\partial T} \, .$$

(5.10)

Integrating over all wavelengths,

$$\frac{\partial V_s}{\partial T} = \frac{\alpha\beta A_o V_n}{\pi\sqrt{ab \, \Delta f_R}} \int_0^\infty \frac{\partial W_\lambda}{\partial T} D*(\lambda) \tau_0(\lambda) \, d\lambda.$$

(5.11)

Using a small signal approximation,

$$\frac{\Delta V_s}{\Delta T} = \frac{\alpha\beta A_o V_n}{\pi\sqrt{ab \, \Delta f_R}} \int_0^\infty \frac{\partial W_\lambda}{\partial T} D*(\lambda) \tau_0(\lambda) \, d\lambda.$$

(5.12)

Rearranging, we get a differential SNR[†] of

$$\frac{\Delta V_s}{V_n} = \Delta T \frac{\alpha\beta\, A_0}{\pi\sqrt{ab\,\Delta f_R}} \int_0^\infty \frac{\partial W_\lambda}{\partial T}\, D^*(\lambda)\, \tau_0(\lambda)\, d\lambda. \tag{5.13}$$

Since we define NETD in terms of unity SNR, let $\Delta V_s/V_n = 1$ and solve for the resulting delta-T which we call NETD:

$$\text{NETD} = \frac{\pi\sqrt{ab\,\Delta f_R}}{\alpha\beta\, A_0 \displaystyle\int_0^\infty \frac{\partial W_\lambda}{\partial T}\, D^*(\lambda)\, \tau_0(\lambda)\, d\lambda}. \tag{5.14}$$

Predictions based on this equation will deviate from reality by the degree to which the basic assumptions are violated.

For some systems it is possible to assume that $\tau_0(\lambda) = \text{constant} = \tau_0$ within $\lambda_1 \leqslant \lambda \leqslant \lambda_2$, and $\tau_0(\lambda) = 0$ elsewhere, so that

$$\text{NETD} = \frac{\pi\sqrt{ab\,\Delta f_R}}{\alpha\beta\, A_0 \tau_0} \; \frac{1}{\displaystyle\int_{\lambda_1}^{\lambda_2} \frac{\partial W_\lambda(T_B)}{\partial T}\, D^*(\lambda)\, d\lambda}. \tag{5.15}$$

If this is not the case, an effective transmission defined by:

$$\tau_0 = \frac{\displaystyle\int_0^\infty \frac{\partial W_\lambda(T_B)}{\partial T}\, \tau_0(\lambda)\, D^*(\lambda)\, d\lambda}{\displaystyle\int_0^\infty \frac{\partial W_\lambda(T_B)}{\partial T}\, D^*(\lambda)\, d\lambda} \tag{5.16}$$

[†] Alternatively, the integral in this expression may be replaced with the equivalent factor:

$$\frac{\partial}{\partial T} \int_0^\infty W_\lambda(T_B)\, D^*(\lambda)\, \tau_0(\lambda)\, d\lambda. \tag{5.17}$$

may be used. This is simply a definition and introduces no loss of accuracy, while it allows the NETD equation to be simplified.

A useful form of the NETD equation results from normalizing the $D^*(\lambda)$ in the integral which appears in equation 5.15 by $D^*(\lambda_p)$, pulling $D^*(\lambda_p)$ out of the integral, and defining an effective change in spectral radiant emittance with temperature as

$$\frac{\Delta W}{\Delta T} \triangleq \int_{\lambda_1}^{\lambda_2} \frac{\partial W_\lambda (T_B)}{\partial T} \frac{D^*(\lambda)}{D^*(\lambda_p)} \, d\lambda. \tag{5.18}$$

Then the simplest possible NETD equation is

$$NETD = \frac{\pi \sqrt{ab} \, \Delta f_R}{\alpha\beta \, A_o \, \tau_o \, D^*(\lambda_p) \, \frac{\Delta W}{\Delta T}}. \tag{5.19}$$

In Section 2.2, we noted that an appropriate approximation in terrestrial thermal imaging is:

$$\frac{\partial W_\lambda (T_B)}{\partial T} \cong \frac{c_2}{\lambda T_B^{\,2}} \, W_\lambda (T_B). \tag{5.20}$$

Making this approximation,

$$NETD = \frac{\pi \sqrt{ab} \, \Delta f_R}{\alpha\beta \, A_o \, \tau_o \, D^*(\lambda_p)} \frac{1}{\left[\dfrac{c_2}{T_B^{\,2}} \displaystyle\int_{\lambda_1}^{\lambda_2} \dfrac{W_\lambda (T_B) \, D^*(\lambda)}{\lambda \, D^*(\lambda_p)} \, d\lambda \right]} \tag{5.21}$$

For the case of photodetectors having theoretical performance such that

$$D^*(\lambda) = \frac{\lambda D^*(\lambda_p)}{\lambda_p} \text{ for } \lambda \leqslant \lambda_p,$$

$$= 0 \text{ for } \lambda > \lambda_p, \tag{5.22}$$

the bracketed expression in equation 5.21 simplifies to

$$\left[\frac{c_2}{T_B{}^2\lambda_p}\int_{\lambda_1}^{\lambda_p} W_\lambda\,(T_B)\,d\lambda\right]\;.$$ (5.23)

This expression is readily evaluated in two steps using a radiation slide rule. For $T_B = 300°K$, the expression becomes:

$$\left[\frac{0.16}{\lambda_p}\int_{\lambda_1}^{\lambda_p} W_\lambda(300°K)\,d\lambda\right]\;.$$ (5.24)

For $\lambda_1 = 8\,\mu m$ and $\lambda_p = 11.5\,\mu m$, the expression equals 1.48×10^{-4} [watt/cm^2 °K].

For an example NETD calculation, consider one possible implementation of the example system introduced in Section 1.5. This system has the following basic parameters:

- Detector lens diameter = 2 cm
- Square detector element dimensions = .005 cm
- Square detector angular subtense = 1.0 mrad
- Frame rate/field rate/interlace = 30/60/2
- Number of active scan lines = 300
- Field of view = 400 mrad by 300 mrad.

We will consider the case shown in Figure 5.4 of a parallel scan system using a 150-element linear gapped array. We will assume that the detectors are not background-limited, so that we do not have to consider the complications of the various detector cold shielding schemes possible. We will further assume that the vertical scan efficiency η_v is 0.8 and that the horizontal scan efficiency η_H is 0.8, giving an overall scan efficiency η_{sc} of 0.64. The system has a spectral bandpass of 8 to 11.5 μm, detectors having per-element $D^*(\lambda_p)'$s of 2×10^{10} $\left[\dfrac{\text{cm Hz}^{1/2}}{\text{watt}}\right]$ and $D^*(\lambda) = \dfrac{\lambda D^*(\lambda_p)}{\lambda_p}$ for $\lambda \leqslant \lambda_p$.

We will assume that the detector noise is white within the system bandpass, and that the effective optical transmission τ_o is 0.8. The dwelltime of the system is given by:

$$\tau_d = \frac{n\,\eta_{sc}\,\alpha\beta}{AB\,\dot{F}} = \frac{(150)\,(.64)\,(1)\,(1)}{(400)\,(300)\,(30)} \; [\text{sec}]$$

$$= 2.67 \times 10^{-5}\;[\text{sec}].$$

The reference bandwidth is:

$$\Delta f_R = \frac{\pi}{2} \frac{1}{2\tau_d} = 29.4 \ [KHz].$$

The appropriate NETD equation to use is:

$$NETD = \frac{4\sqrt{ab \ \Delta f_R}}{\alpha\beta \ D_0^2 \ \tau_o \ D^*(\lambda_p) \int\limits_{8}^{11.5} \frac{\partial W_\lambda}{\partial T} \frac{D^*(\lambda)}{D^*(\lambda_p)} \ d\lambda} \ .$$

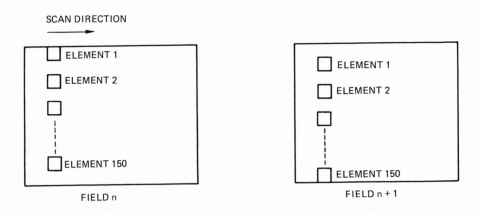

SCAN DIRECTION

ELEMENT 1
ELEMENT 2

ELEMENT 150

FIELD n

ELEMENT 1
ELEMENT 2

ELEMENT 150

FIELD n + 1

Figure 5.4 Scan pattern of the 150 element example system.

For the type of detector described, the value of the integral is 1.48×10^{-4} [watt/cm^2 °K]. Then the NETD is:

$$NETD = \frac{4\sqrt{(0.005 \ [cm])^2 \ (2.94 \times 10^4 \ [Hz])}}{(10^{-3} \ [rad])^2 \ (2[cm])^2 \ (0.8) \left(2.10^{10} \left[\frac{cm \ Hz^{1/2}}{watt}\right]\right)\left(1.48 \times 10^{-4}\left[\frac{watt}{cm^2 °K}\right]\right)}$$

$$= 0.36 [°K]$$

5.4 NETD Equation for BLIP Detectors

A photodetector which exhibits performance at theoretical limits is said to operate in the background-limited infrared photodetection (BLIP) mode. The NETD equations given earlier are valid for BLIP or non-BLIP detectors, but those equations are not suitable for performing design tradeoffs with BLIP detectors. For example, the equation

$$\text{NETD} = \frac{\pi\sqrt{ab\ \Delta f_R}}{\alpha\beta\ A_o\ D^*(\lambda_p)\ \tau_o\ \dfrac{\Delta W}{\Delta T}} \tag{5.25}$$

does not demonstrate the effect on D^* of changing other system parameters. The theoretical limit for $D^*(\lambda)$ for a photoconductor is[†][3]

$$D^*(\lambda) = \frac{\lambda}{2hc}\left(\frac{\eta_q}{Q_B}\right)^{1/2}, \tag{5.26}$$

where η_q is the quantum efficiency and Q_B is the background photon flux density [photons/cm^2 sec] incident on the detector. The Q_B is given by

$$Q_B = \left[\frac{\Omega_{cs}}{\pi}\right]\int_0^{\lambda_p} Q_\lambda(T_B)d\lambda, \tag{5.27}$$

where it is assumed that the detector is a Lambertian receiver with an effective collecting solid angle of π, and where Ω_{cs} is the effective angle to which the detector cold shield reduces reception of background radiation. Substituting for Q_B in the D^* equation gives

$$D^*(\lambda) = \frac{\lambda}{2hc}\left[\frac{\pi}{\Omega_{cs}}\right]^{1/2}\frac{\eta_q^{1/2}}{\left[\displaystyle\int_0^{\lambda_p} Q_\lambda(T_B)\,d\lambda\right]^{1/2}}. \tag{5.28}$$

[†] Multiply by $\sqrt{2}$ for photovoltaics; the λ in the numerator occurs because D^* is defined in units of [cm Hz$^{1/2}$/watt] rather than more physically meaningful units of [(cm Hz$^{1/2}$)/(photon/cm^2 sec)].

A parameter $D^{**}_{BLIP}(\lambda)$ for unity quantum efficiency and for an effective background angle of π steradians is defined by

$$D^{**}_{BLIP}(\lambda) \triangleq \frac{\lambda}{2hc} \frac{1}{\left[\int_{0}^{\lambda_p} Q_\lambda(T_B)\, d\lambda\right]^{1/2}} . \tag{5.29}$$

Then

$$D^*(\lambda) = \left[\frac{\pi}{\Omega_{cs}}\right]^{1/2} \eta_q^{1/2}\, D^{**}_{BLIP}(\lambda). \tag{5.30}$$

Substituting for $D^*(\lambda_p)$ in the NETD equation gives

$$NETD = \frac{\pi^{1/2}\sqrt{ab\,\Delta f_R\,\Omega_{cs}}}{\alpha\beta\, A_o\, \tau_o\, \eta_q^{1/2}\, D^{**}_{BLIP}(\lambda_p)\, \frac{\Delta W}{\Delta T}} . \tag{5.31}$$

Defining a cold shielding efficiency η_{cs} in terms of the actual cold shield solid angle Ω_{cs} used and a theoretically perfect cold shield solid angle Ω_p,

$$\eta_{cs} \triangleq \frac{\Omega_p}{\Omega_{cs}}, \tag{5.32}$$

the NETD becomes

$$NETD = \frac{\pi^{1/2}\sqrt{ab\Delta f_R\Omega_p}}{\alpha\beta\, A_o\, \tau_o\, \eta_{cs}^{1/2}\, \eta_q^{1/2}\, D^{**}_{BLIP}(\lambda_p)\, \frac{\Delta W}{\Delta T}} . \tag{5.33}$$

The perfect cold shield is defined as one which limits reception of background radiation to the cone defined by the optical system F/#. Thus for a circular aperture*

---

*This expression is approximately correct for $F/\# \geqslant 1.5$. Otherwise the exact expression $\Omega_p = \pi \sin^2$ [(arctan $(1/2\, F/\#)$] must be used.

$$\Omega_p = \frac{\pi D_0{}^2}{4 f^2} \tag{5.34}$$

and

$$\text{NETD} = \frac{2 \sqrt{\Delta f_R}}{\sqrt{\alpha\beta} D_0 \, \tau_0 \, \eta_{cs}{}^{1/2} \, \eta_q{}^{1/2} D^{**}_{BLIP}(\lambda_p) \frac{\Delta W}{\Delta T}} \cdot \tag{5.35}$$

Applying the same definitions to D* gives

$$D^* (\lambda) = 2 \, F/\# \, (\eta_{cs}\eta_q)^{1/2} \, D^{**}_{BLIP}(\lambda). \tag{5.36}$$

Both of these equations demonstrate the limitations of design tradeoffs with BLIP detectors.

5.5 Deficiencies of NETD as a Summary Measure

The deficiencies of NETD as a performance measure are numerous:

- Transients often appear on the video which are blanked on the display, but which will cause erroneously large noise readings on the rms voltmeter unless the voltmeter also blanks the transients.
- The peak voltage level V_s (shown in Figure 5.1) may be a matter of interpretation.
- NETD measured at the electronics is not always indicative of the overall system performance because there are other noise sources and spatial filters intervening between the point of NETD measurement and the final image.
- The NETD is a measure of the total in-band noise, whereas the eye does not treat all types and frequencies of noise alike.
- Designing for a low measured NETD without considering the effects on image quality can be disastrous for overall system performance. For example, it is often possible to increase the laboratory-measured NETD by using a wide spectral bandpass, at the expense of poor field performance. One possibility is that the wide spectral bandpass could make the system susceptible to sun glints. Another is that if the detectors are BLIP, the high SNR's seen in the laboratory will be lost in the field when the signal response at the band extremes is attenuated by the atmosphere, leaving only noise-producing background.

- NETD does not account for the effects of emissivity differences on SNR.
- NETD is almost useless for comparison of systems operating in different spectral regions because of atmospheric transmission differences.

In addition, the NETD criterion does not take into account the following facts:

- The visual system acts like a low-pass filter which rejects high spatial frequency noise.
- The visual system consists of many parallel narrowband filters which are activated separately. Thus a target with a narrow frequency spectrum may excite only one filter, so that noise rejection outside that filter's band is achieved.

Therefore one must be very careful in designing a system to be sure that one does not excessively filter noise to achieve a good NETD at the expense of a good MTF. NETD is best used as a measure of the day-to-day performance of a single system and as a diagnostic test. Of course, NETD does indicate gross sensitivity. While the difference between two otherwise identical systems with NETD's of $0.2°C$ and $0.25°C$ is insignificant, the difference between a $0.2°C$ and a $0.4°C$ NETD is significant.

5.6 NETD Trade-offs

Quite often one would like to know the effects on sensitivity of changing a single parameter. For that reason, it is useful to recast the NETD equation into a form using fundamental system parameters. The equation

$$\text{NETD} = \frac{\pi \sqrt{ab} \; \Delta f_R}{\alpha \beta A_o D^*(\lambda_p) \, \tau_o \dfrac{\Delta W}{\Delta T}} \tag{5.37}$$

is usually cast into forms which more specifically represent the system to be described. A common case is that of a parallel scanning sensor with an n-element non-BLIP detector array which exhibits white noise.

For a clear circular unobscured collecting aperture of diameter D_o, and a linear scan having no overscan and producing frame rate $\dot F$, field-of-view A by B, and a combined vertical and horizontal scan efficiency of η_{sc},

$$A_o = \pi D_o^2/4 \tag{5.38}$$

$$\tau_d = \frac{n\alpha\beta\eta_{sc}}{AB\dot F} \tag{5.39}$$

and

$$\Delta f_R = \frac{\pi}{2} \frac{1}{2\tau_d} .$$

(5.40)

Then

$$NETD = \frac{2\sqrt{\pi \text{ ab } AB \dot{F}/n} \, \eta_{sc}}{(\alpha\beta)^{3/2} D_o^2 \, D^*(\lambda_p) \, \tau_o \frac{\Delta W}{\Delta T}} .$$

(5.41)

This equation allows $\alpha\beta$ to be varied explicitly and independently for constant ab, while f is varied implicitly.

As $\alpha\beta = \text{ab}/f^2$, another version is

$$NETD = \frac{2\sqrt{\pi \, AB \dot{F}/n} \, \eta_{sc} \, f^3}{\text{ab } D_o^2 \, D^*(\lambda_p) \, \tau_o \frac{\Delta W}{\Delta T}} .$$

(5.42)

This allows f to be varied explicitly for constant ab while $\alpha\beta$ is varied implicitly.

As $\text{ab} = \alpha\beta \, f^2$, we have

$$NETD = \frac{2\sqrt{\pi \, AB \dot{F}/n} \, \eta_{sc} \, f}{\alpha\beta \, D_o^2 \, D^*(\lambda_p) \, \tau_o \frac{\Delta W}{\Delta T}} .$$

(5.43)

Here, $\alpha\beta$ is varied explicitly with constant f, and ab is varied implicitly.

Finally, for rectilinear non-overlapped but interlaced azimuth scanning with a vertical array, $B = n\beta I$, where I is the interlace factor, and

$$NETD = \frac{2\sqrt{\pi \text{ ab } A \dot{F} I \beta/\eta_{sc}}}{(\alpha\beta)^{3/2} \, D_o^2 \, D^*(\lambda_p) \, \tau_o \frac{\Delta W}{\Delta T}}$$

(5.44)

showing the square root dependence of sensitivity on interlace.

A common application of any such equation involves deciding which spectral region to use for a particular problem. For example, if aperture size is fixed due to some vehicular or material constraint, and field of view and frame rate are fixed, the other parameters can be adjusted to give the most cost effective or least complex problem solution. A trivial example is the aperture diameter comparison for two spectral regions 1 and 2, say 3 to 5 μm and 8 to 14 μm, with all system parameters except $D^*(\lambda)$ and τ_O being equal, and for equal NETD's. In that case,

$$
\frac{D_{O1}}{D_{O2}} = \left\{ \frac{D_2^* \, \tau_{O2} \left(\frac{\Delta W}{\Delta T} \right)_2}{D_1^* \, \tau_{O1} \left(\frac{\Delta W}{\Delta T} \right)_1} \right\}^{1/2}
\tag{5.45}
$$

NETD is a good measure of sensor performance, and a good sensitivity diagnostic test. However, it is not a good image quality summary measure. The following section discusses and derives a more useful parameter.

5.7 Minimum Resolvable Temperature Difference

We found in Chapter Four that four factors heavily influence subjective image quality: sharpness, graininess, contrast rendition, and interference by artifacts. An objective correlate exists for each of these factors which describes its effects when the other three factors are held constant or are not significant. A problem arises in that the four factors are usually simultaneously influential, and since the equivalences of degradations in the correlates are not yet known in detail, it is difficult to predict system performance for a multiplicity of tasks.

This is a common problem in electro-optical devices, but it is accentuated in thermal systems because thermal image contrasts usually don't look quite like the contrasts of a visible scene, so a mental set of standards such as those we apply to commercial TV is difficult to acquire. We are very sensitive to changes in home television reproduction and in newspaper halftone photographs, but it is possible to overlook relatively serious degradations in FLIR image quality because of the somewhat unnatural appearance of even good imagery and the wide variations in system design which are customary.

It is thus desirable to have a summary measure which includes all four elements of image quality, that is, one which is a unified system-observer performance criterion. There are many possible approaches to such a performance criterion based on observer threshold perceptual limitations caused by random noise, clutter, contrast, spatial frequency content, or observation time. In FLIRs, spatial resolution and thermal sensitivity dominate performance. This section concentrates on a widely accepted measure for the signal-to-noise ratio limited thermal sensitivity of a system as a function of spatial frequency.

Schade[5] derived such a measure for photographic, motion picture, and television systems. It is the image signal-to-noise ratio required for an observer to resolve a three-bar standard Air Force chart that is masked by noise. R. Genoud[6] and R.L. Sendall[7] recognized that Schade's formulation was applicable with only minor modifications to thermal imagery, and derived an expression for the bar pattern temperature difference which will produce Schade's SNR threshold for resolution. This function is now generally called the minimum resolvable temperature difference (MRTD), and is defined for four-bar rather than for three-bar targets. The noise-limited case is significant because an infrared imaging system exhibits its ultimate sensitivity when noise is visible to the observer, as it will be when the system gain is set high to compensate for adverse environmental or scene conditions.

The following derivation uses the same assumptions as the preceding NETD derivation, supplemented by the following:

- The effect of temporal integration in the FLIR observer's eye/brain system is approximated by a fixed integration time of 0.2 second. It is assumed that the eye adds signals linearly and takes the root-mean-square value of noise within any 0.2 second interval, although this is clearly not the exact mechanism involved, but merely a convenient fiction.
- The effect of narrowband spatial filtering in the eye in the presence of a periodic square bar target of frequency f_T is approximated by a postulated matched filter for a single bar. This assumed eye filter is

$$\text{sinc}\left(\frac{f}{2f_T}\right) = \sin(\pi f/2f_T)/(\pi f/2f_T). \tag{5.46}$$

Schade and others have demonstrated that the eye/brain functions approximately as a matched filter (or in this case as a synchronous spatial integrator) over small areas and therefore responds to the two-dimensional spatial integral of the displayed irradiance pattern of the target. For simplicity, we will assume that the bars are long enough so that bar end effects may be neglected.

- The electronic processing and monitor are assumed to be noiseless.
- The system is relatively simple with zero overscan, a well-behaved MTF, and a well-behaved noise power spectrum $g^2(f)$.
- The system is operated linearly so that the response to the target is describable by the MTF.
- The image formation is assumed to be spatially invariant (sample-free) in the scan direction.

An operational definition of MRTD is the following. Consider that the system views the largest target shown in Figure 5.5, where the bars and spaces of the target are blackbody radiators of different but uniform temperatures, and the bar height is seven times the bar width. Let the system be adjusted so that noise is clearly visible on the display, and let the initial temperature difference between the bars and spaces be zero. Then increase the temperature difference until the bars can be confidently resolved by an observer viewing the display. The delta-T is the MRTD evaluated at the fundamental spatial frequency of the chart. Repetition at successively higher frequencies yields an MRTD curve of the form shown in Figure 5.6. The steps in arriving at the MRTD at a given frequency are the detection of the test pattern as a square, followed by the accurate establishment of the delta-T at which the operator has a high confidence that he is in fact resolving the individual bars.

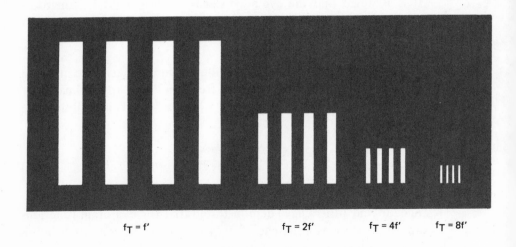

$f_T = f'$ $f_T = 2f'$ $f_T = 4f'$ $f_T = 8f'$

Figure 5.5 MRTD targets.

5.8 MRTD Derivation

It is desirable to preserve historical continuity and to minimize analytical complexity by expressing the MRTD in terms of the NETD. The design tradeoffs become apparent, however, only when the parametric expression for NETD is substituted into the MRTD equation. The following

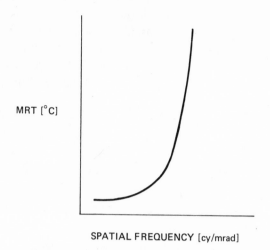

Figure 5.6 A typical shape for an MRTD curve.

quantities in addition to those defined for the NETD derivation are required for the MRTD:

$\widetilde{r}_e(f)$ = electronic amplifier and video processor MTF

$\widetilde{r}_m(f)$ = video monitor MTF

$\widetilde{r}_s(f)$ = overall system MTF

$\overset{\leftrightarrow}{r}_s(f)$ = overall system square wave response

T_e = effective eye integration time [second]

$\dot{F}$ = frame rate [sec^{-1}]

f_T = fundamental target frequency [cy/mrad]

f'_T = fundamental target frequency [Hertz]

We will derive the MRTD by considering first the effects of the system on the SNR in the target image, and then including the effects of the observer. At the output of the monitor, the SNR in the image of one bar is

$$SNR_i = \overset{\leftrightarrow}{r}_s \frac{\Delta T}{NETD_R} \frac{1}{\left[\dfrac{\int_0^\infty g^2(f)\,\widetilde{r}_e^{\,2}(f)\,\widetilde{r}_m^{\,2}(f)\,df}{\Delta f_R} \right]^{1/2}} \qquad (5.47)$$

where the square wave response factor $\overset{\sqcap}{\tau}_s$ describes the modulation loss in the bar image, the factor $\Delta T/NETD_R$ gives the electronic SNR measured at an NETD reference filter for a large target with a particular ΔT, and the bracketed term in the denominator converts from the reference bandwidth to the actual system bandwidth. Since $4/\pi$ is the first harmonic amplitude of a unit square wave, for spatial frequencies greater than or equal to one-third of the nominal cutoff,

$$\overset{\sqcap}{\tau}_s = 4\,\tilde{\tau}_s/\pi. \tag{5.48}$$

For a large class of systems, this is a good approximation for frequencies greater than one-fifth of the cutoff frequency.

When an observer views the target, he perceives a signal-to-noise ratio modified by four perception factors:

1. For this type of target, the eye apparently extracts and operates on mean signal in the image, as demonstrated by Schade[5]. For frequencies greater than one-third of cutoff, the degradation of mean signal level from the peak signal input is given by $2\overset{\sqcap}{\tau}_s/\pi$ or $8\tilde{\tau}_s/\pi^2$, because $2/\pi$ is the average value of the first harmonic of a unit square wave over one-half cycle.

2. SNR is improved by the square root of the number of independent time samples per display point per 0.2 second, that is, by $(T_e\dot{F})^{1/2}$, due to temporal integration.

3. In the vertical bar direction, the eye spatially integrates signal and takes the rms value of noise along the bar, using as a noise correlation length the dimension β, giving a perceived SNR improvement of:

$$(\text{bar length}/\beta)^{1/2} = (7 \times \text{bar width}/\beta)^{1/2} = (7/2\ f_T\beta)^{1/2} \tag{5.49}$$

4. In the horizontal direction, the effect of eye integration is most easily accounted for in the frequency domain by replacing the displayed noise bandwidth

$$\int_0^\infty g^2\,\tilde{\tau}_e^{\,2}\,\tilde{\tau}_m^{\,2}\ df$$

with a noise bandwidth accounting for the eye's matched filter action,

$$\int_0^\infty g^2 \, \tilde{r}_e^{\,2} \, \tilde{r}_m^{\,2} \, \text{sinc}^2 (f/2 \, f_T') \, df \ .$$

This gives a perceived SNR improvement of:

$$\left[\frac{\displaystyle\int_0^\infty g^2 \, \tilde{r}_e^{\,2} \, \tilde{r}_m^{\,2} \, df}{\displaystyle\int_0^\infty g^2 \, \tilde{r}_e^{\,2} \, \tilde{r}_m^{\,2} \, \text{sinc}^2 (f/2 \, f_T') \, df} \right]^{1/2} \ .$$

Combining these four effects with the image SNR_i, the perceived SNR is

$$SNR_p = \frac{8}{\pi^2} \, \tilde{r}_s \, \frac{\Delta T}{NETD_R \left[\dfrac{\displaystyle\int_0^\infty g^2 \tilde{r}_e^{\,2} \tilde{r}_m^{\,2} \, df}{\Delta f_R} \right]^{1/2}} \, [T_e \dot{F}]^{1/2} \left[\frac{7}{2 f_T \beta} \right]^{1/2}$$

$$\times \left[\frac{\displaystyle\int_0^\infty g^2 \, \tilde{r}_e^{\,2} \, \tilde{r}_m^{\,2} \, df}{\displaystyle\int_0^\infty g^2 \, \tilde{r}_e^2 \, \tilde{r}_m^{\,2} \, \text{sinc}^2 (f/2 f_T') \, df} \right]^{1/2}$$

$$= \frac{1.51 \, \Delta T \, [T_e \dot{F}]^{1/2} \tilde{r}_s}{NETD_R \left[\dfrac{\displaystyle\int_0^\infty g^2 \tilde{r}_e^{\,2} \tilde{r}_m^{\,2} \, \text{sinc}^2 (f/2 f_T') \, df}{\Delta f_R} \right]^{1/2} [f_T \beta]^{1/2}}$$

$$(5.50)$$

Letting SNR_p equal a signal-to-noise ratio k in the image of one bar necessary for the observer to detect that bar, solving for ΔT, and defining a bandwidth ratio ρ by

$$\rho^{1/2} \triangleq \left[\frac{\int_0^\infty g^2 \, \tilde{r}_e^2 \, \tilde{r}_m^2 \, \text{sinc}^2 \, (f/2f'_T) \, df}{\Delta f_R} \right]^{1/2} , \tag{5.51}$$

we get

$$\Delta T = \frac{NETD_R \, \rho^{1/2} \, [f_T \beta]^{1/2} \, \dfrac{k}{1.51}}{\tilde{r}_s \, (T_e \dot{F})^{1/2}} . \tag{5.52}$$

Here it must be noted that target frequency appears as f'_T in $\rho^{1/2}$ with units of [Hertz] and as f_T with units [cy/mrad] elsewhere.

The ΔT which is chosen as the MRTD is associated with a perceived signal-to-noise ratio which is in turn associated with a probability of detection. Blackwell found that the threshold of observer confidence is approximately the same as the 90 percent accuracy forced-choice detection threshold. Therefore it seems reasonable to select the constant in the MRTD equation to be consistent with a 90 percent probability of individual bar detection. From Section 4.9.1 we have that the k for $P_d = 0.9$ is approximately 4.5. Letting $k = 4.5$ and $\Delta T = MRTD$, we have that

$$MRTD \, (f_T) = \frac{3 \, NETD_R \, \rho^{1/2} \, (f_T \beta)^{1/2}}{\tilde{r}_s (T_e \dot{F})^{1/2}} . \tag{5.53}$$

The author's experience has been that this equation adequately predicts the measured MRTD of a detector-noise-limited, artifact-free system if all parameters are accurately known.

There are four serious problems with the MRTD as formulated here which make it less than perfect as a performance summary measure. The first is that it does not include the MTF of the eye and magnification effects. The second is that it is a human response measure, and some people are uncomfortable with a man explicitly in the system evaluation process. The third is that the spectral sensitivity is not explicit in the formula. The fourth is that MRTD does not unequivocally correlate with field performance. Thus it is most useful as a diagnostic tool and as a gross indication, per Section 10.5, that field performance will be as desired.

The MRTD concept has been evaluated in the laboratory with many different systems by many different individuals. The correlation with theory has been good, the test is easily performed, and it is repeatable. Furthermore, it is widely accepted because it is an easily grasped, clearly observable concept. The MRTD equation also demonstrates an important point about frame rate and flicker. In the case of a linear scan, the conversion factor from spatial to electrical frequencies along a scan line is proportional to the frame rate. Then Δf_R is proportional to $\dot{F}$, $NETD_R$ is proportional to $\dot{F}^{1/2}$, and the $(\dot{F})^{1/2}$ factor in the denominator removes the apparent $\dot{F}$ dependence of MRTD. This is a significant point because it shows that increasing the frame rate does not reduce the perceived SNR, and in fact improves picture quality by reducing flicker.

If one desires to have an MRTD equation which uses fundamental parameters rather than the NETD, one form derived by substituting NETD is:

$$MRTD = \frac{3\pi\sqrt{ab}\left[\int_0^\infty g^2(f)\,\tilde{r}_e^2\,\tilde{r}_m^2\,\text{sinc}^2\,(f/2f'_T)\,df\right]^{1/2}\,[f_T\beta]^{1/2}}{\alpha\beta A_0\tau_0\,D^*(\lambda_p)\,\frac{\Delta W}{\Delta T}\tilde{r}_s\,(T_e\dot{F})^{1/2}} \qquad (5.54)$$

Finally, the MRTD expression of equation 5.53 can be simplified for most system concepts to allow calculation of MRTD by hand. In many well designed systems, the displayed noise is white within the signal bandpass, or

$$g^2(f)\tilde{r}_e^2\tilde{r}_m^2 = 1. \qquad (5.55)$$

In that case, the expression for $\rho^{1/2}$ reduces to:

$$\rho^{1/2} = \left[\frac{f'_T}{\Delta f_R}\right]^{1/2}, \qquad (5.56)$$

where f'_T has units of Hertz. To permit use of f_T in both places in the equation in units of [cy/mrad] we note that

$$f_T' = \frac{\alpha}{\tau_d}\,f_T \quad . \qquad (5.57)$$

Then

$$MRTD = \frac{3\left(\frac{NETD_R}{(\Delta f_R)^{1/2}}\right) f_T \left[\frac{\alpha}{\tau_d}\right]^{1/2} \beta^{1/2}}{\tilde{r}_s (T_e \dot{F})^{1/2}} .$$ (5.58)

This form has the desirable quality of using the NETD divided by the reference bandwidth. Thus the controversial part of NETD is eliminated and the more or less fixed parameters remain.

For a sample MRTD calculation, consider the example system in its parallel scan implementation. The following MRTD parameters apply:

NETD$_R$ = 0.36°C

Δf_R = 29.4 KHz

$\alpha = \beta$ = 1.0 mrad

τ_d = 2.67 X 10^{-5} sec.

$\dot{F}$ = 30 Hz

MTF's as given in Table 3.5.

Using the MRTD equation for white noise and substituting the above parameters,

$$MRTD = 0.49 \frac{f_T}{\tilde{r}_s} [°C] .$$

The example system MTF and MRT are shown in Figure 5.7.

5.9 Minimum Detectable Temperature Difference

The MRTD concept is a useful analytical and design tool which is indicative of system performance in recognition tasks. However, MRTD does not necessarily correspond in any simple way with practical detection tasks. A function has been proposed by Sendall[7], Hopper[8], and others which should correlate with noise-limited field detection performance. This function is the Minimum Detectable Temperature Difference (MDTD). The MDTD is not at present widely accepted and no conventions for it exist, but it is a useful concept. In deriving an MDTD equation, the usual assumptions about the natures of the system and the observer are made. The target is a square of variable side angular dimension W against a large uniform background. The MDTD is defined as the blackbody temperature difference required for an

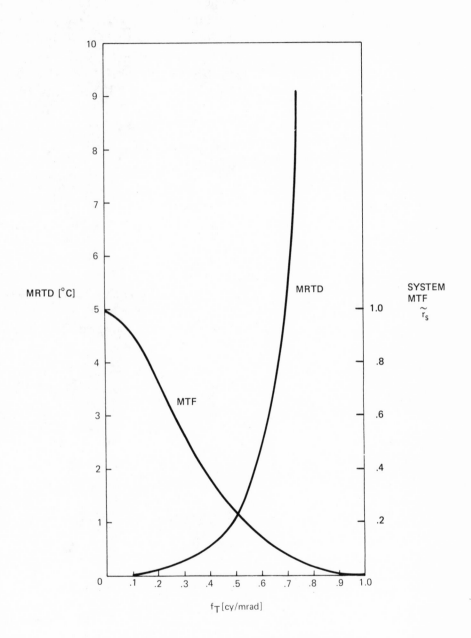

Figure 5.7 MRTD and MTF for the example system.

observer to detect the presence of the square target when he is allowed unlimited time to make a decision and knows where to look for the target.

Let the mean value of the displayed target image (a quantity which we assume the eye computes for spatial integration) be $\overline{I(x,y)}\Delta T$, where $I(x,y)$ is the image of the square target $O(x,y)$ normalized to unity amplitude, and where ΔT is the target to background temperature difference. Then the average signal-to-rms noise ratio per frame per line in the image is:

$$
\text{SNR}_i = \frac{\overline{I(x,y)}\,\Delta T}{\text{NETD}_R \left[\dfrac{\displaystyle\int_0^\infty g^2(f)\,\widetilde{r}_e^{\,2}\,\widetilde{r}_m^{\,2}\,df}{\displaystyle\int_0^\infty g^2(f)\,\widetilde{r}_R^{\,2}\,df} \right]^{1/2}} \quad . \tag{5.59}
$$

Let the system line spread function dimension in the y direction be $|r_y|$. Then for nonoverlapped scanning, the number of independent noise samples across the target image in elevation is approximately $(W + |r_y|)/\beta$.

Assuming the eye spatially integrates in the x direction over the dimension W, the perceived signal-to-noise ratio is

$$
\text{SNR}_P = \text{SNR}_i (T_e \dot{F})^{1/2} \left[\frac{W + |r_y|}{\beta} \right]^{1/2} \left[\frac{\displaystyle\int_0^\infty g^2(f)\,\widetilde{r}_e^{\,2}\,\widetilde{r}_m^{\,2}\,df}{\displaystyle\int_0^\infty g^2(f)\,\text{sinc}^2(Wf)\,\widetilde{r}_e^{\,2}\,\widetilde{r}_m^{\,2}\,df} \right]^{1/2} \tag{5.60}
$$

Substituting for SNR_i, simplifying, setting $\text{SNR}_P = k$, solving for ΔT and letting $\Delta T = \text{MDTD}$ yields

$$
\text{MDTD}(W) = k \, \frac{\left[\dfrac{\displaystyle\int_0^\infty g^2(f)\,\text{sinc}^2(Wf)\,\widetilde{r}_m^{\,2}\,\widetilde{r}_e^{\,2}\,df}{\displaystyle\int_0^\infty g^2(f)\,\widetilde{r}_R^{\,2}\,df} \right]^{1/2}}{\overline{I(x,y)}\,(T_e \dot{F})^{1/2} \left[\dfrac{W + |r_y|}{\beta} \right]^{1/2}} \, \text{NETD}_R \quad . \tag{5.61}
$$

This may be cast as a frequency domain function by defining a fundamental target frequency,

$$f_T = \frac{1}{2W},$$ (5.62)

so that

$$\text{MDTD}\left(f_T = \frac{1}{2W}\right) = \frac{k \left[\dfrac{\displaystyle\int_0^\infty g^2(f)\,\text{sinc}^2\left(\dfrac{f}{2f_T}\right)\tilde{r}_m^2\,\tilde{r}_e^2\,df}{\displaystyle\int_0^\infty g^2(f)\tilde{r}_R^2\,df}\right]^{1/2} \text{NETD}_R}{\overline{I(x,y)}\,(T_e\dot{F})^{1/2}\left[\dfrac{\dfrac{1}{2f_T} + |r_y|}{\beta}\right]^{1/2}} \cdot$$ (5.63)

Dryden[9] has noted that the MDTD can be related to MRTD by assuming that the square dimension W is the same as the bar width in an MRTD target and by rewriting the MDTD equation as:

$$\text{MDTD} = \frac{k\,\text{NETD}_R\,\rho^{1/2}\left(\dfrac{\beta}{1/(2f_T) + |r_y|}\right)^{1/2}}{\overline{I(x,y)}\,(T_e\dot{F})^{1/2}}$$ (5.64)

Neglecting the term $|r_y|$, we get

$$\text{MDTD}\left(f_t = \frac{1}{2W}\right) = \frac{\sqrt{2}\,k\,\text{NETD}_R\,\rho^{1/2}\,(f_T\beta)^{1/2}}{\overline{I(x,y)}\,(T_e\dot{F})^{1/2}}$$ (5.65)

Comparing this to the MRTD equation, we see that

$$\text{MDTD}\left(f_T = \frac{1}{2W}\right) = \frac{\tilde{r}_s \, 1.5\sqrt{2}}{\overline{I(x,y)}} \, \text{MRTD}\left(f_T = \frac{1}{2W}\right) . \qquad (5.66)$$

The difficulty of accurately predicting MDTD arises from the necessity to calculate $\overline{I(x,y)}$, which is the average value of a convolution integral. For targets much smaller than the detector solid angular subtense, $\overline{I(x,y)}$ reduces to the ratio of the target solid angular subtense to detector subtense. The MTDT equation for this case is useful for predicting detectability of targets which are effectively point sources.

5.10 Noise Equivalent Emissivity

The noise equivalent temperature difference is an idealization which allows one to calculate the video signal-to-noise ratio for a large blackbody target against a uniform blackbody background. Since scene apparent temperature differences may arise from emissivity variations as well as from contact temperature variations, it is useful to define an analog of the NETD which uses as the target a uniform emissivity difference in a broad area of an otherwise blackbody surface. Since the greybody area radiates less than its blackbody background at the same temperature, the video signal produced in the system by the greybody target will be negative relative to the background level. Thus, the parameter we seek is the emissivity ϵ which produces a video signal-to-noise ratio of -1 against a blackbody background. We may call this quantity the noise equivalent emissivity (NEE). As defined here, 1/NEE gives the video SNR produced by a particular NEE.

NEE is derivable by paralleling the NETD derivation. Assume that the target is shaped as shown in Figure 5.8. The difference between the target spectral radiant emittance and that of the background is:

$$\Delta W_\lambda = \epsilon W_\lambda (T_B) - W_\lambda (T_B)$$

$$= (\epsilon - 1) W_\lambda (T_B) . \qquad (5.67)$$

The change in spectral power incident on the detector is:

$$\Delta P_\lambda = \frac{(\epsilon - 1) W_\lambda (T_B)}{\pi} A_o \, \alpha\beta \, \tau_o(\lambda) \qquad (5.68)$$

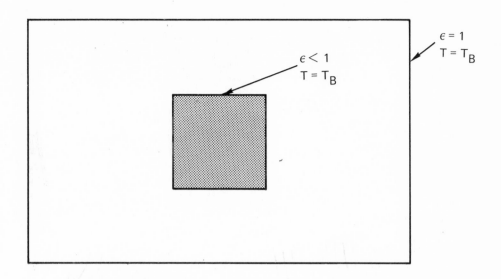

Figure 5.8 Noise equivalent emissivity notation and target pattern.

and the change in signal voltage is:

$$\Delta V_s (\lambda) = \frac{(\epsilon - 1)\, W_\lambda\, (T_B)}{\pi} \frac{A_o\, \alpha\beta\, V_n}{\sqrt{ab\, \Delta f}\ _R}\ \tau_o\, (\lambda)\, D^*(\lambda)\ . \tag{5.69}$$

Integrating over λ, solving for $\Delta V_s/V_n$, and setting it equal to -1 yields:

$$-1 = (\epsilon - 1)\, \frac{A_o\, \alpha\beta}{\pi\sqrt{ab\, \Delta f_R}} \int_0^\infty W_\lambda\, (T_B)\, D^*(\lambda)\, \tau_o(\lambda)\, d\lambda. \tag{5.70}$$

Solving for ϵ and identifying it as NEE,

$$NEE = 1 - \frac{\pi\sqrt{ab\, \Delta f_R}}{A_o\, \alpha\beta \int_0^\infty W_\lambda\, (T_B)\, D^*(\lambda)\, \tau_o\, (\lambda)\, d\lambda}\ . \tag{5.71}$$

Converting to the form of the NETD equation allows us to write

$$
NEE = 1 - NETD \left[\frac{\displaystyle\int_0^\infty \frac{\partial W_\lambda \, (T_B)}{\partial T} \frac{D^*(\lambda)}{D^*(\lambda_p)} \, d\lambda}{\displaystyle\int_0^\infty W_\lambda(T_B) \frac{D^*(\lambda)}{D^*(\lambda_p)} \, d\lambda} \right].
\tag{5.72}
$$

Hawkins[2] has defined a similar quantity called noise equivalent fractional emissivity.

5.11 Sensor Performance

It is difficult to derive a single sensor performance factor which is meaningful for all types of designs and component characteristics because general performance factors do not always translate into subjective image quality or into information extraction capability. To simplify the problem somewhat, Sendall[7] makes the important distinction between those performance figures which are used for system synthesis and those which are used for system analysis. Whereas the NETD and MRTD measures previously derived are analytical, Sendall has suggested quantities which are useful for arriving at an efficient system design.

The following derivation parallels that of Sendall for the broad class of systems whose sensitivity is determined by white detector-generated noise and whose resolution is largely determined by the detector angular subtense. It further assumes a circular clear aperture, contiguous non-overlapped scanning, single-RC low pass filter electronic processing, and square detectors.

The equations appropriate to those assumptions are

$$
NETD = 4\sqrt{ab \, \Delta f_R} \left/ \left[\alpha\beta \, D_0^2 \int_0^\infty \frac{\partial W_\lambda \, (T_B)}{\partial T} D^*(\lambda) \, \tau_0(\lambda) d(\lambda) \right] \right. ,
\tag{5.73}
$$

$$
\Delta f_R = \frac{\pi}{2} \frac{1}{2\tau_d} ,
\tag{5.74}
$$

and

$$
\tau_d = n\alpha\beta\eta_{sc} / AB\dot{F} \text{ (for the case of a parallel scan system).}
\tag{5.75}
$$

Then

$$\Delta f_R = \pi \, AB\dot{F}/4n\alpha\beta\eta_{sc} \; , \tag{5.76}$$

and

$$NETD = \frac{2\sqrt{\pi \, ab \, AB\dot{F}/n\alpha\beta\eta_{sc}}}{\alpha\beta D_o^2 \displaystyle\int_0^\infty \frac{\partial W_\lambda (T_B)}{\partial T} \, D^*(\lambda) \, \tau_o(\lambda) \, d\lambda}$$

$$= \frac{2\sqrt{\pi AB\dot{F}} \, F/\#}{\alpha\beta\sqrt{n \, \eta_{sc}} \, D_o \displaystyle\int_0^\infty \frac{\partial W_\lambda (T_B)}{\partial T} \, D^*(\lambda) \, \tau_o(\lambda) \, d\lambda} \; . \tag{5.77}$$

Now define the sensor performance P as

$$P = \frac{(AB\dot{F})^{1/2}}{\alpha\beta \, NETD} = \frac{\sqrt{n \, \eta_{sc}} \, D_o \displaystyle\int_0^\infty \frac{\partial W_\lambda(T_B)}{\partial T} \, D^*(\lambda) \, \tau_o(\lambda) \, d\lambda}{2\sqrt{\pi} \, F/\#} \; . \tag{5.78}$$

This is a good summary parameter because $NETD/\dot{F}^{1/2}$ is indicative of thermal sensitivity, $(AB)^{1/2}$ is an equivalent square field of view dimension, and $\alpha\beta$ is indicative of resolution performance for many systems. Thus reducing $NETD/\dot{F}^{1/2}$, increasing $(AB)^{1/2}$, and reducing $\alpha\beta$ increase performance.

For a BLIP detector with theoretically perfect cold shielding[†]

$$D^* = 2 \, D_{BLIP}^{**} \, F/\# \, (\eta_q \, \eta_{cs})^{1/2} \tag{5.79}$$

and we can write

$$P = \frac{\sqrt{n \, \eta_{sc}}}{\sqrt{\pi}} \, D_o \, (\eta_{cs} \, \eta_q)^{1/2} \int_0^\infty \frac{\partial W_\lambda (T_B)}{\partial T} \, D_{BLIP}^{**}(\lambda) \, d\lambda . \tag{5.80}$$

[†] Assuming that the detector operates at the same focal ratio as the optic, i.e., that there is no pupil migration with scanning.

If we now let $\tau_0(\lambda) = \tau_0$ and define

$$\text{Sensor design complexity} = C \overset{\Delta}{=} \frac{\sqrt{n}\, D_0}{\sqrt{\pi}} \tag{5.81}$$

$$\text{Sensor efficiency} = E \overset{\Delta}{=} \tau_0\, \eta_{cs}\, \eta_{sc}\, \eta_q)^{1/2} \tag{5.82}$$

and

$$\text{Sensor radiation function} = R \overset{\Delta}{=} \int_0^\infty D^{**}_{BLIP}(\lambda)\, \frac{\partial W_\lambda (T_B)}{\partial T}\, d\lambda, \tag{5.83}$$

then $P = CER$. The interpretations of E and R are straightforward, and C shows that after E and R have been optimized, further improvement in performance can be achieved only at the expense of a wider aperture or more detectors.

If the detector angular subtense does not dominate the sensor MTF, it is desirable to substitute for α the equivalent resolution r_x in the scan direction

$$r_x = \left[2 \int \tilde{r}_x^2\, df \right]^{-1}, \tag{5.84}$$

where $\tilde{r}_x$ is the system MTF in the scan direction. Similarly defining r_y, the performance equation becomes

$$P = \frac{(A\, B\, \dot{F})^{1/2}}{r_x r_y\, \text{NETD}}. \tag{5.85}$$

As a final remark, it should be noted that the above definitions of performance clearly emphasize the constancy of performance with increasing frame rate.

5.12 Spectral Region Selection

The most desirable spectral region for FLIR operation is the one which maximizes performance for a given cost. For two FLIRs identical in all respects except for spectral region, their NETD's are roughly indicative of relative performance except for the effects of atmospheric absorption. The best possible case for both spectral regions is that of BLIP detectors with NETD given by:

$$\text{NETD} = \frac{2\sqrt{\Delta f_R}}{D_o \sqrt{\alpha\beta}\, \eta_{cs}\, \eta_q\, D^{**}_{\text{BLIP}}(\lambda_p)\, \tau_o\, \frac{\Delta W}{\Delta T}} \qquad (5.86)$$

where

$$\frac{\Delta W}{\Delta T} = \int_{\lambda_1}^{\lambda_2} \frac{\partial W_\lambda(T_B)}{\partial T} \frac{D^*(\lambda)}{D^*(\lambda_p)}\, d\lambda.$$

From Planck's distribution for long wavelengths and terrestrial temperatures,

$$\frac{\partial W_\lambda(T_B)}{\partial T} \cong \frac{c_2}{\lambda T_B^2} W_\lambda(T_B) \quad . \qquad (5.87)$$

Since for perfect photodetectors,

$$D^*(\lambda) = \frac{\lambda D^*(\lambda_p)}{\lambda_p} \quad \text{for } \lambda \leqslant \lambda_p$$

$$= 0 \text{ for } \lambda > \lambda_p, \qquad (5.88)$$

we have that the integral in equation 5.86 reduces to

$$\frac{\Delta W}{\Delta T} = \frac{c_2}{\lambda T_B^2} \int_{\lambda_1}^{\lambda_p} W_\lambda\, d\lambda. \qquad (5.89)$$

Then assuming that all other parameters are equal the ratio of the NETD's in two spectral regions is given by

$$\frac{\text{NETD}(\lambda_1 \rightarrow \lambda_2)}{\text{NETD}(\lambda_3 \rightarrow \lambda_4)} = \frac{\lambda_2 \ D^{**}_{BLIP}(\lambda_4) \int_{\lambda_3}^{\lambda_4} W_\lambda d\lambda}{\lambda_4 \ D^{**}_{BLIP}(\lambda_2) \int_{\lambda_1}^{\lambda_2} W_\lambda d\lambda} , \tag{5.90}$$

and one may now compare two theoretically perfect systems which are identical except for spectral region. For BLIP operation in the 3.5 to 5 and the 8 to 14 regions, theoretical D^{**}_{BLIP}'s are:

$$D^{**}_{BLIP}(14 \ \mu m) = 3.27 \times 10^{10} \ [cm \ Hz^{1/2}/watt],$$

and

$$D^{**}_{BLIP}(5 \ \mu m) = 1.09 \times 10^{11} \ [cm \ Hz^{1/2}/watt].$$

The values $\int_{\lambda_1}^{\lambda_p} W_\lambda d\lambda$ are obtained from a radiation slide rule and yield

$$\frac{\text{NET (3.5-5)}}{\text{NET (8-14.5)}} = \frac{5 \times 3.27 \times 10^{10} \times 1.72 \times 10^{-2}}{14 \times 1.09 \times 10^{11} \times 5.56 \times 10^{-4}} = 3.65.$$

This result shows that a perfect FLIR operating in the 8 to 14 μm region will perform better than one operating in the 3.5- to 5-μm region by a factor of 3.65 when all other design parameters are equal, and when the atmospheric transmissions are the same.

5.13 Spectral Bandpass Optimization

The choice of the approximate spectral range in which to operate is normally dictated by economic constraints; one conceives the possible systems which perform the desired function, and compares their estimated costs. Equivalence of performance of the candidate systems may depend strongly on the atmospheric transmissions expected under the anticipated conditions of use, and that effect may dominate the sensor cost, as for example in the case of 3 to 5-μm versus 8 to 14-μm sensors for moderate range usage.

Once the coarse spectral region has been selected, the problem of determining the functional form of the spectral transmission is separable into two cases. In the first case, the sensor noise is not background-limited, but is dominated by generation-recombination, thermal, and excess noises. The problem then is to choose a spectral response which suppresses undesirable spectral elements such as sunlight or expected countermeasures, or which emphasizes desirable target characteristics. In general, this is not an involved procedure.

In the second case, the sensor is background-limited, so that the detector background and the atmospheric transmission determine the detector noise. Thus, in peculiar atmospheres, or in severely signal-to-noise-ratio limited applications, it may prove desirable to maximize the sensor SNR by using a cold filter which immediately precedes the detector and which has an appropriate spectral response.

We recall from previous sections that

$$\text{SNR} \propto \int_0^\infty \tau_a(\lambda)\, \tau_0(\lambda)\, D^*(\lambda) \frac{\partial W_\lambda(T_B)}{\partial T}\, d\lambda, \tag{5.91}$$

where $\tau_a(\lambda)$ is the atmospheric transmission. If a cooled spectral filter is used, it is useful to separate the optical transmission into an uncooled part $\tau_{0u}(\lambda)$ and a cooled part $\tau_{0c}(\lambda)$. Then

$$\text{SNR} \propto \int_0^\infty \tau_a(\lambda)\, \tau_{0u}(\lambda)\, \tau_{0c}(\lambda)\, D^*(\lambda) \frac{\partial W_\lambda(T_B)}{\partial T}\, d\lambda. \tag{5.92}$$

For BLIP performance, we recall that

$$D^*(\lambda) = D^{**}_{BLIP}(\lambda)\, \eta_q^{1/2}(\lambda) \left[\frac{\pi}{\Omega_{cs}} \right]^{1/2}. \tag{5.93}$$

The BLIP limit for photoconductors is

$$D^{**}_{BLIP}(\lambda) = \frac{\lambda}{2hcQ_B^{1/2}}, \tag{5.94}$$

where when a cooled filter is used,

$$Q_B = \int_0^\infty \tau_{0c}(\lambda)\, Q_\lambda\, d\lambda. \tag{5.95}$$

Combining all of these relations,

$$\text{SNR} \propto \int_0^\infty \tau_a(\lambda)\, \tau_{ou}(\lambda)\, \tau_{oc}(\lambda)\, \frac{\lambda\, \eta_q^{1/2}(\lambda)\, \dfrac{\partial W_\lambda(T_B)}{\partial T}\left[\dfrac{\pi}{\Omega_{cs}}\right]^{1/2}}{2hc\left[\displaystyle\int_0^\infty \tau_{oc}(\lambda)\, Q_\lambda\, d\lambda\right]^{1/2}}\, d\lambda \qquad (5.96)$$

As

$$Q_\lambda = \frac{\lambda\, W_\lambda}{hc}\ , \qquad\qquad\qquad (5.97)$$

and dropping constant factors,

$$\text{SNR} \propto \frac{\displaystyle\int_0^\infty \tau_a(\lambda)\, \tau_{ou}(\lambda)\, \tau_{oc}(\lambda)\, \eta_q^{1/2}(\lambda)\, \frac{\partial W_\lambda(T_B)}{\partial T}\, d\lambda}{\left[\displaystyle\int_0^\infty \tau_{oc}(\lambda)\, \lambda\, W_\lambda(T_{BC})\, d\lambda\right]^{1/2}}, \qquad (5.98)$$

where T_{BC} is the absolute temperature of the cooled background.

The problem of maximizing the ratio is now reduced to appropriate selection of the cold filter $\tau_{oc}(\lambda)$. Kleinhans[10] found the necessary (but not sufficient) conditions which maximize the SNR, and showed that the optimum cooled spectral filter is binary at each wavelength.

Kleinhans' analysis is applicable to systems for which

$$\text{SNR} \propto \frac{\displaystyle\int_0^\infty \tau_{oc}(\lambda)\, \Phi_T(\lambda)\, d\lambda}{\left[\displaystyle\int_0^\infty \tau_{oc}(\lambda)\, \Phi_B(\lambda)\, d\lambda\right]^{1/2}} = \frac{S}{N}\ , \qquad (5.99)$$

where $\Phi_T(\lambda)$ and $\Phi_B(\lambda)$ are target and background spectral quantities, respectively. Using variational principles to maximize $(S/N)^2$, using the constraint $0 \leqslant \tau_{oc}(\lambda) \leqslant 1$, and denoting the optimum filter by $\tau'_{oc}(\lambda)$ such that the optimum SNR is

$$\frac{S_o}{N_o} \triangleq \frac{\int_0^\infty \tau_{oc}(\lambda)\, \Phi_T(\lambda)\, d\lambda}{\left[\int_0^\infty \tau'_{oc}(\lambda)\, \Phi_B(\lambda)\, d\lambda\right]^{1/2}}, \tag{5.100}$$

Kleinhans proved that

$$\tau'_{oc}(\lambda) = 1 \text{ when} \frac{\Phi_T(\lambda)}{\Phi_B(\lambda)} > \frac{S_o}{2N_o} \tag{5.101}$$

and

$$\tau'_{oc}(\lambda) = 0 \text{ when } \frac{\Phi_T(\lambda)}{\Phi_B(\lambda)} < \frac{S_o}{2N_o}. \tag{5.102}$$

In our case we have

$$\Phi_T(\lambda) = \tau_a(\lambda)\, \tau_{ou}(\lambda)\, \eta_q^{1/2}(\lambda)\, \frac{\partial W_\lambda(T_B)}{\partial T}, \tag{5.103}$$

and

$$\Phi_B(\lambda) = \lambda\, W_\lambda(T_B). \tag{5.104}$$

The numerical integrations required to find the optimum filter may be prohibitively time-consuming unless the form of the filter is approximately dictated by such effects as sharp spectral cut-on's of cut-off's due to the atmosphere, the optical materials, or the detector. In that case, the filter may take a bandpass form, and the solution for the optimum filter is relatively simple. Since Kleinhans' proof does not assure sufficiency, it is possible that the filter resulting from blind application of the technique may be less optimal than a conventionally-chosen filter.

5.14 Infrared Optics Collection Efficiency Factors

Often it is useful to define and calculate various optical efficiency factors for use in sensitivity equations. The fundamental property of a lens system is the transmission function $T(x,y,\theta,\lambda)$ which describes the attenuation of a ray of light of wavelength λ making an angle θ with the optical axis and directed at lens coordinates (x,y). We will assume that $T(x,y,\theta,\lambda)$ is separable into a real (as opposed to complex) pupil function $p(x,y)$ and a spectral transmission $\tau_0(\theta,\lambda)$,

$$T(x,y,\lambda) = p(x,y)\, \tau_0(\theta,\lambda) \ . \tag{5.105}$$

The effective collecting area A_0 of the optic is

$$A_0 = \iint\limits_{-\infty}^{\infty} p(x,y) dx\, dy \tag{5.106}$$

and average spectral transmission $\overline{\tau}_0$ for a particular field angle θ may be defined by

$$\overline{\tau}_0(\theta) = \frac{1}{\lambda_2 - \lambda_1} \int_{\lambda_1}^{\lambda_2} \tau_0(\theta,\lambda) d\lambda. \tag{5.107}$$

An effective spectral transmission may be defined by

$$\tau_0 = \frac{\displaystyle\int_{-\infty}^{\infty} \tau_0(\lambda)\, D^*(\lambda)\, \frac{\partial W_\lambda(T_B)}{\partial T} d\lambda}{\displaystyle\int_{-\infty}^{\infty} D^*(\lambda)\, \frac{\partial W_\lambda(T_B)}{\partial T} d\lambda} \ . \tag{5.108}$$

However, for a background limited detector, τ_0 is not a good figure of merit because atmospheric attenuation removes many spectral signal components, thereby reducing the signal-to-noise ratio.

A more appropriate quantity is the spectral efficiency η_s:

$$\eta_s = \frac{\int_{-\infty}^{\infty} \tau_o(\lambda)\, \tau_a(\lambda)\, D^*(\lambda)\, \frac{\partial W_\lambda(T_B)}{\partial T}\, d\lambda}{\int_{-\infty}^{\infty} \tau_a(\lambda)\, D^*(\lambda)\, \frac{\partial W_\lambda(T_B)}{\partial T}\, d\lambda}.$$

(5.109)

5.15 Relating Performance Parameters to Operational Effectiveness

Specifiers and designers of thermal imaging systems must be able to select system parameters which will satisfy the performance requirement at hand. The intent of this section is to outline a procedure which will reasonably ensure the specifier that his performance requirements will be met. The experiences of military thermal imaging developments are enlightening in this respect.

Many rules of thumb which were approximately valid in the early days of military thermal imaging do not now apply to the full range of current applications. For example, when most system designs followed the same convention, as in parallel-processed systems, the detector angular subtense and the display element equivalent angular subtense had approximately the same value and dominated the system MTF.

Consequently one could speak of a "resolution", meaning detector angular subtense, of one milliradian, and imply a great deal about the system's performance. The systems were relatively coarse in resolution and possible recognition ranges were short. Consequently if one knew that a system with a resolution of so many milliradians recognized an object at so many meters then one could scale up or down in resolution and confidently infer the resulting recognition capabilities. Implicit in this practice were the assumptions that thermal sensitivity was unchanged, atmospheric absorption was unchanged, and atmospheric MTF effects were negligible.

When ranges were extended as systems improved, and as rigid design conventions were dropped, the latter two assumptions were violated, recognition failed to scale with resolution and resolution in the old sense of the term became much less meaningful. Consequently, new performance parameters and more complex performance theories had to be developed.

The explosion of FLIR technology has immensely complicated the problem, so that thermal systems are now as elusive to define and standardize as are thermal scenes and observers. Two nominally equivalent systems may have different signal transfer functions, raster effects, frame rates, and magnifications. Performance is sensitive to these variations, and it is commonplace to find in a competition that one system is preferable to another even though both are nominally the same.

These uncertainties render absurd such seemingly accurate statements as "with these performance specifications and parameters the system will distinguish a man from a bush at one hundred meters in moderate clutter on a clear warm night". Another observer on another night using a similarly specified system under similar conditions probably will not duplicate that performance. However, users frequently must express their needs in terms of such statements and suppliers must then translate the ambiguities into specific systems parameters. The danger of this practice is that it is easy to overdesign by specifying system parameters to an accuracy greater than that to which the problem is describable.

Any thermal imaging problem statement is affected by the forty or more factors discussed in Section 10.1 which influence search performance. Some uncontrollable variables which exhibit wide variations are target and background characteristics such as operating state, clutter, context, contrast, atmospheric transmission and MTF, and radiation exchange between the target and its environment. In addition, those variables which are controllable by the observer will vary around nominal positions and affect performance. Some examples are surround luminance, sensor and display focus, and brightness and contrast settings. Other factors which are less under the control of the observer and which have enormous impact on performance are motivation, frame of mind, and information extraction skill.

Numerous procedures exist for predicting the field performance of thermal imaging systems. However, given the strong variables in any tactical situation, it is the author's opinion that no thermal imaging model will ever accurately predict *a priori* the results of a field test. Successful *ex post facto* theoretical explications of field test results should not obscure the fact that target acquisition is too complex a phenomenon to measure completely, much less to predict accurately. This is quite clear when the variations in the predictions of two widely differing models fall within the envelope of the measured performance variations.

System specifiers, nonetheless, must know what kind of system is needed to perform a tactical function, and cannot always confidently rely on a paper analysis of a problem. Specifiers also understand from experience approximately how sensor variations affect field performance, and may be reluctant to use theoretical constructs rather than their own judgment. The practical solution to this dilemma is to perform a field test under the conditions of interest with a sensor having characteristics similar to those which are suspected will do the job, and to extrapolate up or down in sensor performance to get the desired performance.

This approach is practical because standard imagery and empirical data are readily available which indicate the effects of differential performance changes. This is not to say that we should not continue searching for simple performance summary measures which correlate with field performance, but that we should not rely on analysis alone.

Basing system design on extrapolations of field data is impractical, however, when all system parameters continue to be treated as design variables. This makes it difficult to assess the effect of varying one or two parameters when the others are not constant. Those who specify systems can have simple design criteria only if some variables are fixed. The following conventions are suggested to make it possible:

1. fix the per frame, per resolution element, displayed NETD at 0.2°C except for the most severe applications.
2. require image uniformity at least as good as commercial television quality, and preferably with no raster and no fixed or moving pattern defects.
3. fix the effective frame rate at 30 fps.
4. fix the magnification at an optimum value of σM = constant.
5. do not allow a thermal maximum dynamic range of less than 20°C above ambient.
6. require system MTF's to be better than or equivalent to a gaussian MTF with a sigma of one-half of the detector angular subtense.

These conventions are in some cases already inherent in good FLIR practice.

5.16 Summary of Primary FLIR Image Quality Analysis Equations

It is useful at this point to summarize what has been proposed so far about FLIR image quality measures. Two functions and one parameter are generally believed to provide a good first-order estimate of thermal imaging system quality. The functions are the minimum resolvable temperature difference (MRTD) and the modulation transfer function (MTF), and the parameter is the noise equivalent temperature difference (NETD). The MTF and summary measures derived from it correlate well with the subjective impression of image sharpness in noise-free imagery. The MRTD describes four-bar resolving power in noisy imagery and is related to target recognition capability. The NETD is a convenient summary measure of broad area thermal sensitivity, but is not related in a simple way to any practical task or subjective correlate of image quality

A typical FLIR has five component contributions to the overall system MTF: the optical blur, the detector geometry, the detector electrical frequency response, the electronic signal processing response, and the display blur. The optics in most FLIR designs are diffraction-limited on-axis, and usually have circular unobscured apertures. For this case, and for entrance pupil diameter D_0 and mean wavelength λ, the MTF is:

$$\tilde{r}_0 = \frac{2}{\pi}\left\{ \arccos\left(\frac{f}{f_c}\right) - \left(\frac{f}{f_c}\right)\left[1 - \left(\frac{f}{f_c}\right)^2\right]^{1/2} \right\}, \tag{5.110}$$

where $f_c = D_0/\lambda$ is the absolute cutoff frequency in cycles/radian.

The detector responsivity contour is usually approximated by a rectangular function of angular dimension α in the scan direction. The MTF is:

$$\tilde{r}_d = \frac{\sin(\pi\alpha f)}{\pi\alpha f}. \tag{5.111}$$

The frequency response of FLIR detectors usually behaves like that of a single-RC circuit with characteristic frequency f_τ given by $1/(2\pi\tau_R)$, where τ_R is the responsive time constant. The MTF is:

$$\tilde{r}_\tau = \left[1 + \left(\frac{f}{f_\tau}\right)^2\right]^{-1/2}. \tag{5.112}$$

Electronic processing MTF's are composed of contributions from preamplifiers, post amplifiers, boost circuits, and video processors. For analysis purposes one usually assumes an overall electronics processing MTF of unity within the bandpass, where any rolloffs have been compensated by the boost circuit.

The video monitor is usually a TV-type monitor and is adequately represented by a gaussian MTF with a monitor spread function standard deviation σ_m related to the shrinking-raster line separation s by

$$\sigma_m = 0.54\ s. \tag{5.113}$$

The MTF is

$$\tilde{r}_m = \exp\left(-2\pi^2\sigma_m^2 f^2\right). \tag{5.114}$$

The reference $NETD_R$ for the case of a clear circular aperture and a rectangular detector is given by

$$NETD_R = \frac{4\sqrt{ab}\ \Delta f_R}{\sqrt{n}\ \alpha\beta\ D_o^2\ D^*(\lambda_p)\int_0^\infty \tau_o(\lambda)\frac{\partial W_\lambda(T_B)}{\partial T}\frac{D^*(\lambda)}{D^*(\lambda_p)}d\lambda} \qquad (5.115)$$

where a and b are the detector linear dimensions, α and β are the detector angular dimensions, D_o is the optic diameter, Δf_R is the reference electronic equivalent noise bandwidth, and n is the number of integrated detectors in the case of a serial scanner. For a parallel processor, n = 1. The noise reference bandwidth Δf_R is given by

$$\Delta f_R = \int_0^\infty g^2\ (f)\ \tilde{r}_R^2\ d\ f, \qquad (5.116)$$

where g(f) is the noise voltage spectrum into a reference filter having an MTF of $\tilde{r}_R$.

The MRTD is predictable by many similar and equivalent equations. One of these is

$$MRTD = \frac{3\ NETD_R\left[\frac{\int_0^\infty g^2(f)\tilde{r}_e^2\ \tilde{r}_m^2\ \text{sinc}^2\ (f/2f_T)df}{\Delta f_R}\right]^{1/2}\left[f_T\beta\right]^{1/2}}{\tilde{r}_s\ (T_e\ \dot{F})^{1/2}} \qquad (5.117)$$

where f_T is the MRTD target frequency in appropriate units, T_e is the eye integration time, and $\dot{F}$ is the frame rate.

For the special case of a gaussian system MTF and component MTF's such that there is white noise on the display, this MRTD equation reduces to

$$MRTD = 3\frac{NETD_R}{(\Delta f_R)^{1/2}}\ (f_T\ [cy/mr])\ (\alpha/\tau_d)^{1/2}\beta^{1/2}$$

$$\exp\ (2\pi^2\ \sigma_s^2\ f_T^2)\ (T_e F)^{-1/2} \qquad (5.118)$$

where τ_d is the detector dwelltime.

The primary usefulness of the MRTD function is that it is an excellent summary measure which indicates both low frequency thermal sensitivity and high frequency limiting resolution. MRTD provides a good method for comparing systems because it includes the effects of noise, resolution, temporal integration, spatial integration, and spurious image defects. In addition, MRTD should correlate well with practical tasks such as target recognition and identification in noisy imagery. MTF, on the other hand, correlates with recognition and identification performance in noise-free imagery. Depending on the task, the integral of the MTF, the integral of the square of the MTF, or the MTF 5 percent modulation frequency may correlate well with performance.

REFERENCES

[1] W.K. Weihe, "Classification and Analysis of Image-Forming Systems", Proc. IRE, 47, pp 1593-1604, September 1959.

[2] J.A. Hawkins, "Generalized Figures of Merit for Infrared Imaging Systems", Report DRL-TR-68-12, Defense Research Laboratory, University of Texas at Austin, February 1968.

[3] R.D. Hudson, *Infrared System Engineering*, Wiley-Interscience, 1969.

[4] N.F. Soboleva, "Calculation of the Sensitivity of an IR Scanner", Soviet Journal of Optical Technology, 37, pp 635-637, October 1970 (English Translation).

[5] O.H. Schade, Sr., "An Evaluation of Photographic Image Quality and Resolving Power", JSMPTE, 73, pp 81-120, February 1964.

[6] R. Genoud, personal communication, Hughes Aircraft Company, Culver City, California.

[7] R.L. Sendall, personal communications, Xerox Electro-Optical Systems, Pasadena, California.

[8] G.S. Hopper, personal communication, Texas Instruments, Inc., Dallas, Texas.

[9] E. Dryden, personal communication, Aerojet Electro-Systems Corporation, Azusa, California.

[10] W.A. Kleinhans, "Optimum Spectral Filtering for Background-Limited Infrared Systems", JOSA, 55, pp 104-105, January 1965.

CHAPTER SIX — OPTICS

6.1 Optical Elements

The simplest categorization of an optical system is the distinction between a refractive and a reflective system. The selection of a reflector, a refractor, or a combination of both (a catadioptric or catoptric) for a particular requirement is complicated by the fact that each type has merits and faults.

For a specific lens diameter and focal length, the reflection and absorption losses of a refractor will be greater than the absorption losses of a reflector, but the overall collection efficiency of the reflector may or may not be better depending on the amount of central obscuration. A reflector tends to be lighter in weight and less expensive than an equivalently sized refractor, but the reflector may not have the same image quality. For example, spherically-surfaced reflectors cannot be corrected for off-axis aberrations without use of a refractive corrector plate, whereas spherically-surfaced refractors facilitate aberration-balanced optical designs. Since spherical surfaces are less expensive to produce than aspherics, this is an important consideration. On the other hand, refractors suffer from chromatic aberrations, while reflectors do not. The prevailing bias in FLIR design is in favor of refractive systems since so many ingenious concepts exist which enable good quality to be achieved in a compact package. However, the cost advantages of reflectors make it necessary to re-evaluate that bias each time a new optical system is conceived.

Beyond this simple distinction it is difficult to discuss optics independently of scan mechanisms because the two are closely related, the type of scanner chosen dictating the optical design to a great extent, and *vice versa*. Thus a brief discussion of scanners is in order. There are two types of scanners: convergent beam scanners and parallel beam scanners.

A convergent beam scanner operates in a part of the optical system where rays are converging, or approaching a focus. A convergent beam scanner typically is placed between the inner-most focusing refractor and the detector array. A parallel beam scanner operates in a part of the optical system where the rays are parallel or collimated. If the scanner is located in front of the optical system, the term "object plane scanner" applies. A parallel beam scanner may also be placed between a relay lens pair. Two examples are shown in Figure 6.1.

Which type of scanner is preferred for a particular application depends on what trade-offs can be made. A convergent beam scanner generally requires fewer elements in its optical system but the scanner introduces aberrations and distortions into the convergent beam which must be corrected.

A parallel beam scanner generally requires a more complicated optic, but image quality in a parallel beam scanner is not as sensitive to deviations from scan mirror flatness as is image quality in a convergent beam. An object plane scan element must be as large as the maximum projected dimension of the entrance pupil, so it is usually desirable to confine parallel scanning to the interior of an optical system. With these distinctions in mind, we may proceed to the discussion of optical systems.

6.2 Basic Optical Processes

There are two idealized processes of interest to us here: refraction at the interface between two nonconducting transparent media as in a refractive lens, and reflection at the surface of a perfect conductor as in a mirror. Consider Figure 6.2, which shows an interface between two nonconductors. Each material is characterized by a parameter called refractive index n, the ratio of the speed of light in the material to the speed of light in a vacuum. For nonmagnetic materials, n is the square root of the dielectric constant evaluated at the optical frequency of interest. The refractive indices of the useful thermal imaging materials are given in Section 6.8.

Figure 6.2 demonstrates Snell's Law of refraction, which is that for angles of incidence θ and of refraction θ' measured from the surface normal, and for refractive indices n and n'

$$\frac{\sin \theta'}{\sin \theta} = \frac{n}{n'} \tag{6.1}$$

or

$$\theta' = \arcsin \left(\frac{n}{n'} \sin \theta \right). \tag{6.2}$$

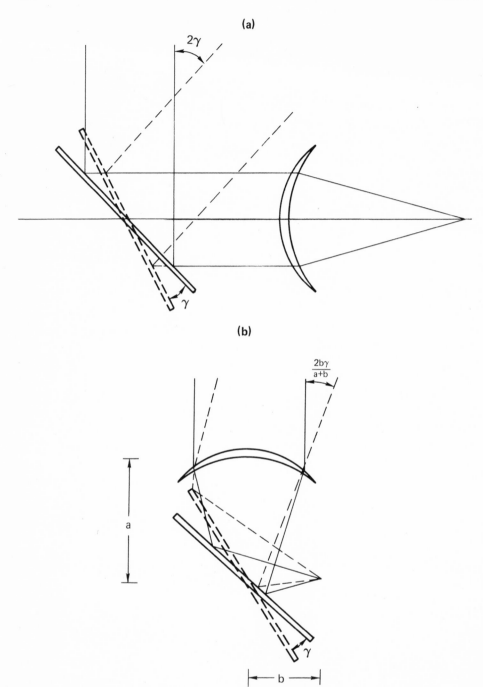

Figure 6.1 (a) A parallel beam and (b) a convergent beam scanner.

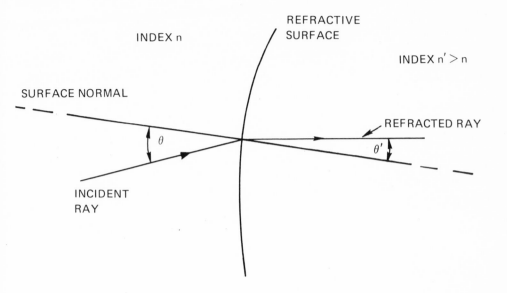

Figure 6.2 Refraction.

A simple way to remember this effect is to note that the refracted ray is bent toward the normal when $n' > n$.

Reflection by a perfect conductor (a mirror) is demonstrated in Figure 6.3. The law of reflection is that

$$\theta = \theta'. \tag{6.3}$$

6.2.1 The Wedge

A useful optical component is the refractive wedge depicted in Figure 6.4. The wedge apex angle is W, the angle of the incident ray is φ_1, the normal to the incident surface is N_1, the angle of the exitant ray is φ_2, and the normal to the exitant surface is N_2. The indices from left to right are n_1, n_2, n_3. The ray deviation produced by a wedge is derived as follows. With reference to the symbols defined by Figure 6.4, using Snell's law at the first surface gives

$$\frac{\sin(\phi_2 - N_1)}{\sin(\phi_1 - N_1)} = \frac{n_1}{n_2}, \tag{6.4}$$

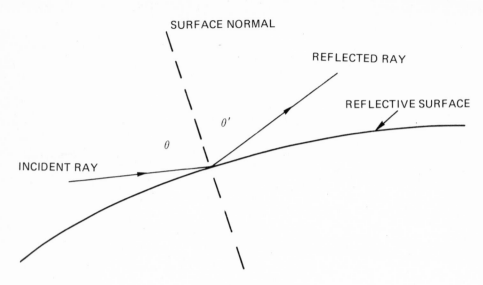

Figure 6.3 Reflection.

and the second surface gives

$$\frac{\sin (\phi_3 - N_2)}{\sin (\phi_2 - N_2)} = \frac{n_2}{n_3}.$$ (6.5)

Combining equations 6.4 and 6.5,

$$\sin (\phi_3 - N_2) = \frac{n_2}{n_3} \sin (\phi_2 - N_2) = \frac{n_2}{n_3} \sin (\phi_2 - N_1 - W)$$

$$= \frac{n_2}{n_3} \sin [(\phi_2 - N_1) - W]$$

$$= \frac{n_2}{n_3} \sin \left\{ \arcsin \left[\frac{n_1}{n_2} \sin (\phi_1 - N_1) \right] - W \right\}.$$ (6.6)

Then

$$\phi_3 = \arcsin \left\{ \frac{n_2}{n_3} \sin \left\{ \arcsin \left[\frac{n_1}{n_2} \sin (\phi_1 - N_1) \right] - W \right\} \right\} + N_2.$$ (6.7)

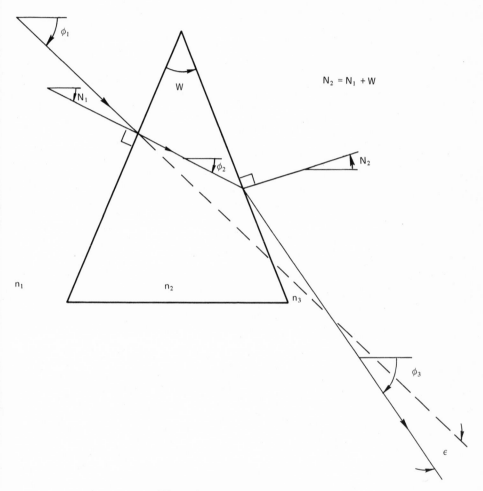

Figure 6.4 Refraction by a wedge.

For small $(\phi_1 - N_1)$, for $n_2 > n_1$, and for small W,

$$\phi_3 \cong \frac{n_1}{n_3} (\phi_1 - N_1) - \frac{n_2}{n_3} W + N_2.$$ (6.8)

For $n_1 = n_3 = 1$,

$$\phi_3 \cong \phi_1 - W (n_2 - 1),$$ (6.9)

and the ray deviation ϵ is

$$\epsilon = \phi_3 - \phi_1 \simeq -W (n_2 - 1).$$ (6.10)

For germanium with $n_2 = 4.003$,

$$\phi_3 = \arcsin \left\{ 4.003 \sin \left\{ \arcsin \left[\frac{1}{4.003} \sin (\phi_1 - N_1) \right] - W \right\} \right\} + N_2.$$

For W = 10 degrees and an equilateral wedge,

$N_1 = -5$ degrees, $N_2 = +5$ degrees, and

$$\phi_3 = \arcsin \left\{ 4.003 \sin \left\{ \arcsin [.2498 \sin (\phi_1 + 5°)] - 10° \right\} \right\} + 5°$$

$$\cong \phi_1 - 10° (3.003) = \phi_1 - 30.03°.$$

Selected values for this case are:

ϕ_1	ϕ_3 exact	ϕ_3 approximate
0	-32.526°	-30.03°
5°	-26.564	-25.03
10°	-21.026	-20.03
15°	-15.840	-15.03

Thus the approximation is not very accurate even for these small angles.

A wedge is a useful component because rotation of the wedge with a selected exitant ray as the axis causes the incident ray in object space to scan in a circle. This property may be used in scanners, as discussed in Section 7.5.

6.2.2 The Simple Lens

The simplest possible refractive focusing optical system consists of a single spherically surfaced lens defined as shown in Figure 6.5. The simple lens is characterized by four numbers: the radii of curvature R_1 and R_2, the refractive index n, and the thickness t. The usual sign conventions are that the sign of the radius of a surface is positive when the center of curvature is to the right of the surface, and that the sign of the radius is negative when the center is to the left of the surface. R_1 is the radius of the leftmost surface and R_2 is the radius of the rightmost surface.

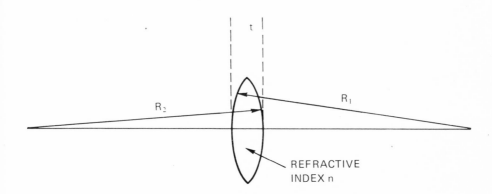

Figure 6.5 A spherically-surfaced simple lens.

The simple lens is called a thin lens when the thickness t is negligible. We are interested in the three modes of imaging of the thin lens which are shown in Figure 6.6. In the first mode, parallel rays of light from an infinitely distant point source are focused to an image at a distance f behind the lens. The quantity f is called the focal length, and to a first order approximation f determines the shift s in the image position of an infinitely distant object due to a small angular beam shift σ,

$$s = \sigma f, \tag{6.11}$$

as shown in Figure 6.7.

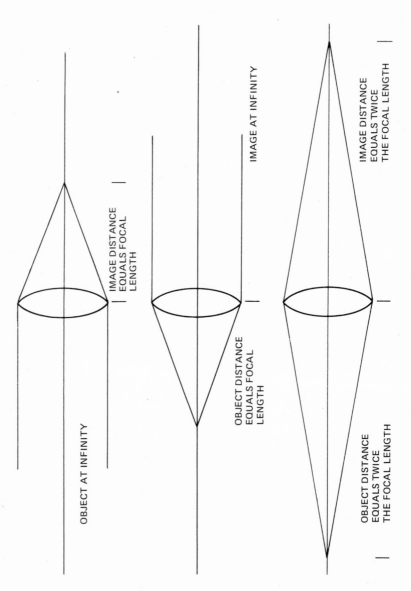

Figure 6.6 Imaging modes of a simple lens.

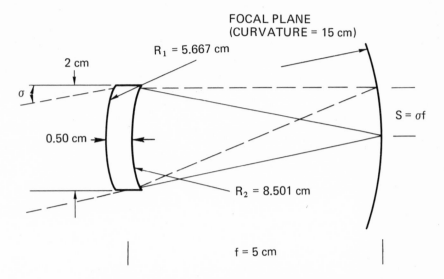

Figure 6.7 Focal property of a practical infrared lens.

In the second mode, a point source at a distance f in front of the lens is converted to a beam of parallel rays leaving the lens. In the third mode, a point object at a distance O neither at infinity nor at f is imaged as a point at distance i given by the thin lens equation

$$\frac{1}{O} + \frac{1}{i} = \frac{1}{f} \, , \tag{6.12}$$

where

$$1/f \overset{\Delta}{=} (n - 1) \left(\frac{1}{R_1} - \frac{1}{R_2} \right) \tag{6.13}$$

or

$$f = \frac{R_1 R_2}{(n - 1)(R_2 - R_1)} \, . \tag{6.14}$$

This relation is derived in Section 6.4.

The importance of the magnitude of the refractive index of a potential lens material is evident from this equation. The higher the index, the larger the radii may be for a particular focal length. This makes it easier to manufacture and to uniformly antireflection coat the lens, and makes the lens thinner, giving less absorption loss.

Figure 6.8 shows all of the possible cases of a thin lens where a lens is considered positive or convergent when an incident parallel beam is focused, or as negative and divergent when an incident parallel beam diverges. Note that two of these lenses are reversals of two others, so that there are only six thin lenses. The focal length of a thick lens with thickness t is

$$f = \frac{R_1 \, R_2}{(n-1) \left[(R_2 - R_1) + \left(\frac{n-1}{n} \right) t \right]}. \tag{6.15}$$

The change Δf in lens focal length from the zero thickness focal length f_o with increasing thickness is readily shown to be

$$\Delta f = - \frac{f_o}{(R_2 - R_1) \frac{t}{n} + 1}. \tag{6.16}$$

The lens law for the thick lens is derived in Section 6.4.

6.3 Generalized Lens Systems

Thick lenses and combinations of lenses are conveniently described by using an equivalent fictional system in which all refractions occur at two planes, called the first and second principal planes, shown in Figure 6.9. The intersections of the principal planes with the optical axis define first and second principal points H and H′. A collimated beam incident from the right comes to a focus at the first focal point F, and a collimated beam from the right focuses at the second focal point F′. The distances from the focal points to the principal points are the first and second focal lengths f and f′ as shown in Figure 6.10.

When object space and image space have the same index, and if object distance O is measured from H, and image distance i is measured from H′, the lens law remains

$$\frac{1}{O} + \frac{1}{i} = \frac{1}{f'}. \tag{6.17}$$

$$f = \frac{R_1 \, R_2}{(n-1)(R_2 - R_1)} \, , \, |R_1| = r_1 \, , \, |R_2| = r_2$$

For $R_1 > 0$, $R_2 < 0$

$$f = \frac{r_1 \, r_2}{(n-1)(r_1 + r_2)} \; > 0 \; \text{(converging)}$$

double convex    or   plano convex or

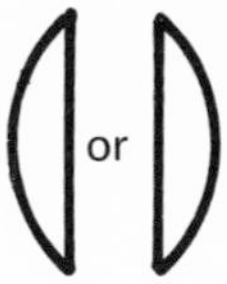

For $R_1 > 0$, $R_2 > 0$

$$f = \frac{r_1 \, r_2}{(n-1)(r_2 - r_1)}$$

$r_2 > r_1$, $f > 0$ (converging)

meniscus    or

$r_1 > r_2$, $f < 0$ (diverging) .

meniscus    or   plano concave

For $R_1 < 0$, $R_2 > 0$

$$f = -\frac{r_1 \, r_2}{(n-1)(r_1 + r_2)} \, , \, f < 0 \quad \text{(diverging)}$$

double concave    or   plano concave

Figure 6.8 Simple lenses.

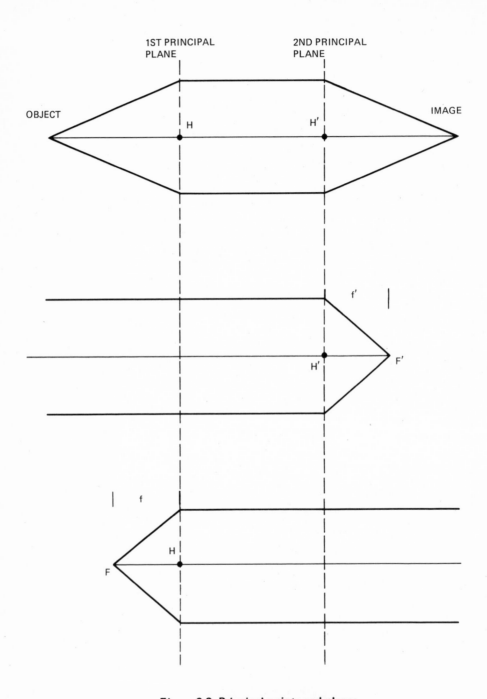

Figure 6.9 Principal points and planes.

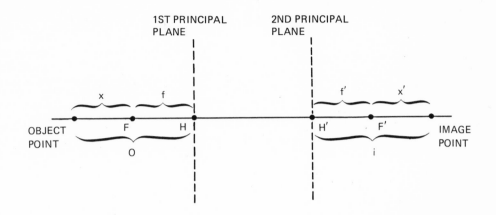

Figure 6.10 Image and object distances for a generalized lens system.

When object distance is denoted by x and is measured from F, and image distance is denoted by x' and measured from F', the lens law becomes

$$f\, f' = x\, x'. \tag{6.18}$$

The locations of the principal planes are calculated as described in Section 6.4.

An optical system typically is characterized by three parameters: the field of view, the effective clear aperture diameter, and the effective focal length. The field of view of an optical system consists of the shape and angular dimensions of the cone in object space within which object points are imaged by the system. For example, a field of view (FOV) may be described as being rectangular with dimensions of 4 degrees wide by 3 degrees high. The FOV is usually defined by a field stop, an opaque aperture placed somewhere in the optical system to block rays which originate from outside of the desired FOV. Figure 6.11 shows two typical field stops. The FOV of a FLIR system, however, is not always determined by the FOV of its lens, because the scan angles or the portion of the video selected for display may further limit the field coverage.

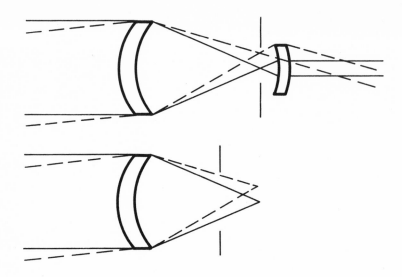

Figure 6.11 Typical field stop arrangements.

The entrance pupil consists of the shape and linear dimensions of the area over which the optical system accepts radiation. It may be a real aperture or a projection of an aperture stop inside the system. The ratio of focal length to clear aperture diameter D_0 is defined as focal ratio or F/number,

$$F/\# = f/D_0 \ . \tag{6.19}$$

If a small radiation-detecting element of dimension "a" is placed at the focal surface, the angular projection α of this detector through the optical system is given by

$$\alpha = \frac{a}{f} \ . \tag{6.20}$$

6.4 Matrix Representation of Lens Systems

The principle of operation of an optical system may be perceived rather quickly by performing a geometrical raytrace using the matrix representation summarized here. Klein[1] is a good source for this technique. Any ray is completely characterized at a point (x,y,z) by prescribing the projections onto the x, y, and z axes of a unit vector tangent to the ray at the point. These projections are called direction cosines because they are the cosines of the angles the unit vector makes with the axes. The direction cosines are the components of this unit vector and are denoted by (l,m,n), as shown in Figure 6.12.

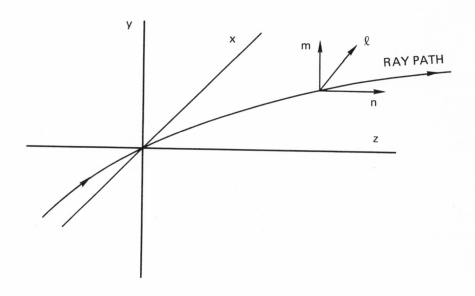

Figure 6.12 Direction cosines.

Raytracing is the process of following an incident ray specified by $(x, y, z; \ell, m, n)$ through a system until it exits as$(x', y', z'; \ell', m', n')$. An all-refractive system is easy to analyze when attention is confined to meridional rays, those rays contained by a plane which includes the optical axis. Then any incident ray can be considered in a coordinate system in which the ray motion is confined to two directions. Since all the ray changes consist of discrete refractions and of translations, the two directions are independent of each other. Thus we may consider ray motion in x together with its direction cosine ℓ, y with m, and z with n. In a reflective system, however, x and ℓ depend on y, z, m, and n because a reflection couples the coordinates.

Figure 6.13 shows a ray translation from a point y_1 to a point y_2 with constant direction cosine. This translation is represented by

$$\begin{pmatrix} y_2 \\ m_2 \end{pmatrix} = \begin{pmatrix} 1 & d_{12} \\ 0 & 1 \end{pmatrix} \begin{pmatrix} y_1 \\ m_1 \end{pmatrix}, \tag{6.21}$$

or

$$y_2 = y_1 + d_{12}\, m_1 \qquad\qquad (6.22)$$

and

$$m_2 = m_1\ . \qquad\qquad (6.23)$$

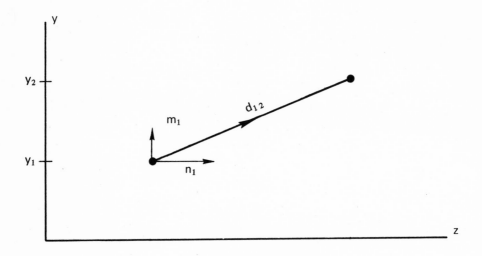

Figure 6.13 Ray translation.

Refraction at a surface of radius R going from index n to n′ was shown in Figure 6.2 and is given by

$$\begin{pmatrix} y' \\ m' \end{pmatrix} = \begin{pmatrix} 1 & 0 \\ \dfrac{-p}{n'} & \dfrac{n}{n'} \end{pmatrix} \begin{pmatrix} y \\ m \end{pmatrix} \qquad\qquad (6.24)$$

where p is called the power of the surface and is given by

$$p = \frac{n' - n}{R}\ . \qquad\qquad (6.25)$$

Refraction by a thin lens going from index n_1 to lens index n_2 to index n_3 as shown in Figure 6.14 is described by:

$$\begin{pmatrix} y_2 \\ m_2 \end{pmatrix} = \begin{pmatrix} 1 & 0 \\ -p & \dfrac{n_1}{n_3} \\ \dfrac{}{n_3} & \end{pmatrix} \begin{pmatrix} y_1 \\ m_1 \end{pmatrix}, \tag{6.26}$$

where p is called the power of the lens and is given by

$$p = \frac{n_2 - n_1}{R_1} + \frac{n_3 - n_2}{R_2} = \frac{1}{f} . \tag{6.27}$$

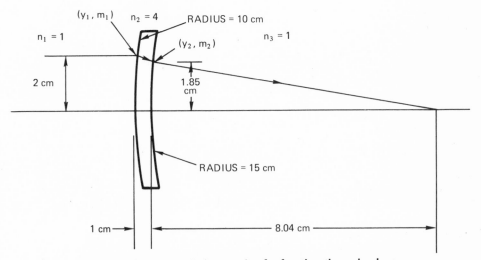

Figure 6.14 A numerical example of refraction through a lens.

The action of a thin lens on rays terminating a distance d to the right of the lens is*

$$\begin{pmatrix} y' \\ m' \end{pmatrix} = \begin{pmatrix} 1 & d \\ 0 & 1 \end{pmatrix} \begin{pmatrix} 1 & 0 \\ -p & 1 \end{pmatrix} \begin{pmatrix} y \\ m \end{pmatrix}$$

$$= \begin{pmatrix} y(1 - dp) + dm \\ m - py \end{pmatrix} . \tag{6.28}$$

---

*Recall that matrix multiplication is associated but not commutative, so the sequence of the matrices is significant.

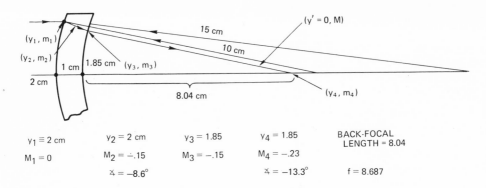

Figure 6.15 A ray trace calculation for an on-axis object at infinity.

For a ray parallel to the optical axis (that is, from a source at infinity), $m = 0$ and $y' = y - dpy$. The image ray intersects the axis ($y' = 0$) when $d = 1/p$. This distance is called the focal length f and is given by

$$\frac{1}{f} = \frac{n_2 - n_1}{R_1} + \frac{n_3 - n_2}{R_2} .$$ (6.29)

In air this becomes

$$\frac{1}{f} = \left(n - 1\right)\left(\frac{1}{R_1} - \frac{1}{R_2}\right) .$$ (6.30)

Figure 6.15 shows a numerical example of a simple lens ray trace.

The action of a simple lens may be derived from the basic processes of refraction and of translation as follows. As shown in Figure 6.16, let the light be transmitted from left to right, passing through indices n_1, n_2, and n_3. Then the matrix equation is

$$\begin{pmatrix} y_3 \\ m_3 \end{pmatrix} = \begin{pmatrix} 1 & 0 \\ \dfrac{-p_2}{n_3} & \dfrac{n_2}{n_3} \end{pmatrix} \begin{pmatrix} 1 & t \\ 0 & 1 \end{pmatrix} \begin{pmatrix} 1 & 0 \\ \dfrac{-p_1}{n_2} & \dfrac{n_1}{n_2} \end{pmatrix} \begin{pmatrix} y_1 \\ m_1 \end{pmatrix}$$ (6.31)

where

$$p_1 = \frac{n_2 - n_1}{R_1} \text{ and } p_2 = \frac{n_3 - n_2}{R_2}.$$

(6.32)

This reduces to

$$\begin{pmatrix} y_3 \\ \\ m_3 \end{pmatrix} = \begin{pmatrix} 1 - \dfrac{p_1 t}{n_2} & \dfrac{n_1}{n_2} t \\ \\ -\dfrac{p_2}{n_3} - \dfrac{p_1}{n_3} + \dfrac{p_1 p_2 t}{n_2 n_3} & \dfrac{n_1}{n_3} - \dfrac{n_1 p_2 t}{n_2 n_3} \end{pmatrix} \begin{pmatrix} y_1 \\ \\ m_1 \end{pmatrix}.$$

(6.33)

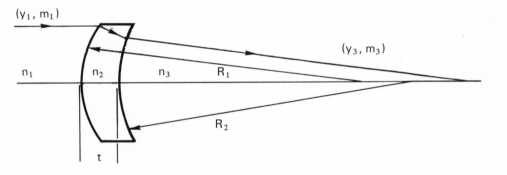

Figure 6.16 Notation for the derivation of the matrix equation
for a simple lens.

Thus the power p_L of the lens is given by

$$-p_L = -\frac{p_2}{n_3} - \frac{p_1}{n_3} + \frac{p_1 p_2 t}{n_2 n_3}$$

(6.34)

and the focal length f by

$$\frac{1}{f} = \frac{p_1}{n_3} + \frac{p_2}{n_3} - \frac{p_1 p_2 t}{n_2 n_3}$$

$$= \frac{n_2 - n_1}{n_3 R_1} - \frac{n_2 - n_3}{R_2} + \frac{(n_2 - n_1)(n_2 - n_3)t}{R_1 R_2 n_2 n_2}. \qquad (6.35)$$

If the lens is in air, $n_1 = n_3 = 1$ and

$$\frac{1}{f} = (n_2 - 1) \left[\frac{1}{R_1} - \frac{1}{R_2} + \frac{(n_2 - 1)t}{R_1 R_2} \right]. \qquad (6.36)$$

When the thickness is negligible,

$$\frac{1}{f} = (n_2 - 1) \left(\frac{1}{R_1} - \frac{1}{R_2} \right), \qquad (6.37)$$

as before.

Klein[1] shows that the thick lens may be treated as a thin lens by finding the principal planes, translating rays into and away from these planes, and using between the principal planes the matrix

$$\begin{pmatrix} 1 & 0 \\ \dfrac{-1}{fn_3} & \dfrac{n_1}{n_3} \end{pmatrix} \qquad (6.38)$$

where f is as given before. The first principal plane is located a distance Δ_1 from the apex of the first surface given by

$$\Delta_1 = \frac{n_1 (n_3 - n_2)}{n_2 n_3} \frac{tf}{R_2} - f \left(\frac{n_3 - n_1}{n_3} \right), \qquad (6.39)$$

where $\Delta_1 > 0$ indicates location to the right and $\Delta_1 < 0$ means location to the left. The second principal plane is located a distance Δ_2 from the second surface by

$$\Delta_2 = - \frac{(n_2 - n_1)tf}{R_1 n_2}. \qquad (6.40)$$

In air

$$\Delta_1 = -\frac{n-1}{n} \frac{tf}{R_2} \cdot \qquad (6.41)$$

and

$$\Delta_2 = -\frac{n-1}{n} \frac{tf}{R_1} \cdot \qquad (6.42)$$

The typical locations of the principal planes for the six simple lenses are shown in Figure 6.17.

Two thin lenses separated by a distance d (or two simple lenses whose principal planes are separated by d) have an effective focal length f_{12} given by:

$$\frac{1}{f_{12}} = \frac{1}{f_1} + \frac{1}{f_2} - \frac{d}{f_1 f_2} \cdot \qquad (6.43)$$

The principal planes are located at

$$\Delta_1 = -\frac{f_{12} d}{f_2} , \qquad (6.44)$$

and

$$\Delta_2 = \frac{f_{12} d}{f_1} \cdot \qquad (6.45)$$

The matrix for a two-lens telescope is:

$$M = \begin{pmatrix} -f_2/f_1 & f_1 + f_2 \\ 0 & -f_1/f_2 \end{pmatrix} \cdot \qquad (6.46)$$

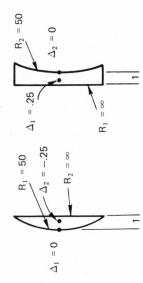

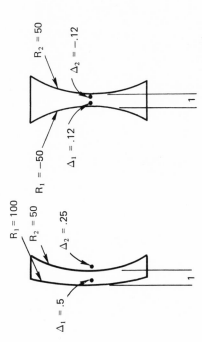

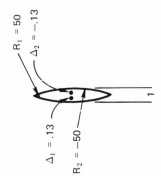

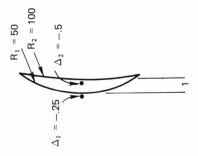

Figure 6.17 Principal planes for 6 simple lenses.

6.5 Matrix Representation of Plane Mirrors

The action of a plane mirror when the direction cosines of the normal pointing into the mirror are (L, M, N) is described by

$$
\begin{pmatrix} \ell_2 \\ m_2 \\ n_2 \end{pmatrix} = \begin{pmatrix} 1-2L^2 & -2LM & -2LN \\ -2LM & 1-2M^2 & -2MN \\ -2LN & -2MN & 1-2N^2 \end{pmatrix} \begin{pmatrix} \ell_1 \\ m_1 \\ n_1 \end{pmatrix} \qquad (6.47)
$$

As an example, consider the case of a mirror whose centroid is the y axis and which is rotated about its centroid parallel to y, normal to z, and with x = 0, as shown in Figure 6.18. Then

$$L = \cos\theta,$$
$$M = 0,$$
and $N = \cos(\pi/2 - \theta) = \sin\theta.$

The direction cosines after reflection are

$$
\begin{pmatrix} \ell \\ m \\ n \end{pmatrix} = \begin{pmatrix} 1-2\cos^2\theta & 0 & -2\sin\theta\cos\theta \\ 0 & 1 & 0 \\ -2\sin\theta\cos\theta & 0 & 1-\sin^2\theta \end{pmatrix} \begin{pmatrix} \ell_o \\ m_o \\ n_o \end{pmatrix}
$$

or

$$
\begin{aligned}
\ell &= (1-2\cos^2\theta)\ell_o - 2(\sin\theta\cos\theta)n_o, \\
m &= m_o \\
\text{and } n &= (2\sin\theta\cos\theta)\ell_o + (1-2\sin^2\theta)n_o.
\end{aligned}
$$

Because the mirror couples the direction cosines, it is necessary to treat the system using coordinate and cosine pairs up to the point of a refraction, to treat the refraction using all the cosines, and then to resume using the simpler notation. This gets so involved, however, that if the aim is a general understanding of the system operation it is much simpler to do a graphical raytrace. For the same reason it is usually clearer to perform a geometrical analysis of a mirror system if one wishes to derive equations for the system.

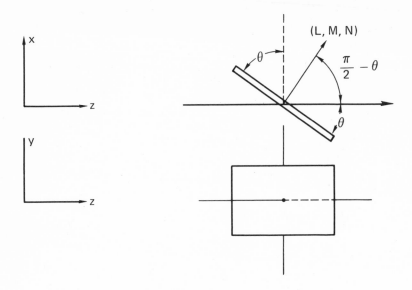

Figure 6.18 Reflection by a plane mirror.

6.6 A Detector Lens for the Example System

We will consider only refractive optics here because their use provides greater design flexibility and because they are more frequently used than reflectors. Spherical and flat surfaces are used almost exclusively because they are easier and less expensive to fabricate than aspherics. Aspherics tend to be used only in optics such as zoom systems where the use of spherically-surfaced-refractors would result in excessive complexity and transmission loss.

An ideal lens converts a converging or diverging spherical wave into a converging or diverging spherical wave, but all real lenses produce wavefront aberrations which cause deviations from the ideal. The primary wavefront aberrations are focusing errors, the Seidel aberrations (spherical aberrations, astigmatism, field curvature, distortion and coma), and chromatic aberrations. Detailed discussions of these and higher order aberrations and of designing for minimum aberration are treated extensively in many texts on optics, and in the context of infrared systems by the standard infrared texts. Attention here is confined to consideration of a few simple concepts.

The example system requires a lens having an aperture of 2 cm and a focal length of 5 cm. The first choice necessary is the selection of the type of lens to be used. For minimum complexity, it is obviously desirable to make the lens of a single element. We will discover some conditions for which this is possible.

6.6.1 Effects of Spherical Aberration and of Coma

The type of single element lens which is allowable is determined primarily by the aberrations which are tolerable. First, consider longitudinal spherical aberration. The deviations of the wavefront from a perfect spherical converging wavefront emanating from the exit pupil are given by the aberration function W. The aberration function produced by spherical aberration at a radius r in the exit pupil is expressable[1] as:

$$W = r^4 \, _0C_{40} \tag{6.48}$$

where $_0C_{40}$ is the spherical aberration coefficient. The maximum value of spherical aberration occurs at the pupil edge for $r = r_0$. For a thin lens of focal length f and index n which produces an image at an image distance i,

$$_0C_{40} = -\frac{1}{4f^3} \frac{1}{8n(n-1)} \left[\frac{n+2}{n-1} q^2 + 4(n+1)pq \right.$$

$$\left. + (3n+2)(n-1)p^2 + \frac{n^3}{n-1} \right], \tag{6.49}$$

where

$$q = \frac{R_1 + R_2}{R_2 - R_1}, \tag{6.50}$$

and

$$p = 1 - \frac{2f}{i} \tag{6.51}$$

If we use germanium for the lens, the mean wavelength in the 8 to 12 band is n = 4.004 and

$$_0C_{40} = -\frac{1}{4f^3} [0.0208q^2 + 0.200 \, pq + 0.4209p^2 + 0.2137]. \tag{6.52}$$

We are interested in an infinity focus for which i = f so that

$$p = -1 \text{ and } _0C_{40} = -\frac{1}{4f^3} [0.0208q^2 - 0.2q + 0.6346].$$

The aberration coefficient is approximately minimized when

$$q = 5 \text{ and } _0C_{40} = -\frac{1}{4f^3}(0.155).$$ (6.53)

Now note that the wavefront deviation $W = r_0^4 \, _0C_{40}$ becomes $W = -0.155 \, D_0/(64\,(F/\#)^3)$. For our detector lens, $D_0 = 2_{cm}$, $F/\# = 2.5$, and $W = 3.1 \times 10^{-4}$ cm.

This is a negligible amount of aberration (less than one-half wave). If it had been significant, the spherical aberration could have been partially compensated by a linear focal shift in the direction opposite to the spherical aberration. Obviously longitudinal spherical aberration cannot be totally eliminated in a distantly-focused thin lens of germanium because the aberration function never goes to zero in the long wavelength band. The value $q = 5$ implies $R_2 = 1.5 \, R_1$.

Next consider coma. According to Born and Wolf[3], the coma of a thin lens is eliminated when the following equations are satisfied:

$$\frac{1}{R_1} = \frac{2n+1}{n+1}\frac{1}{O} + \frac{n^2}{n^2-1}\frac{1}{f},$$ (6.54)

and

$$\frac{1}{R_2} = \frac{2n+1}{n+1}\frac{1}{O} + \frac{n^2-n-1}{n^2-1}\frac{1}{f},$$ (6.55)

where O is the object distance. For $O = \infty$ and $n = 4.004$,

$$R_1 = \frac{n^2-1}{n^2} f = 0.938 \, f \text{ and } R_2 = \frac{n^2-1}{n^2-n-1} f = 1.363f.$$

Thus

$$R_2 = 1.454 \, R_1,$$

which fortuitously is approximately the same solution as the radii which minimize spherical aberration.

If the example system consisted only of a detector lens and a scanner, the detector lens shown in Figure 6.19 satisfying the above equations would suffice. Here a center thickness of 0.50 cm has been allowed for structural rigidity. However, if auxiliary optics such as magnifying telescopes are to be used, this design may not suffice because the best tradeoff between aberration balance and overall complexity may dictate a two-element detector lens.

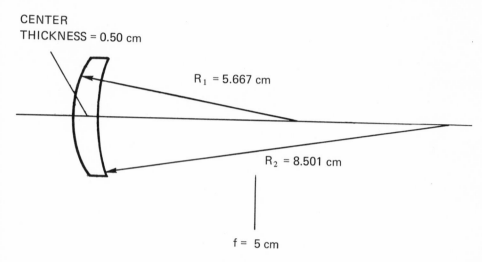

CENTER
THICKNESS = 0.50 cm

R_1 = 5.667 cm

R_2 = 8.501 cm

f = 5 cm

Figure 6.19 An example detector lens with minimum coma and spherical aberration.

6.6.2 Effect of Chromatic Aberration

The longitudinal chromatic aberration, or the difference in focus between the two extreme wavelengths in the bandpass, is shown in Figure 6.20. It is found by calculating the difference between the focal lengths for the two wavelengths:

$$\Delta f = f(n = 12 \ \mu m) - f(n = 8 \ \mu m).$$

First, define

$$k = \frac{R_1 R_2}{R_1 + R_2},$$

$$n = \frac{n(8) + n(12)}{2},$$

and

$$\Delta n = n(8) - n(12).$$

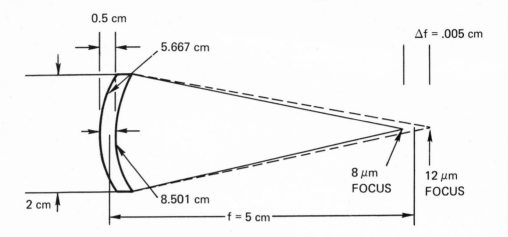

Figure 6.20 **Longitudinal chromatic aberration of the
example detector lens.**

Then the separation of the two foci is:

$$\Delta f = \frac{k}{n - \dfrac{\Delta n}{2} - 1} - \frac{k}{n + \dfrac{\Delta n}{2} - 1}$$

$$= k \left[\frac{\Delta n}{n^2 - 2n - \left(\dfrac{\Delta n}{2}\right)^2 + 1} \right]$$

$$\simeq f \frac{\Delta n}{n - 1}, \tag{6.56}$$

where f is defined for the mean index n. A dispersion index V (also called the Abbé number) is defined by

$$V = \frac{n - 1}{\Delta n}, \tag{6.57}$$

so that

$$\Delta f \simeq \frac{f}{V}. \tag{6.58}$$

The shift on either side of a best focus is

$$\Delta i = \frac{f}{2V} .$$
(6.59)

For the example detector lens of germanium,

$$V (8 \rightarrow 12 \ \mu m) = \frac{4.0038 - 1}{4.0053 - 4.0023} = 1001.27,$$

$$\Delta f = \frac{(5 \ cm)}{1001.27} \cong 5 \times 10^{-3} \ cm$$

and

$$\Delta i \cong \frac{\Delta f}{2} = 2.5 \times 10^{-3} \ cm.$$

This amount of linear defocus produces a dimensionless quality parameter Δ (see Section 6.6.4) of

$$\Delta = 2 \ \sin^2\left(\frac{D}{2f}\right)\frac{\Delta i}{\lambda}$$

$$= 0.197,$$
(6.60)

which produces negligible MTF loss.

Had this chromatic aberration not been acceptable, a technique of combining two or more separationless thin lenses to eliminate chromatic aberration would have to be used. Hertzberger and Salzberg[4] showed that materials are available in the thermal infrared which permit 2 lenses to match focus for 2, and possibly 3 wavelengths. In theory, 5 matched wavelengths imply perfect color correction, and in practice four wavelengths suffice. Such a perfectly achromatized lens assembly is called a superachromat. Some examples of achromats are shown in Figure 6.21.

6.6.3 Effect of Field Curvature

The surface of best focus for an optical system may have a variety of three-dimensional shapes. When the optical system is completely free of

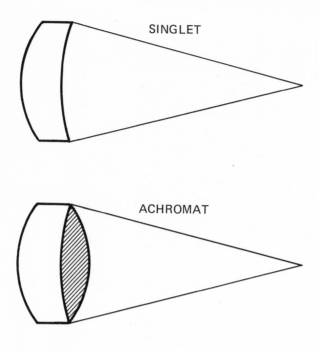

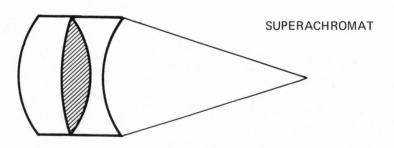

Figure 6.21 Converging beam scan optics.

aberrations, the curved image plane remaining is called the Petzval surface. The Petzval radius R_p of this field curvature is given by

$$R_p = \left[\sum_i \frac{1}{n_i f_i}\right]^{-1}, \tag{6.61}$$

where the n_i and the f_i are the indices and focal lengths of the individual lenses in the system. Thus an unaberrated lens produces a focal surface which is a section of a spherical shell, as shown in Figure 6.22. Obviously a vertical detector array contained in a plane and having its center at the optical axis will be increasingly out of focus as the array extremes are approached. This is shown geometrically in Figure 6.23.

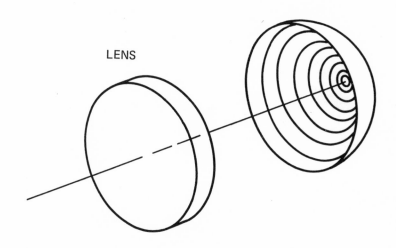

LENS

Figure 6.22 Petzval curvature.

The distance x from the lens to the Petzval surface as a function of the field angle θ may be derived as follows. By construction,

$$R_p^2 = (R_p - f)^2 + x^2 + 2x (R_p - f) \cos\theta.$$

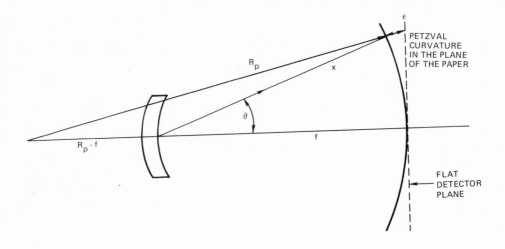

Figure 6.23 A section of the Petzval surface in the plane of the paper.

Solving for x we find

$$x = \sqrt{R_p^2 - \sin^2 \theta \ (R_p - f)^2} - (R_p - f) \cos \theta. \tag{6.62}$$

The defocus $\epsilon(\theta)$ from a flat detector plane is given by

$$\epsilon(\theta) = \frac{f}{\cos \theta} - x(\theta). \tag{6.63}$$

The geometrical blur diameter ω is shown in Figure 6.24, and is given by

$$\omega = \frac{y}{f + \epsilon} = \frac{D \, \epsilon}{2f \, (f + \epsilon)} \ .$$

This problem is exaggerated in a parallel scan system where an extended detector array is necessary for simultaneous total vertical field coverage by all detectors. A serial scanner having a short array and operating with a parallel beam scanner does not have this difficulty. Many solutions to the problem exist, among them designing the optical system to have a flat focal surface, fabricating the detectors on a continuously curving focal plane assembly, or constructing the array of flat canted segments as shown in Figure 6.25.

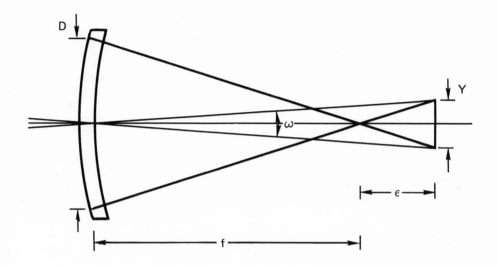

Figure 6.24 Blur due to focus error.

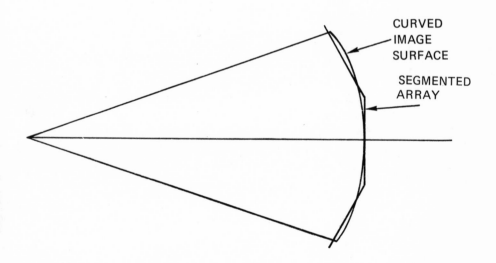

Figure 6.25 Segmented detector array.

For our example system with its 5 cm focal length detector lens constructed of germanium of index 4.004, the radius of field curvature is 20.02 cm. If the example system is implemented as a parallel scanner with 150 elements 0.005 cm high separated by 0.005 cm, the array height will be 1.495 cm. The field of view half-angle A/2 subtended by half of the array including interlace is:

$$\frac{A}{2} = \arctan\left(\frac{0.75 \text{ cm}}{5 \text{ cm}}\right) = 8.531°.$$

For $\theta = A/2$,

$$x = \sqrt{(20.02)^2 - (20.02 - 5)^2 \sin^2(8.531°)} - (20.02 - 5) \cos 8.531°$$

$$= 5.046 \text{ cm}.$$

The focus error at the array edge will be:

$$\epsilon = \frac{5 \text{ cm}}{\cos(8.531°)} - 5.046 = 0.01 \text{ cm}.$$

The geometrical blur resulting is

$$\omega = \frac{(2 \text{ cm})(.01 \text{ cm})}{2(5 \text{ cm})(5 \text{ cm} + 0.01 \text{ cm})} = 0.000399 \text{ radian} \simeq 0.4 \text{ mrad}.$$

Thus for the example system edge defocus is not a problem.

One optical solution to the problem of field curvature consists of redesigning the detector lens as a doublet having zero curvature such that two conditions are satisfied:

$$f_{12} = \frac{f_1 f_2}{f_1 + f_2} \tag{6.64}$$

and

$$\frac{1}{n_1 f_1} + \frac{1}{n_2 f_2} = 0, \text{ or } f_2 = \frac{-n_1}{n_2} f_1. \tag{6.65}$$

Thus the requirements are

$$f_1 = -\frac{n_2 - n_1}{n_1} f_{12},$$ (6.66)

and

$$f_2 = \frac{n_2 - n_1}{n_2} f_{12},$$ (6.67)

and a flat field with a doublet requires that two different lens materials be used. The other optical solution to field curvature is the addition of a field flattening lens near the image plane.

6.6.4 Effect of Defocus in a Diffraction-Limited Optic

A very practical problem concerns the loss of image quality due to a defect of focus such as could arise from thermal effects in the optic or by human error in mechanical focusing. Consider the simple case shown in Figure 6.26, where a lens of diameter D and focal length f images an infinitely distant object.

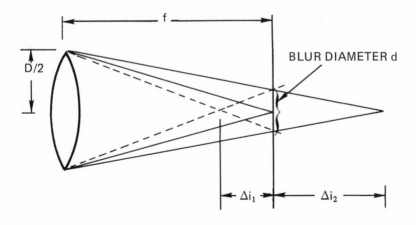

Figure 6.26 Blur due to defocus.

The construction of Figure 6.26 shows the effect upon an optic focused on a point at infinity of an error in the placement of the focal plane. Two cases, error Δi_1 to the left of, and an error Δi_2 to the right of the focal plane, are shown which result in the same geometric blur diameter d. By similar triangles

$$\frac{d/2}{\Delta i_1} = \frac{D/2}{f - \Delta i_1} \text{ or d } \frac{D\Delta i_1}{f - \Delta i_1} \simeq \frac{D\Delta i_1}{f} , \tag{6.68}$$

and

$$\frac{d/2}{\Delta i_2} = \frac{D/2}{f + \Delta i_2} \text{ or d } = \frac{D\Delta i_2}{f + \Delta i_2} \simeq \frac{D\Delta i_2}{f} . \tag{6.69}$$

Note that a specific defocus on either side of the focal plane produces approximately the same geometrical blur, $d = D\Delta i/f$. The corresponding angular geometric blur diameters δ are obtained by dividing d by f,

$$\delta = \frac{D\Delta_i}{f^2} . \tag{6.70}$$

Hopkins[5] derived an expression for the OTF as a function of the focus error Δi, the F/#, and the wavelength, for the case of a diffraction-limited circular aperture illuminated by a monochromatic plane wave. Levi and Austing[6] simplified Hopkins' equation to

$$\text{MTF } (\nu_r, \Delta) = \frac{4}{\pi a} \int_{\nu_r}^{1} (1 - x^2)^{1/2} \cos [a(x - \nu_r)] dx \tag{6.71}$$

where

$$a = 2\pi \nu_r \Delta \tag{6.72}$$

$$\Delta = 2 \sin^2 \left(\frac{D}{2f}\right) \Delta i/\lambda \tag{6.73}$$

Δi = linear defocus in the same units as λ

ν_r = relative spatial frequency in dimensionless units of
      f/f_c, $0 \leqslant \nu_r \leqslant 1$, $f_c = D/\lambda$.

They calculated and tabulated values of MTF (ν_r, Δ) for values of ν_r from 0 to 50. Figure 6.27 graphs some of those results. Geometrical optics predicts an MTF degradation of Bessinc (df), but according to Hopkins, this prediction is accurate only for frequencies up to about $f/f_c = .1$, regardless of the focus error. At higher frequencies, the geometrical optics prediction overestimates MTF, regardless of focus error.

To demonstrate the use of these analyses, consider the example system used without a telescope. The detector lens focal length is 5 cm, the aperture is 2 cm in diameter, and the detector element size is 0.005 cm by 0.005 cm. If the detector lens is diffraction-limited, its MTF cutoff frequency f_c for a perfect focus is

$$f_c = \frac{D_d}{\lambda} = \frac{2 \text{ cm}}{10^{-3} \text{ cm}} = 2 \text{ cy/mrad.}$$

The detector angular subtense is 1 mrad, giving an MTF first-zero frequency of 1 cy/mrad. The resulting MTF's and their product are shown in Figure 6.28.

Let us assume for the sake of argument that a defocus which causes the product $\tilde{r}_o \tilde{r}_d$ to go to zero near 0.5 cy/mrad is severe enough to be objectionable. Referring to Figure 6.27, we see that $\Delta = 2.5$ is sufficient to cause this degradation. For our example system,

$$\Delta = 2 \sin^2 \left(\frac{D}{2f}\right) \frac{\Delta i}{\lambda}$$

$$2.5 = 2 \sin^2 \left(\frac{2 \text{ cm}}{2 (5 \text{ cm})}\right) \frac{\Delta i}{10^{-3} \text{ cm}}$$

or $\Delta i = 0.0316$ cm.

Since

$$\Delta i = \frac{df}{D} \text{ or } d \simeq \frac{D \Delta i}{f},$$

$$d \simeq \frac{2 \text{ cm} (.0316 \text{ cm})}{5 \text{ cm}} = 0.0127 \text{ cm}$$

and

$$\delta \simeq \frac{d}{f} = \frac{0.0127 \text{ cm}}{5 \text{ cm}} = 0.00253 \text{ radian} = 2.53 \text{ mrad.}$$

The tolerable defocus blur is thus about 2-1/2 mrad.

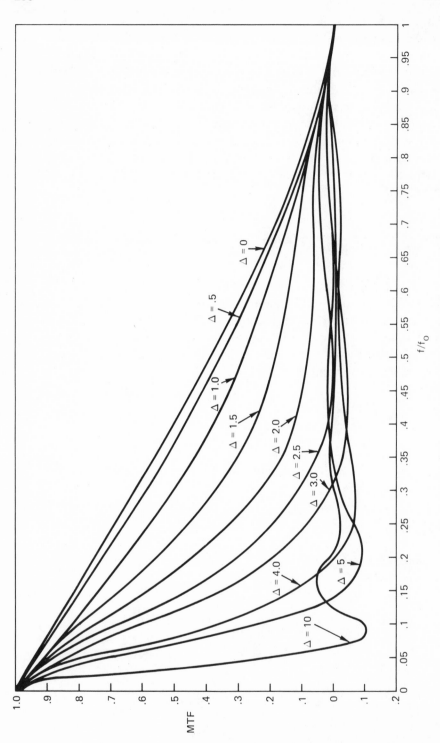

Figure 6.27 MTF's for defocussed diffraction - limited optics adapted from reference 6.

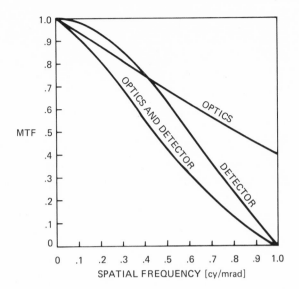

Figure 6.28 MTF's of the example system for perfect optics and detectors.

6.7 Auxiliary Optics

The detector lens may be augmented by the use of a magnifying telescope to increase the resolution. Magnification decreases the field of view, increases the angular resolution, and keeps the ratio of detector subtense to clear aperture diameter constant so that thermal sensitivity does not change. Figure 6.29 shows some possible astronomical (image inverting) telescopes. The first telescope consists of a positive single-element objective lens, a real image plane at which a field-of-view-defining stop is located, and a positive single element collimator or eyepiece lens.

The second telescope shows a weakly negative element used to reduce chromatic aberrations in the objective. The material of this chromatic corrector must exhibit a higher dispersion than the material of the first element in order to correct the shift of focus with wavelength. The focal length of this objective is given by

$$f_{12} = \frac{f_1 f_2}{f_1 + f_2 - d} .$$

(6.74)

Zero longitudinal chromatic aberration in the objective requires that

$$f_{12} (\lambda_1) = f_{12} (\lambda_2).$$

(6.75)

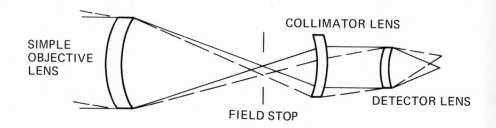

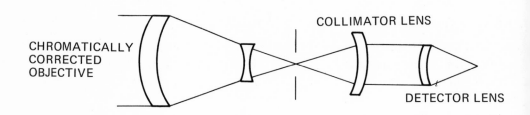

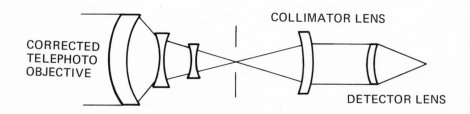

Figure 6.29 Astronomical telescopes for parallel beam scanning systems.

Solving for d and substituting the relations

$$f_{12}(\lambda) = \frac{f_1(\lambda)\, f_2(\lambda)}{f_1(\lambda) + f_2(\lambda) - d} \tag{6.76}$$

and

$$f = \frac{k}{n(\lambda) - 1} \tag{6.77}$$

expressed for the two wavelengths λ_1 and λ_2 yields

$$d = \frac{f_1}{1 + \dfrac{\Delta n_1\,(n_2 - 1)}{\Delta n_2\,(n_1 - 1)}} + \frac{f_2}{1 + \dfrac{\Delta n_2\,(n_1 - 1)}{\Delta n_1\,(n_2 - 1)}}. \tag{6.78}$$

In equation 6.78,

$$\Delta n_1 = n_1(\lambda_2) - n_2(\lambda_1), \tag{6.79}$$

$$\Delta n_2 = n_2(\lambda_2) - n_2(\lambda_1), \tag{6.80}$$

$$n_1 = n_1\left(\frac{\lambda_1 + \lambda_2}{2}\right), \tag{6.81}$$

$$n_2 = n_2\left(\frac{\lambda_1 + \lambda_2}{2}\right). \tag{6.82}$$

Since

$$V = \frac{n - 1}{\Delta n},$$

$$d = \frac{f_1}{1 + \dfrac{V_2}{V_1}} + \frac{f_2}{1 + \dfrac{V_1}{V_2}}. \tag{6.83}$$

The combined focal length then is expressed by

$$f_{12} = \frac{f_1 \, f_2 \, (V_1 + V_2)}{f_1 \, V_2 + f_2 \, V_1}. \tag{6.84}$$

As an example, if the first lens is made of germanium, $V_1 = 1001.27$ from 8 to 14 μm, and if the second lens is of zinc selenide, $V_2 = 34.45$. Then

$$d = 0.9667f_1 + 0.03324f_2.$$

The parameters may be selected for the best aberration balance. The chromatic correction equation for a cemented doublet of thin lenses having focal length

$$\frac{1}{f} = \frac{1}{f_1} + \frac{1}{f_2} \tag{6.85}$$

is found by setting d = 0 and solving to find

$$f_1 = \frac{-V_2}{V_1} \, f_2 \, , \tag{6.86}$$

$$f_1 = \frac{-V_1}{V_2 - V_1} \, f \, , \tag{6.87}$$

and

$$f_2 = \frac{V_2 - V_1}{V_2} \, f. \tag{6.88}$$

The third telescope shown in Figure 6.29 is an astronomical telescope with a telephoto objective. The first element has a higher power than the first element of the simpler astronomical telescope, followed by a negative second element. This produces a shorter overall length for a particular F/#.

Figure 6.30 shows a Galilean telescope. The first element is positive, the second element is negative, and the separation is the sum of the focal lengths. There is no image plane, and for a compact design the exit pupil is inside the telescope and the exciting beams diverge. Therefore, Galileans are most useful when placed in front of the detector lens in a converging beam scanner.

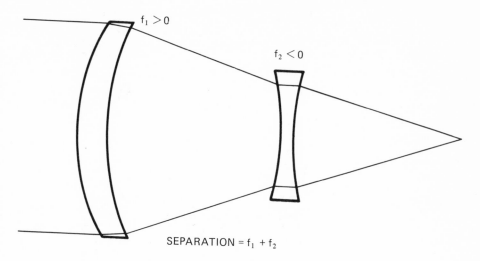

Figure 6.30 A Galilean telescope.

A dual field of view may be implemented in a variety of ways. Some typical telescopic systems are shown in Figure 6.31. Finally, Jamieson[7] has shown that workable continuous-zoom infrared optical systems can be designed using no more than two or three elements. Jamieson discusses the image quality of such simple zoom systems and presents some designs.

6.8 Thermal Imaging Optical Materials

Three classes of optical materials commonly are used in thermal imaging lenses and windows: semiconductor (single crystal or polycrystalline) Si and Ge, hot-pressed polycrystalline II-IV compounds such as the IRTRANs;* chemical vapor deposited ZnSe and ZnS; and the chalcogenide glasses such as TI 1173.† Silicon and germanium are used for the majority of applications because their high refractive indices and excellent mechanical strengths and hardnesses simplify optical design.

Ideally the refractive index n of a thermal imaging lens material should be high, should have zero change with temperature, and zero change with wavelength (zero dispersion). The material should have zero coefficient of thermal expansion, high surface hardness, high mechanical strength, compatibility with antireflection (AR) coatings, zero solubility in water, and low infrared absorptance. High refractive index is necessary to minimize lens curvatures, lens thicknesses, and the number of lens elements in a given F/# system. Low thermal coefficient of refractive index is necessary to

---

*Trademark of the Eastman-Kodak Corporation
†Trademark of Texas Instruments, Incorporated.

(a)

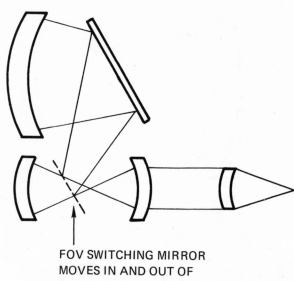

FOV SWITCHING MIRROR
MOVES IN AND OUT OF
PLANE OF THE PAPER

(b)

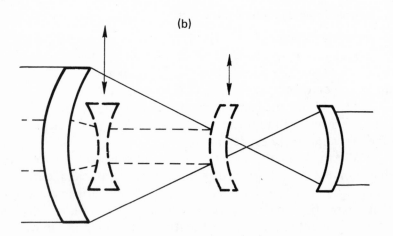

Figure 6.31 Dual field telescopes: (a) two separate objectives selected
by a switching mirror, (b) a coaxial two-step zoom telescope.

prevent aberration unbalancing and focal length changes as the lens temperature changes. Low dispersion is necessary to minimize chromatic aberration and the number of lenses necessary to compensate for it. Zero change of lens dimensions with temperature is necessary to avoid aberration-unbalancing and focal length changes, and to simplify lens mountings.

High surface hardness prevents scratching and abrasion, and high mechanical strength allows thin lenses (high diameter-to-thickness ratio) to be used. Compatibility with AR coatings is absolutely essential because lens surface reflection losses increase with increasing n. Nonsolubility in water is necessary to assure lens integrity in rain and high humidity. Low radiation absorptance minimizes absorption losses and allows efficient system design.

No known material possesses all of these desirable qualities, but silicon, germanium, zinc selenide, zinc sulfide, and TI 1173 come the closest to perfection. They have high indices, hardnesses, and strengths, are compatible with AR coatings, are insoluble, and have low absorptance. Silicon and germanium are the most frequently used thermal lens materials. The optically important properties of the useful materials are summarized in Table 6.1.

Herzberger and Salzberg[4] found that the dispersion of infrared optical materials is accurately described by the equation

$$n = A + BL + CL^2 + D\lambda^2 + E\lambda^4,$$

where

$$L = (\lambda^2 - 0.028)^{-1}.$$

The coefficients they derived are given in Table 6.2. Such an equation is useful for interpolating between measured index values. The dispersion of several materials is shown in Figure 6.32. References 8 and 9 through 12 contain the detailed discussions from which this section was abstracted.

6.9 Thermal Effects in Optics

Most thermal imaging refractive lens materials, including germanium, have refractive indices which change significantly with temperature at terrestrial temperatures. Many applications of thermal imaging systems are used in environments where temperature extremes of -20°C to +40°C are not uncommon, and in other applications such as high altitude aircraft, high speed aircraft, or space vehicles, the extremes may be greater. Thus the effects on lens parameters due to thermal index changes cannot be ignored.

TABLE 6.1
Properties of the Useful Insoluble Infrared Optical Materials

Material Name	Composition	Useful Infrared Wavelength Range [μm]	Refractive Indices							
			n(3μm)	n(4μm)	n(5μm)	n(8μm)	n(9μm)	n(10μm)	n(11μm)	n(12μm)
Germanium	Low resistivity n-type	2 – 23	4.049	4.0244	4.0151	4.0053	4.0040	4.0032	4.0026	4.0023
Silicon	Low resistivity p- or n-type	1.5 – 15	3.4324	3.4254	3.4221	3.4184	3.4180	3.4177	3.4177	–
IRTRAN 1	Hot pressed (HP) magnesium fluoride ($Mg\ F_2$)	0.5 – 9.0	1.3640	1.3526	1.3374	1.2634	1.2269	–	–	–
IRTRAN 2	HP zinc sulfide (Zn S)	0.4 – 14.5	2.2558	2.2504	2.2447	2.2213	2.2107	2.1986	2.1846	2.1689
Zinc Sulfide	Chemical vapor deposited (CVD)	0.4 – 14.5	Same as IRTRAN 2							
IRTRAN 3	HP calcium fluoride (CaF_2)	0.4 – 11.5	1.4179	1.4097	1.3990	1.3498	1.3269	1.3002	1.2694	–
IRTRAN 4	HP zinc selenide (Zn Se)	0.5 – 22	2.440	2.435	2.432	2.418	2.413	2.407	2.401	2.394
Zinc Selenide	CVD	0.5 – 22	Same as IRTRAN 4							
IRTRAN 5	HP magnesium oxide (MgO)	0.4 – 9.5	1.6920	1.6684	1.6368	1.4824	1.406	–	–	–
IRTRAN 6	HP cadmium telluride (Cd Te)	0.9 – 31	2.695	2.688	2.684	2.677	2.674	2.672	2.669	2.666
Cadmium Telluride	CVD Cd Te	0.9 – 15	–	–	2.67	–	–	2.56	–	–
Cadmium Sulfide	CVD Cd S	1 – 14	–	–	2.27	–	–	–	–	–
TI 1173	Amorphous $Ge_{28}\ Sb_{12}\ Se_{60}$	1 – 14	2.6263	2.620	2.6165	2.6076	2.604	2.600	2.596	2.592
TI 20	Amorphous $Ge_{33}\ As_{12}\ Se_{55}$	0.8 – 16	–	–	2.49	–	–	3.135	3.045	–
Gallium Arsenide	Ga As	0.9 – 11	–	–	–	3.34	–	~3.1	–	–
Arsenic Trisulfide	$As_2\ S_3$ Glass	1 – 11	2.4168	2.4118	2.4077	2.3937	2.3878	2.3811	2.3735	2.3650

TABLE 6.1 (Continued)

Material Name	$\frac{\partial n}{\partial T}$ at 300°K [10^{-6} °C^{-1}]	$a = \frac{1}{\ell}\frac{\partial \ell}{\partial T}$ $10^{-6}\frac{cm}{cm\,°C}$	Hardness [Knoop]	Approximate Absorption Coefficient at 300°K and at 3, 4 and 5 μm [cm^{-1}]	Approximate Absorption Coefficient at 300°K and at 8, 9, 10 and 11 μm [cm^{-1}]	Young's Modulus 10^{-6} [psi] at 25°C	Dispersion Index 3–5 μm	Dispersion Index 8–11 μm
Germanium	+280 to 300	5.5 to 6.1	692–850	0.0047, .0048, .0051	0.015, .018, .021, .029	15	88.44	1112.59
Silicon	+162 to 168	4 to 4.15	1150	—	—	19	235.65	3454.36
IRTRAN 1	—	10.4 to 12.0	576	—	—	16.6	13.18	—
IRTRAN 2	+51	6.9 to 7.4	250	.22, .12, .07	.07, .08, .11, .45	14	112.64	32.78
Zinc Sulfide	—	—	250	Same as IRTRAN 2	Same as IRTRAN 2		112.64	32.78
IRTRAN 3	—	18.9 to 20	200	.16, .12, .06	—	14.3	21.61	3.851
IRTRAN 4	+48 to +58	7.4 to 8	100	—	.14, .19, .13, .13	10.3	179.5	82.91
Zinc Selenide	+100	8.53	100	<.003	<.003	9.75	179.5	82.91
IRTRAN 5	—	10.4 to 12	640	.24, .18, .16	—	48.2	12.04	—
IRTRAN 6	96	5.3 to 6	45	—	.24, .25, .29, .27	5.2	153.6	209.1
Cadmium Telluride	+93 to +107	5.9	45	~.005 at 10μm	.002 at 10 μm	3.2	—	—
Cadmium Sulfide	—	3	120		.03 at 10 μm	—	—	—
TI 1173	+80	15 to 16	150		.059 at 11 μm	3.1	165.45	138.09
TI 20	~+80	13	171		.075 at 10 μm .091 at 11 μm	3	—	—
Gallium Arsenide	+149	5.7 to 6	750		~.02 at 10 μm	12	—	—
Arsenic Trisulfide	-8.6 to -10	24.6	109	—	—	2.3	155.2	68.5

TABLE 6.2

Thermal imaging optical materials dispersion coefficients, from reference 4

Material	Wavelength Range [μm]	A	B	C	D	E
Ge	2.0 − 13.5	3.99931	0.391707	0.163492	−.0000060	.000000053
Si	1.3 − 11.0	3.41696	0.138497	0.013924	−.0000209	.000000148
IRTRAN 1	1.0 − 6.7	1.37770	0.001348	0.000216	−.0015041	.00000441
IRTRAN 2	1.0 − 13.5	2.25698	0.032586	0.000679	−.0005272	.000000604
IRTRAN 3	1 − 10	1.4278071	2.2806966×10^{-3}	$-9.1939015 \times 10^{-5}$	$-1.1165792 \times 10^{-3}$	$-1.5949659 \times 10^{-6}$
IRTRAN 4	1 − 10	2.4350823	5.1567472×10^{-2}	2.4901923×10^{-3}	$-2.7245212 \times 10^{-8}$	$-9.8541275 \times 10^{-8}$
IRTRAN 5	1 − 10	1.7200516	5.6119400×10^{-3}	$-1.0986148 \times 10^{-5}$	$-3.0994558 \times 10^{-3}$	$-9.6139613 \times 10^{-6}$
IRTRAN 6	1 − 10	2.682384	1.180290×10^{-1}	3.276801×10^{-2}	-1.202984×10^{-4}	2.177336×10^{-8}

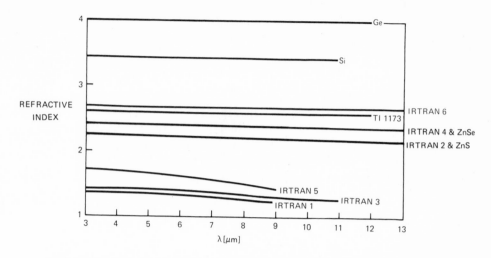

Figure 6.32 Refractive indices of common thermal imaging lens materials.

In a single lens composed of one material and located in air, the lens focal length f is related to the refractive index n by

$$f = \frac{k}{n - 1},$$
(6.89)

where k is a geometrical constant if the lens material does not expand. The partial derivative of f with respect to n is

$$\frac{\partial f}{\partial n} = - \frac{k}{(n - 1)^2},$$
(6.90)

and the partial derivative of n with respect to T is $\partial n/\partial T$. Thus a small focal length shift for a small temperature change T is given approximately by

$$\Delta f = - \frac{k}{(n - 1)^2} \frac{\partial n}{\partial T} \Delta T = - \frac{f}{n - 1} \frac{\partial n}{\partial T} \Delta T.$$
(6.91)

If the focal shift is large, it will be given by

$$f = \frac{k}{n(T_2) - 1} - \frac{k}{n(T_1) - 1}.$$
(6.92)

The thermal coefficients of index of the useful materials were given in Table 6.1. The image of an infinitely distant object shifts by an amount Δf toward the lens for negative $\partial n/\partial T$ and by Δf away from the lens for positive $\partial n/\partial T$. This must be compensated if it exceeds the desired depth of focus.

Since all of the useful materials have significant thermal coefficients, it is necessary to devise a compensation scheme. There are four possible solutions:

1. Provide a manual focus adjustment.
2. Provide an automatic electromechanical focus adjustment.
3. A-thermalize the lens by a combination of lens materials having an effective $\partial n/\partial T$ of zero.
4. A-thermalize the lens by using lens mountings which move passively to compensate the defocus.

The detector lens for the example system at $300°K$ has $f = 5$ cm, $n = 4.004$,

$$\frac{\partial n}{\partial T} = +300 \times 10^{-6} \ °C^{-1},$$

and

$$\Delta f = -4.993 \times 10^{-4} \ \Delta T \ [cm].$$

For

$$\Delta T = +10°C,$$

$$\Delta f = -4.933 \times 10^{-3} \ cm.$$

This produces a defocus dimensionless Δ (see Section 6.6.4) of 0.395, which is a trivial defocus. A temperature shift of $+50°C$, however, would produce a Δ of 1.97, which is a serious defocus which would have to be compensated for by a refocus mechanism.

For small temperature shifts the thermal expansion of any lens dimension R is given by

$$\Delta R = R \, \alpha \Delta T, \tag{6.93}$$

where α is the coefficient of thermal expansion in $[cm/(cm°C)]$. It is trivial to show that the focal length of a single lens changes as

$$\frac{\partial f}{\partial T} = \alpha f. \tag{6.94}$$

The germanium detector lens for the example system has

$$f = 5 \text{ cm and } \alpha = 6.1 \times 10^{-6} \text{ }^\circ\text{C}^{-1}.$$

For $\Delta T = +10^\circ C$, the defocus is

$$\Delta f = (5 \text{ cm}) (6.1 \times 10^{-6} \text{ }^\circ\text{C}^{-1}) (10^\circ C) = 3.05 \times 10^{-4} \text{ cm},$$

which is an insignificant defocus.

This example demonstrates that lens thermal expansion effects tend to be small compared to the effects of thermal refractive index change. However, lens housings also expand and contract to produce defocus. If the example detector lens were held in an aluminum housing having $\alpha = 23 \times 10^{-6}°C^{-1}$, the linear defocus δ produced by the housing expansion alone for a $10°C$ ΔT would be

$$\delta = (5 \text{ cm}) (23 \times 10^{-6}) (10)$$

$$= 1.15 \times 10^{-3} \text{ cm}.$$

All of these effects are accountable by a total defocus given by

$$\delta = \left\{ -\frac{f}{n-1} \frac{\partial n}{\partial T} + \alpha f - \alpha L \right\} \Delta T, \tag{6.95}$$

where L is the overall length of the lens system. The thermal expansion coefficients for some common lens housing materials are given in Table 6.3.

TABLE 6.3

Thermal expansion coefficients of some common lens materials

MATERIAL	APPROXIMATE $\alpha \left[\dfrac{10^{-6} \text{ cm}}{\text{cm}°C} \right]$
Aluminum	23
Magnesium	26
Stainless Steel	11
Titanium	9.5
Beryllium	11.5
Ni 42 (low expansion alloy)	5.7

6.10 Passive Optical Compensation of Thermal Refractive Index Changes

The focal length of a simple lens at temperature T_0 is defined by:

$$f(T_0) = \frac{k}{n(T_0) - 1},$$ (6.96)

where $n(T_0)$ is the index at that temperature. Most useful thermal imaging optical materials exhibit significant thermal coefficients of index. For zero thermal coefficient of expansion, and non-zero coefficient of index, the focal length at $T = T_0 + \Delta T$ is given approximately by:

$$f(T) = \frac{k}{n(T_0) + \Delta T \dfrac{\partial n(T_0)}{\partial T} - 1}$$ (6.97)

if $(\partial^2 n/\partial T^2) = 0$ or for small ΔT.

The change in focal length is:

$$\Delta f = f(T) - f(T_0) = -\frac{k}{n(T_0) - 1} \left[\frac{\Delta T \dfrac{\partial n(T_0)}{\partial T}}{n(T_0) - 1 + \Delta T \dfrac{\partial n(T_0)}{\partial T}} \right]$$

$$\cong - f \, \Delta T \frac{\partial n(T_0)}{\partial T},$$ (6.98)

since $n-1 \gg \Delta T(\partial n/\partial T)$ for infrared materials and for reasonable temperatures.

Passive optical compensation requires that $\partial f/\partial T \cong 0$ over some specified temperature range. Completely successful passive compensation is probably never realizable because the lenses of a system will not always heat up uniformly, so that gradients across lenses and differences from lens to lens will occur.

As an example, consider the doublet shown in Figure 6.33. If the lens separation s is small compared to the product of the two lens focal lengths, the focal length of the combination is:

$$f = \frac{f_1 \, f_2}{f_1 + f_2}.$$ (6.99)

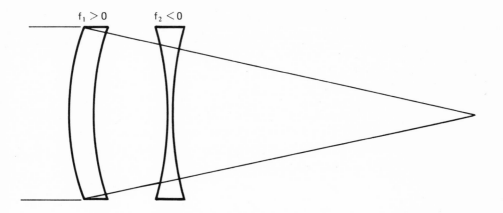

Figure 6.33 A lens system for passive optical compensation of refractive index change with temperature.

The derivative is:

$$\frac{\partial f}{\partial T} = -\frac{f}{f_1 + f_2}\left(\frac{\partial f_1}{\partial T} + \frac{\partial f_2}{\partial T}\right) + \frac{1}{f_1 + f_2}\left(f_1\frac{\partial f_2}{\partial T} + f_2\frac{\partial f_1}{\partial T}\right). \qquad (6.100)$$

Using

$$\frac{\partial n}{\partial T} = -\frac{f}{n-1}\frac{\partial f}{\partial T},$$

one can readily show that:

$$\frac{\partial f}{\partial T} = -f^2\left[\frac{1}{f_1}\frac{\partial n_1/\partial T}{n_1 - 1} + \frac{1}{f_2}\frac{\partial n_2/\partial T}{n_2 - 1}\right]. \qquad (6.101)$$

Thus, for perfect passive compensation, we require the bracketed term to be zero. Defining the relative coefficient of index I as

$$I = \frac{\partial n/\partial T}{n-1}, \qquad (6.102)$$

the requirement is:

$$\frac{I_2}{I_1} = -\frac{f_2}{f_1} .$$

(6.103)

If the lens combination is to be convergent, only three cases are allowed:

CASE 1: $\frac{I_2}{I_1} < 0, f_1 > 0, f_2 > 0$

CASE 2: $\frac{I_2}{I_1} > 1, f_1 > 0, f_2 < 0, |f_2| > |f_1|$

CASE 3: $0 < \frac{I_2}{I_1} < 1, f_1 < 0, f_2 > 0, |f_1| > |f_2|$

As an example, if the first element is of ZnSe and the second is of germanium,

$$\frac{I_2}{I_1} = 2.94,$$

$$f_1 = -0.34 \, f_2 = .66 \, f,$$

$$f_2 = -1.94 \, f.$$

TABLE 6.4

MATERIAL	$\frac{1}{n-1}\frac{\partial n}{\partial T}$ for T = 25°C, $8 \leqslant \lambda \leqslant 11 \mu m$
Germanium	$+100 \times 10^{-6} \,°C^{-1}$
ZnSe	$+34 \times 10^{-6}$
ZnS	Unknown
Tl 1173	$+50 \times 10^{-6}$
$As_2 S_3$	-7.25×10^{-6}
Tl_2 Br I	-171×10^{-6}

One disadvantage of this approach is that the chromatic aberration correction becomes more complicated. The longitudinal chromatic aberration for a separationless doublet of thin lenses is derived as follows:

$$\Delta f = \frac{\partial f}{\partial n_1} \Delta n_1 + \frac{\partial f}{\partial n_2} \Delta n_2 \qquad (6.104)$$

$$\frac{\partial f}{\partial n} = \frac{\partial}{\partial n} \left[f_1 f_2 (f_1 + f_2)^{-1} \right]$$

$$= \frac{\partial f_1}{\partial n} \frac{f_2}{f_1 + f_2} + \frac{\partial f_2}{\partial n} \frac{f_1}{f_1 + f_2} - \left(\frac{\partial f_1}{\partial n} + \frac{\partial f_2}{\partial n} \right) \frac{f_1 f_2}{(f_1 + f_2)^2} \qquad (6.105)$$

$$\frac{\partial f}{\partial n_1} = \frac{\partial f_1}{\partial n_1} \frac{f_2}{f_1 + f_2} - \frac{\partial f_1}{\partial n_1} \frac{f_1 f_2}{(f_1 + f_2)^2} \qquad (6.106)$$

$$\frac{\partial f}{\partial n_2} = \frac{\partial f_2}{\partial n_2} \frac{f_1}{f_1 + f_2} - \frac{\partial f_2}{\partial n_2} \frac{f_1 f_2}{(f_1 + f_2)^2} \qquad (6.107)$$

$$\frac{\partial f}{\partial n} = - \frac{f}{n - 1} \qquad (6.108)$$

and Δf is found to be

$$\Delta f = - f^2 \left[\frac{1}{V_1 f_1} + \frac{1}{V_2 f_2} \right]. \qquad (6.109)$$

In the case cited above, V_2 (germanium) = 1113, V_1 (zinc selenide) = 83, and $\Delta f = -f/73$.

Therefore, the effect in this case is equivalent to that of a single lens using a material more dispersive than ZnSe, and chromatic correction requires use of a third element even more dispersive, such as ZnS.

6.11 Off-Axis Image Quality

The system specifier or designer is frequently faced with the necessity of deciding how much off-axis image degradation to allow relative to the performance in the center of the field.

Section 4.11 indicates how an observer's sensitivity to change in resolution depends on the baseline resolution. One liminal unit of resolution change is typically 5 to 10 percent of the baseline resolution at low resolution, and resolution change becomes harder to detect as resolution improves. Obviously off-axis degradations may fall near one liminal unit with little or no performance loss. Section 10.2 will show that observers searching a field tend to concentrate on the central 9 degrees of the display, paying less attention to the display extremes the larger the angular size of the display becomes. Clearly if image quality is to be sacrificed in some part of the image, it should be the least-used part. Thus resolution at the edges of large displays can be poorer, but not so poor that it becomes distracting or disconcerting. Some resolution loss at the field edges of optical systems is unavoidable due to distortion and lateral chromatic aberration. A factor of two to three in resolution is not objectionable, and is usually not noticed if one is not specifically looking for it. This has been demonstrated conclusively in the operation of circular-scan FLIR's having deliberate resolution variation from the center to the edge of the field[13].

6.12 Image Defects in Fixed-Focus Systems

An optical system which has its focus fixed for a particular object distance O is not necessarily in good focus for all object distances of interest. Consider the simple converging lens with aperture diameter D shown in Figure 6.34 focused at an image distance i for an object at O less than infinity. Figures 6.35 and 6.36 show the foci for objects at O_1 and O_2 other than O. If a thin detector is placed at i, it will see an object at O in perfect focus and objects at O_1 and O_2 imperfectly focused. First-order geometrical optics predicts that the images at i of points O_1 or O_2 will be circles with diameters d_1 and d_2 respectively. The quality of the focus depends in a complicated way upon the blur diameter, as for example the case of the defocused diffraction-limited optic described in Section 6.6.4.

It is common practice to establish a criterion of goodness relative to the detector dimension based on theoretical considerations or practical experience. Whatever the criterion, it is useful to know the limits of object or image movement for which a particular goodness criterion is violated and to know this in terms of the blur diameter d or the associated angular blur diameter δ. Referring to Figures 6.37 and 6.38, the distance $O_2 < O$ for which the criterion is met may be called the near depth of field, and the distance $O_1 > O$ for which the criterion is met may be called the far depth of field. The corresponding shifted image plane focal points may be called the near depth of focus i_2 and the far depth of focus i_1.

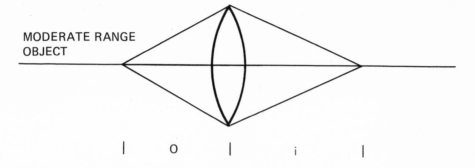

MODERATE RANGE
OBJECT

Figure 6.34 Focus for a distant object not at infinity.

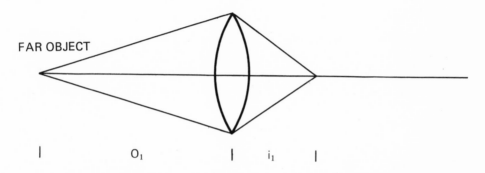

FAR OBJECT

Figure 6.35 Focus for an object more distant than in Figure 6.34.

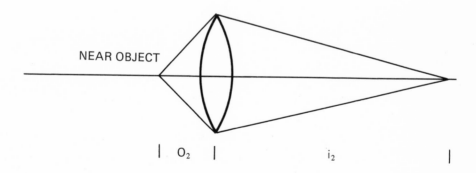

NEAR OBJECT

Figure 6.36 Focus for an object nearer than in Figure 6.35.

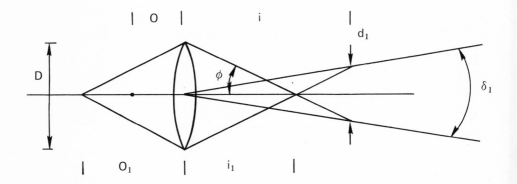

Figure 6.37 Defocus for an object more distant than the in-focus range.

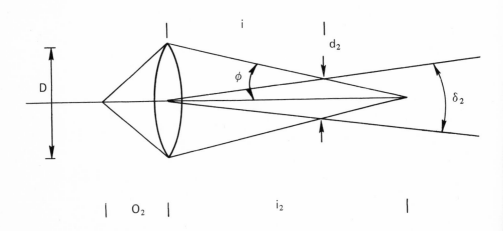

Figure 6.38 Defocus for an object closer than the in-focus range.

Consider first the case of Figure 6.37 where the image at i_1 of an object at O_1 comes to a focus in front of the detector at i. A point at O_1 is imaged as a circular blur of diameter d_1. For sufficiently large $F/\#$,

$$\frac{D}{i_1} = \frac{d_1}{i - i_1} \text{ or } \frac{1}{Di} = \frac{1}{(d_1 + D)i_1} \ . \tag{6.110}$$

Using

$$\frac{1}{i} = \frac{1}{f} - \frac{1}{O} \text{ and } \frac{1}{i_1} = \frac{1}{f} - \frac{1}{O_1},$$

it is easy to show that

$$O_1 = \frac{DOf}{f(D + d_1) - d_1 O} \simeq \frac{DOf}{Df - d_1 O} \tag{6.111}$$

for small defocus and that

$$i_1 \simeq \frac{Df}{D + d_1} \ . \tag{6.112}$$

Next, consider the case shown in Figure 6.38 where the image at i_2 of an object at O_2 imaged as a circular blur of diameter d_2 at i. As before, one can show that

$$O_2 = \frac{DOf}{Df + d_2 (O - f)} \simeq \frac{DOf}{Df + d_2 O} \ , \tag{6.113}$$

and

$$i_2 \simeq \frac{Df}{D - d_2} \ . \tag{6.114}$$

The far depth of field O_1 extends to infinity when $O = Df/d_1$. This in-focus object distance is called the hyperfocal distance O_h. When $O = O_h$, the near depth of field O_2 becomes $Df/2d_2$, or $O_h/2$.

Note that from previous equations, when the blur diameter is small compared to the aperture, $i_1 \simeq i_2 \simeq f$, so that the angular blurs associated with d_1 and d_2 can be written as

$$\delta_1 \simeq \frac{d_1}{f} \text{ and } \delta_2 \simeq \frac{d_2}{f}.$$

Then the depths of field and of focus may be rewritten as

$$O_1 = \frac{DO}{D - \delta_1 O} \tag{6.115}$$

$$i_1 = \frac{Df}{D + f\delta_1} \tag{6.116}$$

$$O_2 = \frac{DO}{D + \delta_2} \tag{6.117}$$

$$i_2 = \frac{Df}{D - f\delta_2} . \tag{6.118}$$

When the object distance is infinity,

$$O_2 = \frac{D}{\delta_2} , \tag{6.119}$$

so that the hyperfocal distance and the near depth of field for the infinity focus case are identical.

Solving for the d's and δ's gives

$$d_1 = Df\left(\frac{1}{O} - \frac{1}{O_1}\right) \tag{6.120}$$

$$d_2 = Df\left(\frac{1}{O_2} - \frac{1}{O}\right) \tag{6.121}$$

$$\delta_1 = D\left(\frac{1}{O} - \frac{1}{O_1}\right) \tag{6.122}$$

$$\delta_2 = D\left(\frac{1}{O_2} - \frac{1}{O}\right) . \tag{6.123}$$

When the object is at infinity,

$$\delta_2 = \frac{D}{O_2} . \tag{6.124}$$

Of related interest is the amount the detector focus may shift due to mechanical wear, to shock, or to vibration, before intolerable blurring results. Consider Figure 6.39 which shows two cases of defocus Δi. For large F/#

$$2\theta = \frac{D}{i} = \frac{d}{\Delta i} ,$$

and

$$\Delta i = \frac{di}{D} = \frac{\delta f i}{D} \cong \frac{\delta}{D} f^2 , \tag{6.125}$$

where δ is the tolerable angular blur,

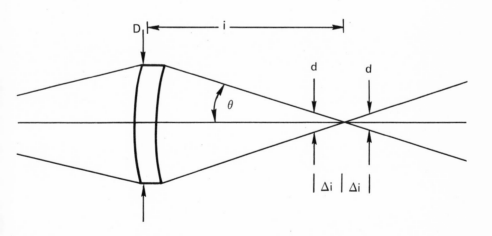

Figure 6.39 Defocus due to shift in image plane.

Finally, consider Figure 6.40, which shows the focus at an image distance f + Δi for an object at O less than infinity. The lens law is

$$\frac{1}{O} + \frac{1}{f + \Delta i} = \frac{1}{f} \qquad\qquad (6.126)$$

or

$$\Delta i = \frac{f^2}{O - f} \cong \frac{f^2}{O} . \qquad\qquad (6.127)$$

Thus the image shifts away from the lens and the infinity focus by this amount as the object distance shifts inward from infinity.

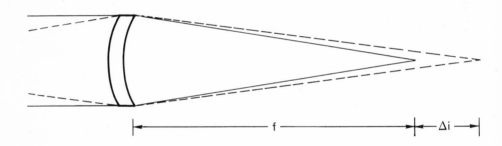

Figure 6.40 Focal plane shift as an object moves in from infinity.

Referring back to Section 6.6.4, the tolerable defocus blur for the example system is about 2-1/2 mrad. If the detector lens is focused at infinity, the near depth of field O_2 for this defocus is

$$O_2 = \frac{D}{\delta_2} = \frac{2 \text{ cm}}{0.0024 \text{ radian}}$$

$$\cong 789 \text{ cm} \cong 8 \text{ meters.}$$

The hyperfocal distance for the allowable defocus is

$$O_h = \frac{D_d}{\delta} = 8 \text{ meters},$$

and the near depth of field for the hyperfocal focus is 4 meters.

This example demonstrates how relatively insensitive the MTF is to defocus. Here a 1 mrad system with a 5 cm focal length has been reduced to a 2 mrad system by a geometrical defocus blur of 2-1/2 mrad with a linear defocus of 32 thousandths of an inch.

6.13 Cold Reflections

Any thermal imaging system which uses refractive optics and cooled detectors can exhibit an image defect resulting from internal retroreflection of cold surfaces onto the detectors. When the detectors sense their own cold surfaces relative to their warm surround, this phenomenon is called the narcissus effect. Narcissus arises whenever at any point in the active scan any part of the cold detector focal plane is reflected by an optical element so that the reflected image is in focus at the array and a warmer background appears at other points of the scan. Two common examples are narcissus from a flat window and from a lens surface concave toward the detector.

As a simple example, consider the case of Figure 6.41, where a convergent meniscus lens is used with a converging beam scanner. The lower part of the figure shows a moving detector equivalent of the system where the effect of defocus due to the scanner is neglected. Although narcissus problems in practice are not so simple, we will use this system as an example because it demonstrates the essential features of any narcissus problem. Assume that the lens focal length is f, the detector dimensions are a and b, the detector focal plane is at a cold temperature T_c, and the focal plane surroundings are at a warmer temperature T_w.

Consider the worst possible case where the lens inner radius of curvature R_2 equals the lens focal length f. Then if the rear lens surface has a reflectance r, that surface acts as a reflective lens with a focal length equal to $R_2/2$. Then any point in the focal plane is reflected to another point in the focal plane as shown in Figure 6.42 and as may be verified by using the lens law. This is an unrealistic example, but it is necessary to understand this case before practical examples may be analyzed.

Figure 6.41 illustrates by means of the moving-detector equivalent that the detector will see the warm surround of the focal plane at the scan extremes, and the cold focal plane itself in the center portion of the scan. The reflections from the lens surface therefore cause the detector to sense not only the scene radiation but also scan-varying radiation from the focal surface.

A CONVERGENT-BEAM MIRROR SCANNER

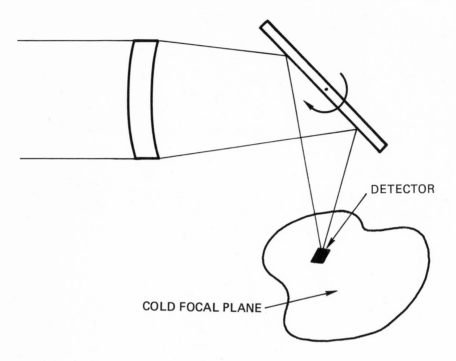

DETECTOR

COLD FOCAL PLANE

MOVING-DETECTOR EQUIVALENT

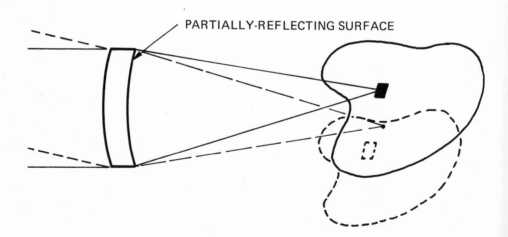

PARTIALLY-REFLECTING SURFACE

Figure 6.41 A convergent beam mirror scanner
and its moving detector equivalent.

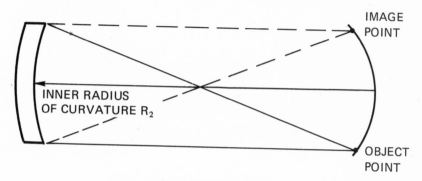

Figure 6.42 Narcissus geometry.

As the detector signals are ac coupled, the large difference signal between the cold focal plane and its warm surroundings will be superimposed on the scene video, and the display will show the scene and a dark image of the focal plane. This typically has the appearance shown in Figure 6.43, where the shape of the narcissus pattern depends on the shape of the focal plane cold area and the extent to which the focal plane is defocused. The magnitude of this negative signal may be calculated as follows: The cold surface at T_c radiates as:

$$W_\lambda (T_c) = \frac{c_1}{\lambda^5 (e^{c_2/\lambda T_c} - 1)}. \tag{6.128}$$

The warm surface at T_w radiates as:

$$W_\lambda (T_w) = \frac{c_1}{\lambda^5 (e^{c_2/\lambda T_w} - 1)}. \tag{6.129}$$

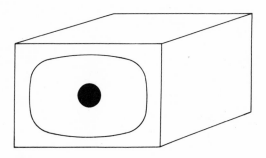

Figure 6.43 Appearance of narcissus.

If we assume that these surfaces are Lambertian radiators, the detector receives "cold" spectral power $P_\lambda (T_c)$,

$$P_\lambda (T_c) = r \ \frac{W_\lambda (T_c)}{\pi} \ \frac{\pi D^2}{4R_2^2} \ \frac{ab}{R_2^2} \ R_2^2 = r \left(\frac{W_\lambda (T_c) D^2 \ ab}{4R_2^2} \right) \qquad (6.130)$$

and "warm spectral power $P_\lambda (T_w)$,

$$P_\lambda (T_w) = r \left(\frac{W_\lambda (T_w) D^2 \ ab}{4R_2^2} \right). \qquad (6.131)$$

The difference in spectral power ΔP_λ is:

$$\Delta P_\lambda = r \frac{ab \ D^2}{4R_2^2} \ [W_\lambda (T_w) - W_\lambda (T_c)] . \qquad (6.132)$$

The total integrated difference power ΔP within a spectral bandpass from λ_1 to λ_2, assuming uniform reflectivity, is:

$$\Delta P = \frac{r \, ab \, D^2}{4 R_2^2} \int_{\lambda_1}^{\lambda_2} \left[\frac{c_1}{\lambda^5 \, (e^{c_2/\lambda T_w} - 1)} - \frac{c_1}{\lambda^5 \, (e^{c_2/\lambda T_c} - 1)} \right] d\lambda. \quad (6.133)$$

To determine the objectionability of this signal, it must be compared to the difference power received by the detector from typical scenes. With reference to Figure 6.44 showing a radiating object subtending the detector solid angle, the change in received power with respect to scene temperature T_s change is:

$$\frac{\partial P_s}{\partial T_s} = \alpha \beta \frac{1}{\pi} \left[\int_{\lambda_1}^{\lambda_2} \frac{\partial W_\lambda \, (T_s)}{\partial T_s} \, d\lambda \right] \frac{\pi D_0^2}{4 \, R^2} \, R^2 \, \tau_0$$

$$= \frac{1}{4} \left[\int_{\lambda_1}^{\lambda_2} \frac{\partial W_\lambda (T_s)}{\partial T_s} \, d\lambda \right] D_0^2 \, \tau_0 \, \alpha \beta. \quad (6.134)$$

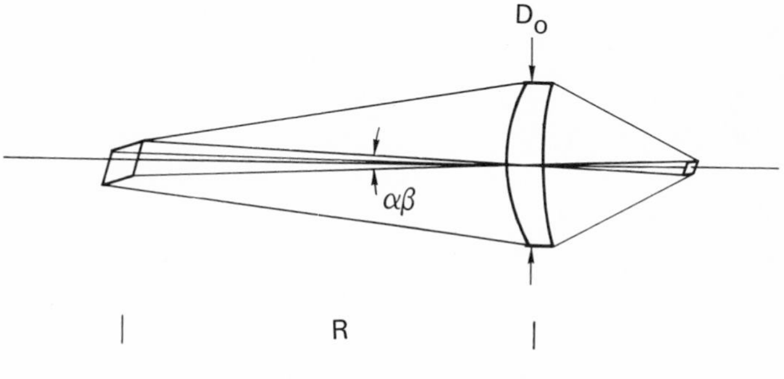

Figure 6.44 Geometry for radiation calculation.

For a small scene temperature difference ΔT_s, the change in received power is:

$$\Delta P_s = \frac{1}{4} \Delta T_s \, \alpha \beta \, \tau_0 \, D_0^2 \int_{\lambda_1}^{\lambda_2} \frac{\partial W_\lambda \, (T_s)}{\partial T_s} \, d\lambda. \quad (6.135)$$

Finally, we can take the ratio of ΔP to ΔP_s,

$$\frac{\Delta P}{\Delta P_s} = \frac{r}{\Delta T_s} \frac{(ab/R_2^2)D^2}{\alpha\beta\,\tau_0\,D_0^2} \frac{\displaystyle\int_{\lambda_1}^{\lambda_2}\left[\frac{c_1}{\lambda^5\,(e^{c_2/\lambda\,T_w}-1)} - \frac{c_1}{\lambda^5\,(e^{c_2/\lambda\,T_c}-1)}\right]d\lambda}{\displaystyle\int_{\lambda_1}^{\lambda_2}\frac{\partial W_\lambda\,(T_s)}{\partial T}\,d\lambda} \quad .$$

$$(6.136)$$

For the simple single lens case, $R_2^2 = f^2$, and $ab/R_2^2 = a\beta$, and also $D = D_0$. To make a sample calculation assume that $T_w = T_s = 300°K$, that $T_c = 77°K$, and that the spectral bandpass is 8 to 14 microns. Then the second integral in the numerator becomes negligible with respect to the first integral, and

$$\frac{\Delta p}{\Delta P_s} = \frac{r}{\Delta T_s\,\bar\tau_0} \frac{\displaystyle\int_{8}^{14}\frac{c_1}{\lambda^5\,(e^{c_2/\lambda T_s}-1)}\,d\lambda}{\displaystyle\int_{8}^{14}\frac{\partial W_\lambda(T_s)}{\partial T_s}\,d\lambda} \quad .$$

$$(6.137)$$

The integral in the numerator has the value 1.72×10^{-2} watt/cm^2 and the denominator integral is 2.62×10^{-4} watt/cm^2. Thus for unity lens transmission,

$$\frac{\Delta P}{\Delta P_s} = 66\,\frac{r}{\Delta T_s} \quad .$$

If the NETD and MDTD of the system are such that broad targets having ΔT_s of $0.01°C$ or better will be detected, and if the lowest possible reflectivity using an antireflection coating is 0.01, then the cold reflection will exceed the NETD by a factor of 66.

The above calculations represent the worst case where a lens surface is confocal with the focal plane. In practice, however, the lens surfaces are usually such that retroreflections are defocused. In general, a confocal surface in a single detector lens will produce

$$\frac{\Delta P}{\Delta P_s} = 66\,\left(\frac{fD}{RD_0}\right)^2\frac{r}{\tau_0\,\Delta T_s} \,,$$

where R is the radius of curvature and D is the diameter of the confocal lens surface.

For the case of the example detector lens, $R = R_2 = 8.501, D = D_0 = 2$ cm, and $f = 5$ cm. Then

$$\frac{\Delta P}{\Delta P_s} = 66 \left(\frac{(5) (2)}{8.501 (2)} \right)^2 \frac{r}{\tau_0 \, \Delta T_s}$$

$$= \frac{23 \, r}{\tau_0 \, \Delta T_s}$$

If $r = 0.01$ and $\tau_0 = 1$,

$$\frac{\Delta P}{\Delta P_s} = \frac{.23}{\Delta T_s} \, ,$$

which is an unacceptable level of narcissus.

Since the cold signal is negative relative to the average scene value, and since the average scene value is usually imaged on the display at a low luminance value, the cold signal will usually drive the display to black. Practical cases where narcissus occurs may involve reflections from an entrance window, from parts of the sensor housing, from the scanner itself, and from lens surfaces anywhere in the system. The essential features of the derivation given above will usually be adequate to analyze any problem if the proper geometrical reflection factors and intervening optical transmissions are used.

Five procedures may be used to reduce cold reflections to an acceptable level. These are:

1. Reduce the focal plane effective radiating cold area by warm baffling.
2. Reduce lens surface reflections by using high efficiency antireflection coatings.
3. Defocus the cold return by designing the optical system so that no confocal surfaces are present.
4. Cant all flat windows.
5. When all else fails, null out the cold reflection with a thermal source or by electronic video signal compensation.

Combination of the first four techniques will usually suffice to eliminate narcissus.

REFERENCES

[1] M.V. Klein, *Optics*, Wiley, 1970.

[2] S. Walles and R.E. Hopkins, "The Orientation of the Image Formed by a Series of Plane Mirrors", Appl. Opt., *3*, pp 1447-1452, December 1964.

[3] M. Born and E. Wolfe, *Principles of Optics*, Pergamon, 1959.

[4] M. Hertzberger and C.D. Salzberg, "Refractive Indices of Infrared Optical Materials and Color Correction of Infrared Lenses", JOSA, *52*, pp 420-427, April 1962.

[5] H.H. Hopkins, "The Frequency Response of a Defocused Optical System", Proc. Roy. Soc., *231*, pp 91-103, 19 July 1955.

[6] L. Levi and R.H. Austing, "Tables of the Modulation Transfer Function of a Defocused Perfect Lens", Appl. Opt., *7*, pp 967-974, May 1968.

[7] T.H. Jamieson, "Zoom Lenses for the 8 to 13μ Waveband", Optical Acta, *18*, pp 17-30, January 1971.

[8] A.R. Hilton, "Nonoxide Chalcogenide Glasses as Infrared Optical Materials", Appl. Opt., *5*, pp 1877-1882, December 1966.

[9] A.R. Hilton and C.E. Jones, "The Thermal Change in the Nondispersive Infrared Refractive Index of Optical Materials", Appl. Opt., *6*, pp 1513-1517, September 1967.

[10] A.R. Hilton, C.E. Jones, and M. Brau, "New High Temperature Infrared Transmitting Glasses – III", Infrared Physics, *6*, pp 183-194, 1966.

[11] C.D. Salzberg and J.J. Villa, "Infrared Refractive Indexes of Silicon, Germanium, and Modified Selenium Glass", JOSA, *47*, pp 244-246, March 1957.

[12] A.R. Hilton, "Infrared Transmitting Materials", J. Elec. Materials, *2*, pp 211-225, 1973.

[13] R.L. Sendall, personal communication, Xerox Electro-Optical Systems, Pasadena, California.

CHAPTER SEVEN — SCANNING MECHANISMS

7.1 Introduction

The function of a scanner in a FLIR is to move the image formed by the optical system in the plane of the detector array in such a way that the detectors dissect the image sequentially and completely. As shown in Figure 7.1, there are two basic types of scanners: parallel beam and converging beam scanners. The parallel beam scanner consists of an optical angle-changing device such as a moving mirror placed in front of the final image forming lens. The converging beam scanner consists of a moving mirror or other scanning device placed between the final lens and the image.

There are seven commonly-used optical scanning mechanisms: the oscillating mirror, the rotating polygonal mirror, the rotating refractive prism, the rotating wedge, the revolving lens, the rotating sensor, and the rotating V-mirror. One-dimensional and two-dimensional scanners may be implemented by various combinations of the basic scanning mechanisms discussed in the following sections.

7.2 Oscillating Plane Mirror

The oscillating mirror shown in Figure 7.2 oscillates periodically between two stops, and may be used as either a parallel beam or a convergent beam scanner. The oscillating mirror produces a ray angular deviation of twice the mirror angle change, as shown in Figure 7.2. This property may be verified by using the law of reflection as follows. Consider Figure 7.3, where two mirror positions M_1 and M_2 separated by an angle γ are shown together with the respective mirror normals N_1 and N_2. Incident rays I_1 and I_2 are shown, both of which exit as E. The angle between I_1 and I_2, $\angle(I_1,I_2)$ as a function of γ is

$$\angle(I_1,I_2) = \angle(E,I_1) - \angle(E,I_2). \tag{7.1}$$

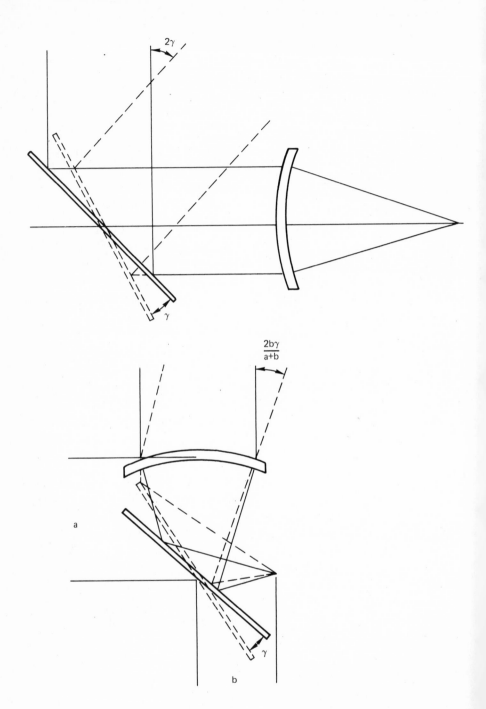

Figure 7.1 A parallel beam and a convergent beam scanner.

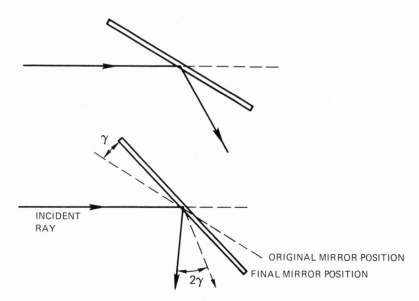

Figure 7.2 Angle doubling by an oscillating mirror.

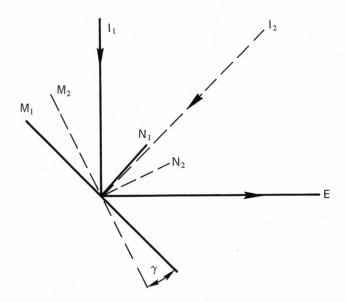

Figure 7.3 Geometry for angle doubling derivation.

From the law of refraction,

$$\angle(I_1, I_2) = 2\angle(E, N_1) - 2\angle(E, N_2). \tag{7.2}$$

Since the mirror normals change by γ,

$$\angle(I_1, I_2) = 2[\angle(E, N_1) - (\angle(E, N_1) - \gamma)] = 2\gamma. \tag{7.3}$$

7.2.1 Converging Beam Scanner

The oscillating mirror convergent beam scanner is such a common mechanism that it is worthwhile determining in detail how the mirror angular motion produces scanning and affects focussing. Consider Figure 7.4, which for simplicity shows only central rays for two mirror positions. Figure 7.5 shows the same deviated and undeviated rays reflected about the deviated mirror position, and shows the ray paths which would result if an equivalent in-line scan mechanism were used. This construction simplifies the derivation. Let the scan angle measured from the optical axis be θ; let the mirror displacement from an arbitrary rest position be γ; let the mirror pivot point be P; and let the detector point be D. Dimensions "a" and "b" are defined as shown.

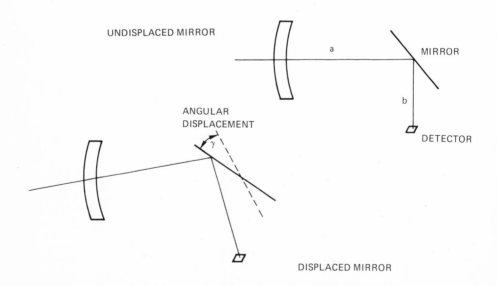

Figure 7.4 Oscillating scan mirror principle.

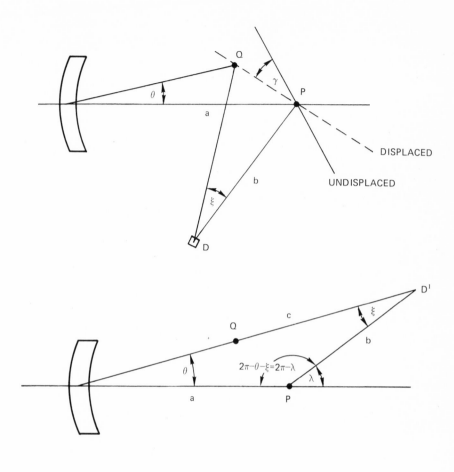

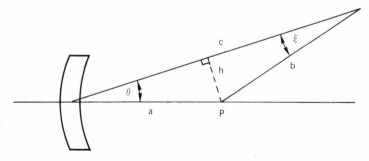

Figure 7.5 Path length change.

The scan angle θ as a function of mirror angular displacement γ may be found from the construction of Figure 7.5. Here the triangle PDQ is reflected through the dotted displaced mirror plane to form triangle PD'Q. The total ray path length in the displaced position is c, and

$$c \leqslant a + b.$$

It is evident from the construction that

$$\lambda = \theta + \xi. \tag{7.4}$$

Since geometric reflection of path PD about the mirror causes an angular shift of double the mirror angle change γ, we have also that

$$\lambda = 2\gamma \tag{7.5}$$

so that

$$\gamma = \frac{\theta + \xi}{2} . \tag{7.6}$$

Now consider the triangle formed by a, b, and c, and extend a perpendicular h to side c from P. Clearly

$$\sin \xi = \frac{h}{b}, \tag{7.7}$$

$$\sin \theta = \frac{h}{a}, \tag{7.8}$$

and therefore,

$$\xi = \arcsin \left(\frac{a \sin \theta}{b} \right). \tag{7.9}$$

Substitution of equation 7.9 into equation 7.6 yields

$$\gamma = \frac{\theta + \arcsin \left(\frac{a \sin \theta}{b} \right)}{2} . \tag{7.10}$$

Manipulation of equation 7.10 yields

$$\tan \theta = \frac{b \sin 2\gamma}{a + b \cos 2\gamma},$$ (7.11)

or

$$\theta = \arctan \left[\frac{b \sin 2\gamma}{a + b \cos 2\gamma} \right].$$ (7.12)

This is the desired relation. For sufficiently small θ, this reduces to

$$\theta = \frac{2 b \gamma}{a + b},$$ (7.13)

or

$$\gamma = \frac{\theta (a + b)}{2b}.$$ (7.14)

These expressions may also be derived from the construction of Figure 7.6. From that figure,

$$\tan \theta = \frac{y}{a + z},$$ (7.15)

$$y = b \sin 2\gamma,$$ (7.16)

and

$$z = b \cos 2\gamma.$$ (7.17)

Then

$$\tan \theta - \frac{\sin 2\gamma}{\frac{a}{b} + \cos 2\gamma},$$ (7.18)

or

$$\theta = \arctan \left[\frac{\sin 2\gamma}{\frac{a}{b} + \cos 2\gamma} \right].$$ (7.19)

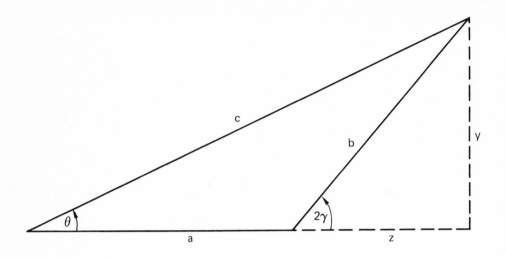

Figure 7.6 Path length construction.

Referring to Figure 7.6, it is evident that the path length c from the lens to the detector is shortened as the mirror causes the chief ray to scan off-axis, or

$$c < a + b = f \tag{7.20}$$

for $\theta \neq 0$. A perfect lens produces an image plane which is spherical with a radius R given by equation 6.61. The distance of this surface from the lens is given by equation 6.62,

$$x(\theta) = \sqrt{R^2 - \sin^2 \theta \, (R - f)^2} - (R{-}f) \cos \theta. \tag{7.21}$$

Figure 7.7 shows that if the field is not curved opposite to the curvature introduced by the changing path length, a defocus will result. The defocus as a function of scan angle θ is

$$\text{defocus} = c \, (\theta) - x \, (\theta). \tag{7.22}$$

Thus zero defocus requires

$$c \, (\theta) = x \, (\theta). \tag{7.23}$$

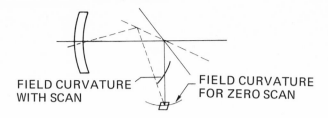

FIELD CURVATURE
WITH SCAN

FIELD CURVATURE
FOR ZERO SCAN

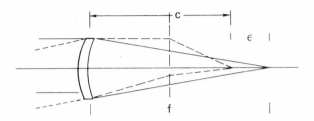

Figure 7.7 Defocus due to path length change.

The path length c may be found from Figure 7.6 by noting that

$$c^2 = a^2 + b^2 - 2ab \cos (\pi - 2\gamma), \tag{7.24}$$

and substituting for γ to get

$$c = \left\{a^2 + b^2 - 2ab \cos \left[\pi - \theta - \arcsin \left(\frac{a \sin \theta}{b}\right)\right]\right\}^{1/2}. \tag{7.25}$$

The construction of Figure 7.6 gives

$$c = \frac{y}{\sin \theta} = \frac{b \sin 2\gamma}{\sin \theta} \tag{7.26}$$

or

$$c = \frac{b \sin \left[\theta + \arcsin \left(\frac{a}{b} \sin \theta\right)\right]}{\sin \theta}. \tag{7.27}$$

Finally, the relation

$$c = \frac{a + z}{\cos \theta} \tag{7.28}$$

gives yet another equivalent form,

$$c = \frac{a + b \cos \left[\theta + \arcsin \left(\frac{a \sin \theta}{b} \right) \right]}{\cos \theta}. \tag{7.29}$$

The construction of Figure 7.8 allows the defocus equation for a flat focal surface to be derived. Here we deal with an axial ray, and consider the deviation of that ray from an ideal flat focal surface. The segment "a" is common to both deviated and undeviated ray paths, and since a mirror deviation γ causes a ray deviation of 2 γ,

$$\epsilon = b' - b = \frac{b}{\cos 2\gamma} - b = b[\sec(2\gamma) - 1] \simeq 2b \gamma^2 \tag{7.30}$$

for small γ, or

$$\epsilon \simeq \frac{\theta^2 (a + b)^2}{2b}. \tag{7.31}$$

The resulting blur diameter d is found by similar triangles from Figure 7.9 to be

$$d = \frac{\epsilon D}{f} = \frac{(f - c)D}{f}, \tag{7.32}$$

and the corresponding angular dimension δ is

$$\delta = \frac{d}{c} = \frac{D(f - c)}{fc}. \tag{7.33}$$

This defocus can be compensated by designing the focal surface to be curved horizontally to compensate for the scan defocus and to be flat vertically to allow a flat detector array. Field lenses in the converging beam near the detector have been used for this purpose. An alternate focus correction scheme mounts the mirror axis on a cam which moves the mirror in and out to prevent defocussing.

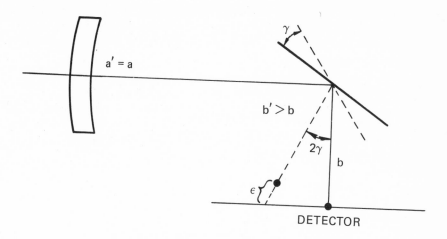

Figure 7.8 Construction of the linear defocus in a convergent beam scanner.

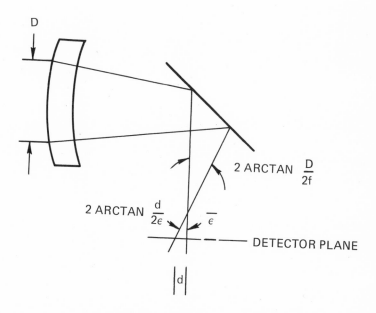

Figure 7.9 Construction of defocus blur diameter in a convergent beam scanner.

7.2.2 Parallel Beam Scanner

The most common use of the oscillating mirror is as a parallel beam scanner located between a telescope and detector lens, as shown in Figure 7.10. The important feature of this type of system is that the mirror is not scanning in object space but near the exit pupil of a telescope. The beams corresponding to specific field angles coincide only in one plane, the exit pupil. Thus if the detector lens is to be kept small, the scan mirror must be located near the exit pupil to keep the spread of the scanned beams small. Such a configuration is shown in Figure 7.11. Figure 7.12 shows the scan geometry in more detail for a non-moving exit beam. The cases shown are those where the initial and final mirror positions correspond to the extremes of the field of view of the telescope.

The required mirror dimension ℓ is determined from Figure 7.12 as follows. To a first order approximation, the beam diameter D for a particular field half angle A/2 is related to the exit pupil diameter P by

$$D = P \cos\left(\frac{A}{2}\right). \tag{7.34}$$

Summing angles around the beam axis gives

$$\frac{A}{2} + \frac{\pi}{2} + K + \gamma + \gamma = \pi \tag{7.35}$$

or

$$\gamma = \frac{\pi}{4} - \frac{A}{4} - \frac{K}{2}. \tag{7.36}$$

As shown in Figure 7.12, the mirror length ℓ is related to the mirror displacement γ and the beam diameter D by

$$\ell = \frac{D}{\sin \gamma}, \tag{7.37}$$

so that

$$\ell = \frac{D}{\sin \left(\frac{\pi}{4} - \frac{K}{2} - \frac{A}{4}\right)}. \tag{7.38}$$

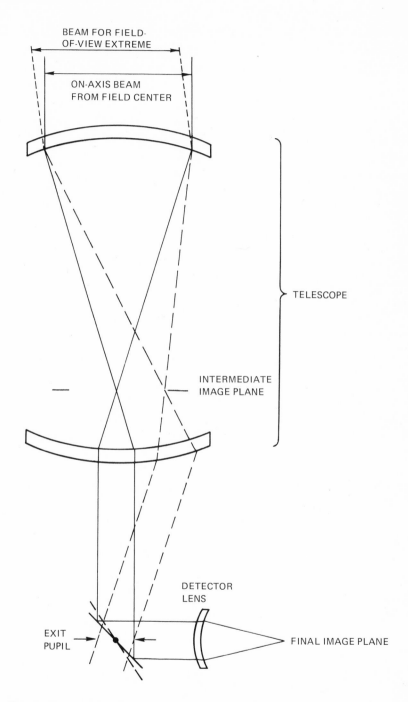

BEAM FOR FIELD-
OF-VIEW EXTREME

ON-AXIS BEAM
FROM FIELD CENTER

TELESCOPE

INTERMEDIATE
IMAGE PLANE

DETECTOR
LENS

EXIT
PUPIL

FINAL IMAGE PLANE

Figure 7.10 A magnified parallel beam oscillating mirror scanner.

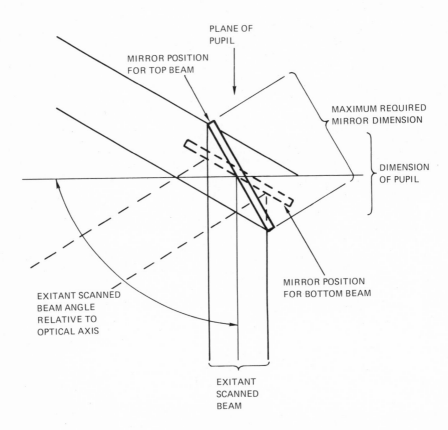

Figure 7.11 Mirror placement for migrationless beam.

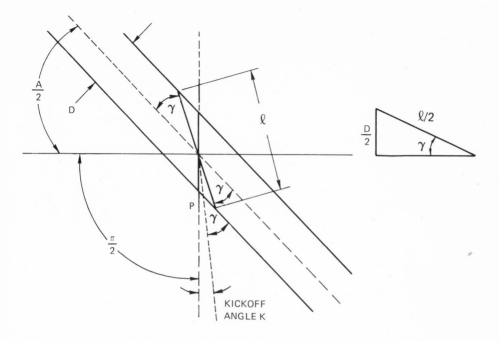

Figure 7.12 Mirror length construction.

This scanner does not scan a parallel beam if the object is not at infinity, as shown in Figure 7.13. When the mirror is displaced by $\sigma/2$, the axial ray shifts by σ and the linear shift in object space is

$$s \simeq B\sigma. \tag{7.39}$$

The angular subtense σ of s at a distance O measured from the lens is

$$\sigma = \frac{s}{O} = \frac{B\alpha}{O} = \sigma\left(\frac{O-A}{O}\right) = \sigma\left(1 - \frac{A}{O}\right). \tag{7.40}$$

Thus the angle scanned in object space changes as a function of the in-focus range. The detector angular subtense also changes from

$$\alpha = \frac{a}{f} \text{ to } \alpha = \frac{a}{i} = \frac{\alpha O}{O+f}. \tag{7.41}$$

Similarly, the motion in the image plane varies with in-focus range, as follows. Figure 7.14 shows how the image of a point closer than infinity moves as the

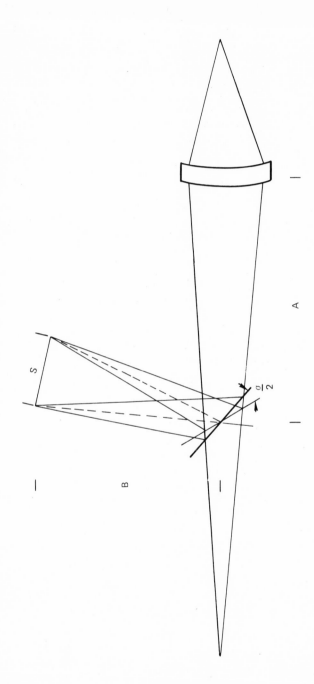

Figure 7.13 Motion in object space for fixed image point.

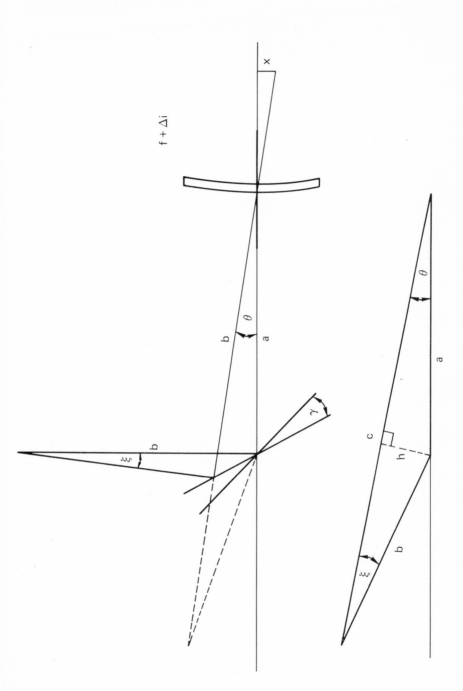

Figure 7.14 Motion in image space for a fixed object point.

mirror is displaced. Reflecting the two ray paths about the displaced mirror position produces the triangle shown in the bottom of the figure. The angle θ by which the central ray shifts is derived as follows:

$$\sin \xi = \frac{h}{b} \tag{7.42}$$

$$\sin \theta = \frac{h}{a} \tag{7.43}$$

$$\xi = \arcsin \frac{a \sin \theta}{b} \tag{7.44}$$

$$\gamma = \frac{\theta + \xi}{2} \quad \text{(see section 7.2.1)} \tag{7.45}$$

$$\simeq \frac{\theta \, (a + b)}{2b} \quad \text{for small } \theta, \tag{7.46}$$

or

$$\theta \simeq \frac{2b \, \gamma}{a + b} . \tag{7.47}$$

The image point moves by a distance x given approximately by

$$x \simeq \theta i = \frac{2b\gamma}{a + b} \, i = \frac{2b\gamma}{a + b} \, (f + \Delta i), \tag{7.48}$$

where

$$\Delta i = \frac{f^2}{O} \quad \text{(see equation 6.127).} \tag{7.49}$$

Using the lens law,

$$\frac{1}{a + b} + \frac{1}{f + \Delta i} = \frac{1}{f} , \tag{7.50}$$

or

$$\frac{1}{a + b} = \frac{\Delta i}{f(f + \Delta i)} , \tag{7.51}$$

and

$$b = \left(f^2 + \Delta i(f-a) \right) / \Delta i. \tag{7.52}$$

Then

$$x = \frac{2\gamma \left(\dfrac{f^2 + \Delta i(f-a)}{\Delta i} \right) \Delta i \, (f+\Delta i)}{f(f + \delta)}$$

$$= 2\dot{\gamma} \left(f + \frac{\Delta i(f - a)}{f} \right) . \tag{7.53}$$

The rate of image motion in the image plane is

$$\frac{dx}{dt} = 2 \frac{d\gamma}{dt} \left[f + \frac{\Delta i(f - a)}{f} \right]. \tag{7.54}$$

The term $2f \, d\gamma/dt$ is the rate for an infinitely distant object and $(2 \, \Delta i \, (f-a)/f)d\gamma/dt$ is the change in rate as the object moves in to a distance $O = f^2/\Delta i$. The percent error is $(\Delta i(f-a)/f^2) \times 100\% = \left((f-a/O) \right) \times 100\%$.

7.3 Rotating Reflective Drum Mirrors

Oscillating mirrors are unsuited for high speed scanner operation because they tend to become unstable near the field edges and require high motor drive power. In such cases, rotating drum mirror assemblies such as that shown in Figure 7.15 are frequently used because their motion is continuous and stable. Consider the polygonal drum mirror shown in Figure 7.16, where the mirrored surfaces are circumscribed by a circle of radius r_o on the outside and are tangent to a circle of radius r_i on the inside. This particular mirror is ten-faceted. The facet angle θ_f measured from the drum axis is

$$\theta_f = \frac{2\pi}{n \text{ facets}} . \tag{7.55}$$

If we let the facet length be ℓ, it is evident from Figure 7.17 that

$$\ell = 2r_o \sin (\theta_f/2) \tag{7.56}$$

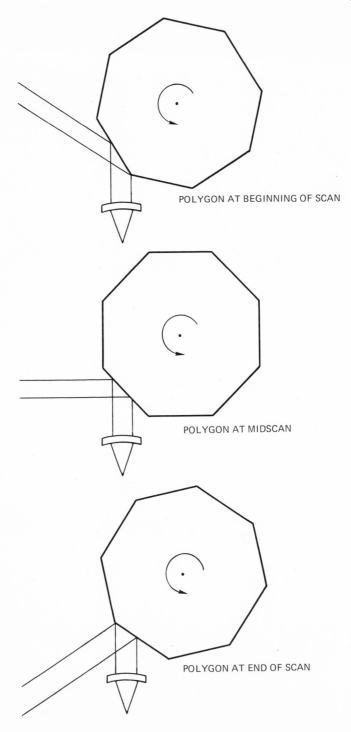

POLYGON AT BEGINNING OF SCAN

POLYGON AT MIDSCAN

POLYGON AT END OF SCAN

Figure 7.15 Rotating polygonal mirror scanner.

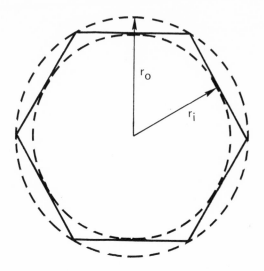

Figure 7.16 Dimensions of a polygonal mirror.

and that

$$\frac{r_o}{r_i} = \left[\cos \left(\frac{\theta_f}{2} \right) \right]^{-1}.$$ (7.57)

The ratios of the facet length to the outer radius and of the outer radius to the inner radius are tabulated in Table 7.1 versus number of facets.

Now consider Figure 7.18, where two drum angular positions are shown with an angular rotational displacement θ_R. The geometrical construction shown demonstrates that the final displacement δ of the center of the facet in the direction normal to the initial undisplaced mirror surface is given by:

$$\delta = r_i (1 - \cos \theta_R)$$

$$= r_o [1 - \sin (\theta_f/2) \tan (\theta_f/4)] [1 - \cos \theta_R].$$ (7.58)

Thus the scanning action of a rotating reflective polygon is similar to the action of an oscillating mirror with the exception that the center of a facet moves in and out at a rate

$$\frac{d\delta}{dt} = r_i \sin (\dot{\theta}_R t) \dot{\theta}_R$$ (7.59)

assuming zero displacement at $t = 0$.

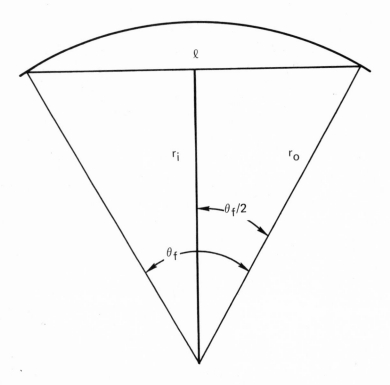

Figure 7.17 Facet length construction.

Table 7.1
Characteristics of a Polygonal Mirror

Number of Facets	Facet Angle	$\dfrac{\ell}{r_o}$	$\dfrac{r_o}{r_i}$
3	120°	1.732	2.000
4	90°	1.414	1.414
5	72°	1.176	1.236
6	60°	1.000	1.155
7	51.43°	0.868	1.110
8	45°	0.765	1.082
9	40°	0.684	1.064
10	36°	0.618	1.051

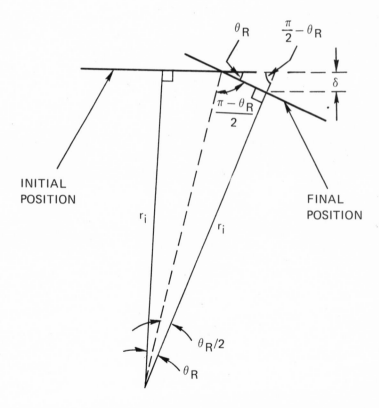

Figure 7.18 Mirror motion with rotation.

If such a scanner were used in a converging beam, this oscillatory motion would change the focal point with time and could cause severe defocusing. Therefore these scanners are used primarily in parallel beam systems. When the angle each mirror makes with the vertical is the same, a non-interlaced rectilinear scan results. If the angles are incremented, interlaced or overscanned patterns can be implemented. For example, the reflective cube scanner can produce interlace factors of one, two, or four. Interlace is accompanied, however, by an increasing image rotation as the face edges are approached due to the vertical mirror angling.

Levi[1] has shown that the minimum outer radius r_0 necessary to keep the beam unvignetted during the scan of a parallel-beam is given by:

$$r_0 = \frac{D}{2 \cos\left(\frac{\delta}{2}\right)\sin\left(\frac{\theta_f - \gamma}{2}\right)},$$
(7.60)

where D is the beam diameter, δ is the beam deviation at the midpoint of the scan, 2γ is the beam deflection angle produced by the scan, θ_f is the facet angle equalling $2\pi/N$. For $\delta = 90°$,

$$R = \frac{D}{\sqrt{2} \sin\left(\frac{\theta_f - \gamma}{2}\right)}.$$
(7.61)

The polygonal scanner is not as effective at reducing beam shift as is the oscillating mirror because the facet motion is not exactly equivalent to that of an oscillating mirror, as is evident from Figure 7.19. Two different initial scan positions are shown in Figure 7.20 to demonstrate the approximate differences in beam shift which result.

Gorog, et al.[2] assert that the maximum rotation rate M_{max} [rps] of a polygonal mirror is limited to the rate at which the mirror will fly apart, and is given by

$$M_{max} = \frac{1}{2\pi r_0} \sqrt{\frac{8T}{\rho(3 + \eta)}}.$$
(7.62)

The mirror material parameters are:

ρ = density
T = ultimate tensile strength
η = Poisson's ratio.

Actually the mirror would become useless due to facet deformation long before mirror disintegration.

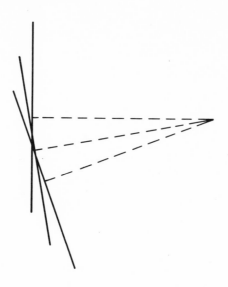

Figure 7.19 An example of facet rotation.

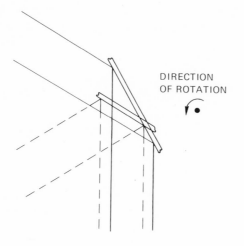

DIRECTION
OF ROTATION

CASE 1 (SHOWN):

MIRROR FACET CENTER AT
PUPIL FOR UPPER BEAM.

THEN THE FACET INTERSECTION
WITH THE BEAM MOVES TOWARD
THE BEAM AND THE EXIT BEAM
SHIFTS TO THE LEFT SLIGHTLY.

CASE 2 (NOT SHOWN):

MIRROR FACET CENTER AT
PUPIL FOR CENTER BEAM.

THEN MIRROR INTERSECTION
MOVES IN AND OUT AND EXIT
BEAM MOVES LEFT AND RIGHT.

Figure 7.20 Beam shift.

Another scanner somewhat similar to the polygonal mirror scanner is the circular-segment-raster scanner shown in Figure 7.21 and referred to as the "soupbowl". This has been used to generate a stable scan in a converging beam. Another mirror arrangement is the "carousel" scanner shown in Figure 7.22 which gives a rectangular raster.

7.4 Rotating Refractive Prisms

The third scanner type is the rotating refractive prism having $2(n+2)$ faces as shown in Figure 7.23 and which is rotated about its centroid. Figure 7.24 depicts the convergent beam configuration, which has the advantage that it allows an in-line system design. It produces the scan by laterally displacing the converging beam, sweeping it across the detector array. The principle of operation is shown for deviated and undeviated rays. First, it is evident that if the incident and exitant beams are in the same medium, the angle both beams make with the horizontal will be the same. This occurs because the refraction at the first surface is reversed by that at the second surface, giving zero angular shift. However, the focus is displaced vertically by y and horizontally by z.

The ray deviation y is derived from the construction of Figure 7.25 as follows, where ϕ_1 is the incident ray angle measured from the horizontal,

$$\frac{a + b}{t} = \tan (\varphi_1 - N) \tag{7.63}$$

$$a + b = t \tan (\varphi_1 - N) \tag{7.64}$$

$$\frac{a}{t} = \tan (\varphi_2 - N) \tag{7.65}$$

$$b = t \left[\tan (\varphi_1 - N) - \tan (\varphi_2 - N) \right] \tag{7.66}$$

$$\frac{y}{b} = \cos \gamma \tag{7.67}$$

$$y = t \cos \gamma \left[\tan (\varphi_1 - N) - \tan (\varphi_2 - N) \right]. \tag{7.68}$$

Since $N = \gamma$,

$$y = t \cos \gamma \left[\tan (\varphi_1 - \gamma) - \tan (\varphi_2 - \gamma) \right] \tag{7.69}$$

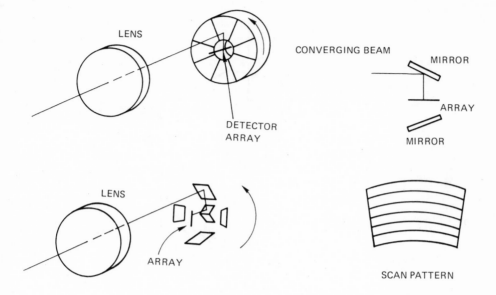

Figure 7.21 Rotating reflective "soupbowl" scanners.

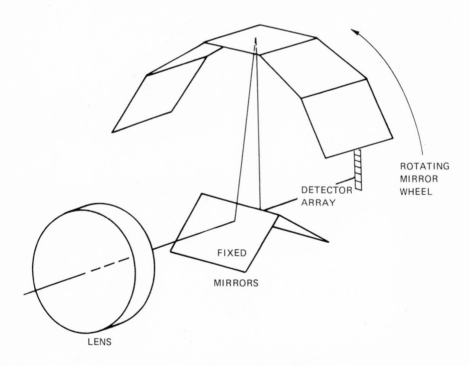

Figure 7.22 Rotating reflective carousel scanner.

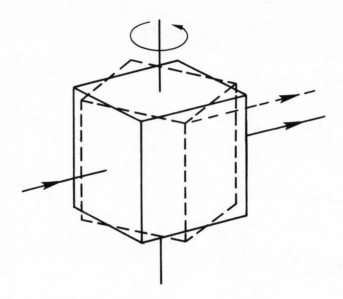

Figure 7.23 Rotating refractive cube scanner.

$$\frac{\sin(\varphi_2 - \gamma)}{\sin(\varphi_1 - \gamma)} = \frac{n_1}{n_2},$$ (7.70)

so

$$\tan(\varphi_2 - \gamma) = \frac{\dfrac{n_1}{n_2}\sin(\varphi_1 - \gamma)}{\cos(\varphi_2 - \gamma)}$$ (7.71)

and

$$\tan(\varphi_2 - \gamma) = \frac{\dfrac{n_1}{n_2}\sin(\varphi_1 - \gamma)}{\sqrt{1 - \sin^2(\varphi_2 - \gamma)}}$$

$$= \frac{\dfrac{n_1}{n_2}(\varphi_1 - \gamma)}{\sqrt{1 - \left[\dfrac{n_1}{n_2}\sin(\varphi_1 - \gamma)\right]^2}}.$$ (7.72)

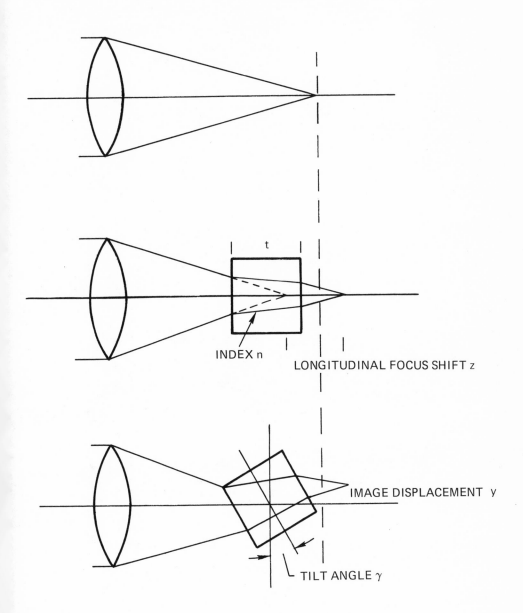

Figure 7.24 Focal shift and image displacement.

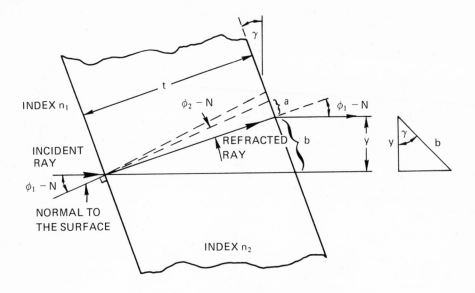

Figure 7.25 Image displacement construction.

Then

$$y = t \cos \gamma \left[\frac{\sin(\varphi_1 - \gamma)}{\cos(\varphi_1 - \gamma)} - \frac{\dfrac{n_1}{n_2} \sin(\varphi_1 - \gamma)}{\sqrt{1 - \left[\dfrac{n_1}{n_2} \sin(\varphi_1 - \gamma) \right]^2}} \right]. \qquad (7.73)$$

In air, $n_1 = 1$ and

$$y = t \cos \gamma \left[\frac{\sin(\varphi_1 - \gamma)}{\cos(\varphi_1 - \gamma)} - \frac{\sin(\varphi_1 - \gamma)}{\sqrt{n^2 - \sin^2(\varphi_1 - \gamma)}} \right]. \qquad (7.74)$$

This has been expressed in various equivalent forms in the literature. For $(\varphi_1 - \gamma)$ small and γ small,

$$y \cong t \cos \gamma \left[\varphi_1 - \gamma - \frac{\dfrac{1}{n}(\varphi_1 - \gamma)}{\sqrt{1 - \dfrac{(\varphi_1 - \gamma)^2}{n^2}}} \right], \qquad (7.75)$$

or

$$y \cong t \cos \gamma \left[\varphi_1 - \gamma - \frac{1}{n}(\varphi_1 - \gamma) \right]$$

$$= t \cos \gamma \, (\varphi_1 - \gamma) \left[\frac{n-1}{n} \right].$$

(7.76)

The ultimate small angle approximation is

$$y = t \, (\varphi_1 - \gamma) \left(\frac{n-1}{n} \right).$$

(7.77)

Thus, over small angles, prism rotation produces an approximately linear scan. Gorog, et al.[2], Lindberg[3], and Kulikovskaya, et al.[4] present these and other design equations. Lindberg also presents designs of refractive cube scanners used in Swedish military and commercial devices.

The differential change in displacement with differential rotation is found by taking $dy/d\gamma$. For a paraxial ray, $\varphi_1 = 0$ and

$$y = -t \sin \gamma \left[1 - \frac{\cos \gamma}{\sqrt{n^2 - \sin^2 \gamma}} \right].$$

(7.78)

Then

$$\frac{dy}{d\gamma} = -t \left[\cos \gamma + \frac{\sin^2 \gamma - \cos^2 \gamma}{(n^2 - \sin^2 \gamma)^{1/2}} - \frac{\sin^2 \gamma \cos^2 \gamma}{(n^2 - \sin^2 \gamma)^{3/2}} \right].$$

(7.79)

For tilt angles $\gamma \ll \pi/2$,

$$y \approx t \, \gamma \left(1 - \frac{1}{n} \right) + \text{terms of order } \gamma^3.$$

(7.80)

For $\gamma = 0$, the shift z in focus away from the lens with increasing ray angle ϕ_1 is, according to Gorog, et al.[2], given by

$$z = t \left[1 - \frac{\cos \phi_1}{n^2 - \sin^2 \phi_1} \right]$$

(7.81)

or for $\phi_1 \ll \pi/2$,

$$z \approx t \left(1 - \frac{1}{n} \right) + \phi_1^2 \frac{t}{n} \left(1 - \frac{1}{2n^2} \right).$$

(7.82)

The central ray is shifted by

$$z = t \left(1 - \frac{1}{n}\right) .$$ (7.83)

They also derived the focal plane shift z for nonzero tilt angle γ for a paraxial ray ($\phi_1 = 0$),

$$z = t \left[\cos \gamma - \frac{\cos 2\gamma}{(n^2 - \sin^2 \gamma)^{1/2}} - \frac{1}{4} \frac{\sin^2 2\gamma}{(n^2 - \sin^2 \gamma)^{3/2}}\right].$$ (7.84)

For $\gamma \ll \pi/2$,

$$z = t \left(1 - \frac{1}{n}\right) + \gamma^2 \frac{t}{n} \left(2 - \frac{n}{2} - \frac{3}{2n^2}\right) .$$ (7.85)

Jenkins and White[5], Levi[1], Smith[6], and Waddell[7] show that such a prism introduces transverse and lateral chromatic aberration, coma, spherical aberration, and astigmatism. The mechanical analysis used for the rotating mirror prism is also applicable for a refractive prism. Lindberg[3] reported that in a system application he removed the edges of a prism to increase mechanical strength and to reduce drag.

Prism scanners have several good features. Their motion is continuous and therefore tends to be stable. Used in a convergent beam, the prism can be small in size, and the prism edges can be removed to reduce inertia and drag for low scan motor power. Mechanical noise can be very low because simple uniformly rotating mechanisms are used. Interlace is readily achieved by canting prism faces, as shown in Figure 7.26. The disadvantages are that scan efficiency is poor, the focal length shifts, and the prism introduces aberrations. Another problem is that refractive materials used for such a prism have high surface reflection losses and thus must be anti-reflection coated. However, these coatings have reduced efficiency for large incidence angles, so near the prism edges, reflection losses will be higher. If the prism temperature differs from the apparent scene temperature, the prism may also spuriously modulate the scene image because of varying self-emission as a function of scan angle.

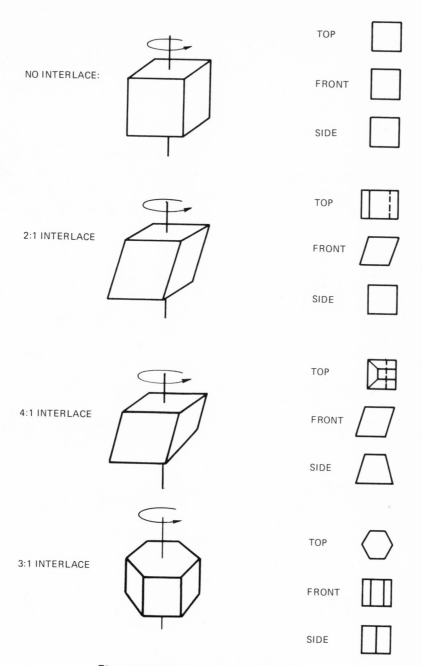

Figure 7.26 Rotating refractive prisms.

7.5 Rotating Refractive Wedges

The rotating refractive wedge is one of the most flexible scan mechanism elements available. As discussed by Rosell[8] and by Wolfe[9], the use of two rotating wedges allows generation of scan patterns which are unidimensional, circular, elliptical, rosette-shaped, or spiral. Wedge scanners must be used in a parallel beam because they introduce severe aberrations into a converging beam. Therefore we will analyze only the parallel beam case shown in Figure 7.27. Let the incident ray be represented by an angle φ_1 having two components φ_{1X} and φ_{1Y}, and let the ray between the wedges be φ_2 with components φ_{2X} and φ_{2Y}. The coordinates (x,y) are chosen so that positive angles are rotated upward from the axis. The exitant ray is parallel to the axis to represent the parallel beam case. The wedge angular rotations around the axis are measured as positive when counter-clockwise and are denoted by θ_1 and θ_2.

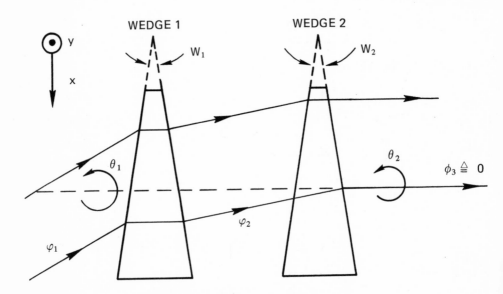

Figure 7.27 Rotating prism-pair scanner.

The motion of the ray φ_1 as θ_1 and θ_2 change may be derived by following φ_3 backward through the system using successive applications of the wedge deviation principle as follows:

1. Find the deviation of φ_3 as it leaves wedge 2 in the coordinates rotated by the wedge rotation θ_2;
2. Resolve these φ_2 ray components into unrotated coordinates and then into the rotated coordinates of wedge 1;

3. Find the deviated ray components leaving wedge 1 in rotated coordinates;
4. Resolve these into unrotated coordinates to find the desired φ_1 components.

Figure 7.28 shows the constructions which are necessary. Working backward through the system, ray φ_3 enters wedge 2 with components

$$\varphi_{3X} = \varphi_{3Y} = 0 \qquad (7.86)$$

in unrotated coordinates. Leaving the wedge the ray angle is the wedge deviation angle δ_2. Resolved into unrotated coordinates, the components are

$$\varphi_{2X} = \delta_2 \cos \theta_2 \qquad (7.87)$$

and

$$\varphi_{2Y} = -\delta_2 \sin \theta_2. \qquad (7.88)$$

Resolving these components into the rotated coordinates of wedge 1 yields

$$\varphi'_{2X} = \varphi_{2X} \cos \theta_1 - \varphi_{2Y} \sin \theta_1 \qquad (7.89)$$

and

$$\varphi'_{2Y} = \varphi_{2Y} \cos \theta_1 + \varphi_{2X} \sin \theta_1. \qquad (7.90)$$

Leaving the wedge, the ray components in rotated coordinates are

$$\varphi'_{1X} = \varphi'_{2X} + \delta_1 \qquad (7.91)$$

and

$$\varphi_{1Y} = \varphi_{2Y}. \qquad (7.92)$$

Finally, resolving these into unrotated coordinates gives

$$\varphi_{1X} = \varphi'_{1X} \cos \theta_1 + \varphi'_{1Y} \sin \theta_1 \qquad (7.93)$$

and

$$\varphi_{1Y} = \varphi'_{1Y} \cos \theta_1 - \varphi'_{1X} \sin \theta_1. \qquad (7.94)$$

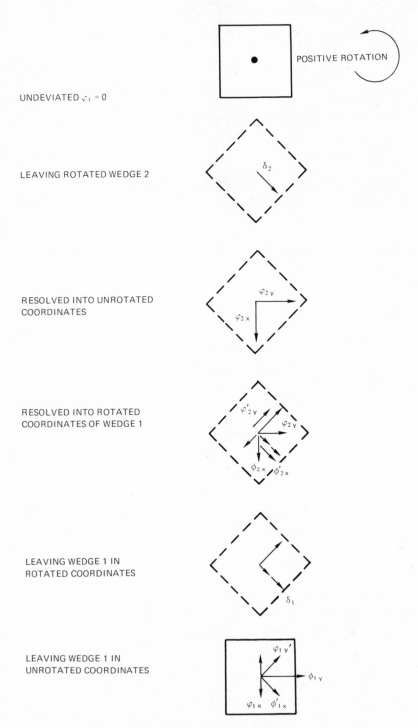

POSITIVE ROTATION

UNDEVIATED $\varsigma_3 = 0$

LEAVING ROTATED WEDGE 2

RESOLVED INTO UNROTATED
COORDINATES

RESOLVED INTO ROTATED
COORDINATES OF WEDGE 1

LEAVING WEDGE 1 IN
ROTATED COORDINATES

LEAVING WEDGE 1 IN
UNROTATED COORDINATES

Figure 7.28 Ray deviation constructions.

Combining these equations gives the desired general equations for a two-wedge scanner:

$$\varphi_{1X} = [\delta_2 \cos(\theta_1 - \theta_2) + \delta_1] \cos\theta_1 + \delta_2 \sin(\theta_1 + \theta_2) \sin\theta_1 \quad (7.95)$$

and

$$\varphi_{1Y} = \delta_2 \sin(\theta_1 - \theta_2) \cos\theta_1 - [\delta_2 \cos(\theta_1 - \theta_2) + \delta_1] \sin\theta_1. \quad (7.96)$$

Table 7.2 gives some of the useful cases of the wedge pair for constant and equal wedge velocities. Spiral and rosette patterns can be generated using unequal velocities. A disadvantage of wedge scanners for line scanning is that the sinusoidal scan leads to a lower scan efficiency than is accomplishable with other scanners. Also, frame-to-frame and line-to-line registration require tight control of the prism angular velocities. When gear drives have been used, gear backlash and wear have been serious problems.

7.6 Other Scanners

The revolving lens is a brute force solution to the problem of scan generation. Figure 7.29 shows the simplest mechanization, a lens revolving back and forth between two stops. A variation of this is the "spin-ball" scanner depicted in Figure 7.30, which was used in an early FLIR. Both systems are impractical for large aperture systems because too much glass must be moved, and the spinball has the added disadvantage of poor scan efficiency.

The dominant scanner models employ rectilinear scan generation, but another interesting though less frequently used class is that of circular scanners. Two approaches have been used: rotating a complete sensor having a radial detector array, or rotating the scene using rotating optical elements. The former approach is illustrated in Figure 7.31. Such scanners have the disadvantages that large sensor masses must be rotated, and that the associated gyroscopic effect may complicate gimballing.

Rotating the scene rather than the sensor is much easier, and one successful method which has been used is the rotating V-mirror of Figure 7.32. Another example of a circular scanner is shown in Figure 7.33. Major advantages of such scanners are that their motion is continuous and that any image rotation introduced by a pointing gimbal is removable by optical derotation. Finally, it is also possible to oscillate the complete electronics assembly as shown in Figure 7.34.

Table 7.2

Scan motions for the four possible dual wedge scanner modes

Case	Conditions	φ_{1X}	φ_{1Y}	Motion
1	$\theta_1 = \theta_2 = \theta$ $\delta_1 = \delta_2 = \delta$	$2\delta \cos\theta$	$-2\delta \sin\theta$	circular
2	$\theta_2 = -\theta_1 = -\theta$ $\delta_1 = \delta_2 = \delta$	$2\delta \cos\theta(1 - \sin^2\theta)$ $\simeq 2\delta \cos\theta$ for small θ	0	unidimensional, bidirectional, approximately cosinusoidal
3	$\theta_2 = -\theta_1 = -\theta$ $\delta_1 \neq \delta_2$	$2\delta_2 \cos^3\theta + (\delta_1 - \delta_2)\cos\theta$	$(\delta_2 - \delta_1)\sin\theta$	approximately elliptical
4	$\theta_2 = \theta_1 + P$ $\theta_1 = \theta$ $\delta_1 = \delta_2 = \delta$	$\delta\,[\cos\theta + \cos\theta \cos P$ $+ \sin\theta \sin(2\theta + P)]$	$-\delta\,[\sin(\theta + P) + \sin\theta]$	same as case 2 except scan is rotated by P

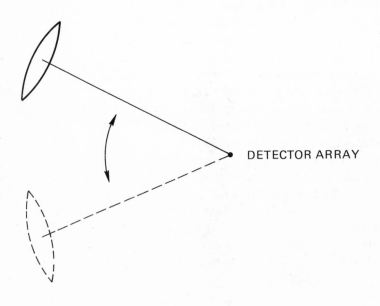

DETECTOR ARRAY

Figure 7.29 Simple nutating lens scanner.

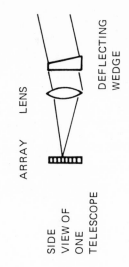

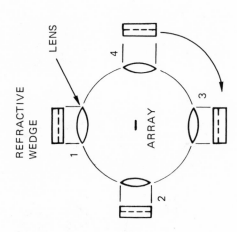

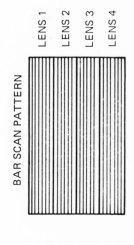

Figure 7.30 Revolving lens ("spin-ball") scanners.

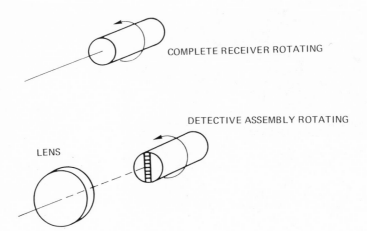

COMPLETE RECEIVER ROTATING

DETECTIVE ASSEMBLY ROTATING

LENS

Figure 7.31 Rotating sensor scanners.

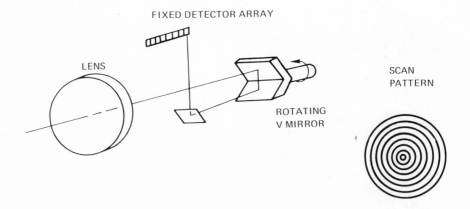

FIXED DETECTOR ARRAY

LENS

ROTATING
V MIRROR

SCAN
PATTERN

Figure 7.32 V-mirror rotating optics scanner.

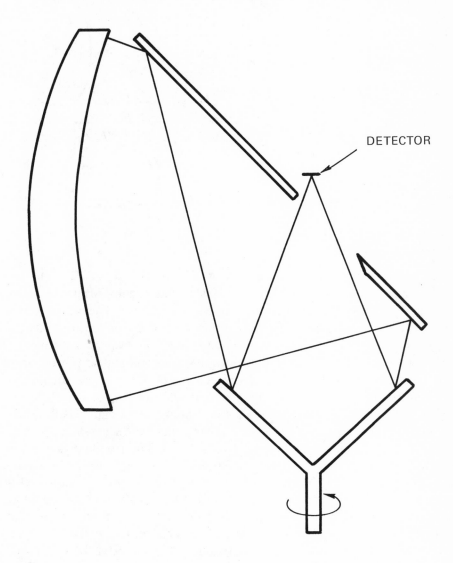

Figure 7.33 Another circular scanner.

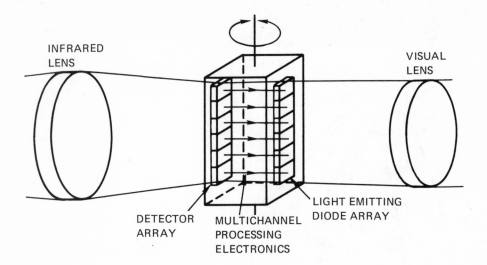

INFRARED
LENS

VISUAL
LENS

DETECTOR
ARRAY

MULTICHANNEL
PROCESSING
ELECTRONICS

LIGHT EMITTING
DIODE ARRAY

Figure 7.34 Nutating focal plane scanner.

7.7 Shading Effects

In certain types of suboptimal FLIR designs the displayed scene has
superimposed on it a broad, slowly-varying spurious signal. This image defect
is called shading, after a similar but unrelated effect in television camera
tubes. Shading is produced whenever either of two effects occurs. First, if
the detector signal from a uniform source in the scene varies as a function
of the scan angle, the displayed image of the source will appear to vary or to
shade. Second, if the component of the detector background signal originating
from sources not in the field of view varies, the ac coupling of the FLIR will
pass the variation into the video and a scene-independent shading will result.

There are two types of shading: signal-dependent and signal-independent.
Signal dependent shading arises from four sources. The first is distortion in
the optical system which results in differential image expansion or contrac-
tion as a function of field angle. This has the effect of locally dimming or
brightening the focal plane signal. Because FLIRs must always be ac coupled,
the changes in background flux due to distortion are emphasized. The second
is vignetting in the optical system which causes a rolloff of signal corresponding to the vignetting pattern. These two sources can often be controlled
satisfactorily by good optical design. The third source is shading in a uniform
target because the optical collecting area projected along the instantaneous
scan angle varies as the cosine of the scan angle measured from the optical
axis.

The fourth source of signal-dependent shading is due to variation in the angle at which a converging beam strikes a detector, and can be a much more serious problem. To a first order approximation, plane detectors are Lambertian receivers; that is, the signal output varies as the cosine of the angle between the input beam and the normal to the detector surface. All convergent beam scanners and all parallel beam scanners with pupil migration produce a change in beam angle, as shown in Figure 7.35. These variations in converging beam angle cause cosine dependent signal shading. In some detectors, increasing reflection losses with angle may also contribute to this effect.

Signal-independent shading may occur whenever the scanner and optics are configured so that housing radiation reaches the detector. Again, convergent beam scanners and shifting-pupil parallel beam scanners are the offenders. A particularly simple case is shown in Figure 7.36, where a converging beam scanner has been modelled by an equivalent oscillating lens scanner for diagrammatic clarity. For any instantaneous scanner position, the detector must receive the radiation from the full solid angle Ω_1 formed by the rays converging from the lens. However, to accommodate all scanning positions, the fixed detector must accept rays over the solid angle $\Omega_2 > \Omega_1$. Thus, the detector will receive radiation not only from the scene, but from the housing as well. The housing is not in focus, but the power reaching the detector from the housing may vary with position, so that spurious stationary patterns are introduced into the video.

In the case of a migrating-pupil scanner, the detector lens must be oversized, with the result that different parts of the housing are seen with changing scan angle, causing shading if housing temperature differences exist. It is important to note that a parallel beam scanner which scans in only one direction unavoidably has beam shift in the non-scanned direction. In itself, such shading may not be too objectionable if it is limited to a few percent of the maximum displayed scene luminance. However, if the FLIR is to be used to provide inputs to an automatic target motion tracker or to some automatic target classification device, shading must be minimized. When optical design alone cannot eliminate shading, electronic compensation and optical baffling can be employed.

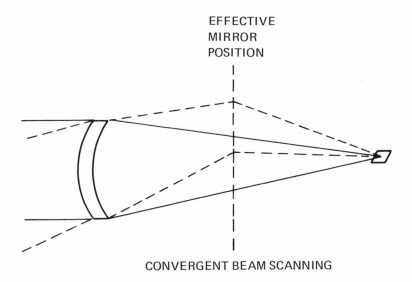

EFFECTIVE
MIRROR
POSITION

CONVERGENT BEAM SCANNING

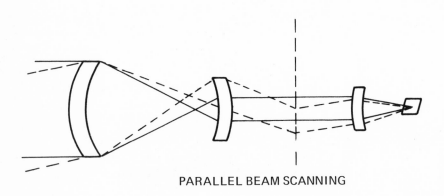

PARALLEL BEAM SCANNING

Figure 7.35 Beam shift due to scanning.

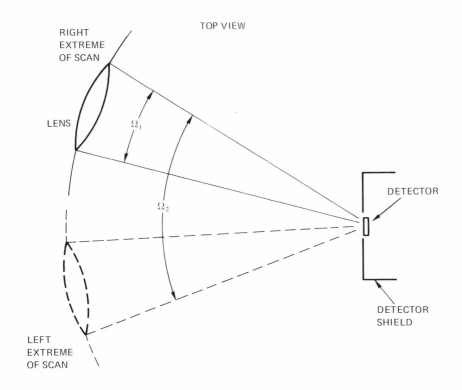

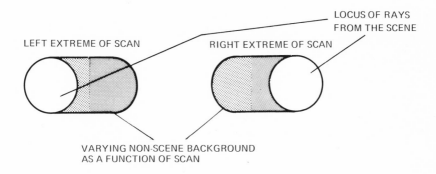

Figure 7.36 An example of aperture scanning.

REFERENCES

[1] L. Levi, *Applied Optics,* Wiley, 1968.

[2] I. Gorog, J.D. Knox and P.V. Goedertier, "A Television-Rate Laser Scanner. 1. General Considerations", RCA Rev., *33*, pp 623-666, December 1972.

[3] P.J. Lindberg, "A Prism Line-Scanner for High Speed Thermography", Optica Acta, *15*, pp 305-316, July-August 1966.

[4] N.I. Kulikovskaya, L.A. Gal'Perin, and V.M. Kurysheva, "Optical Characteristics of Thermal Infrared Scanning Systems", Soviet Journal of Optical Technology, *37*, pp 646-648, Oct. 1970.

[5] F.A. Jenkins and H.E. White, *Fundamentals of Optics,* McGraw-Hill, 1957.

[6] W.J. Smith, *Modern Optical Engineering,* McGraw-Hill, 1966.

[7] J.H. Waddell, "Rotating Prism Design for Continuous Image Compensation Cameras", Applied Optics, *5*, pp 1211-1223, July 1966.

[8] F.A. Rosell, "Prism Scanner", JOSA, *50*, pp 521-526, June 1960.

[9] W.L. Wolfe, editor, *Handbook of Military Infrared Technology,* USGPO, 1965.

CHAPTER EIGHT — THERMAL IMAGING SYSTEM TYPES

8.1 Introduction

The classified nature of most advanced thermal imaging development has prevented much discussion of device characteristics in the open scientific literature. The few examples are the thermographs described in references 1 through 7, the commercial FLIR's in references 8 and 9, and the military FLIR's in references 10 through 12. From the viewpoint of complexity, the single element system considered briefly in Chapter One is the simplest conceivable system. However, fundamental limits or the detector state of the art may not allow such a system to have a sufficient thermal sensitivity, so that the per frame, per resolution element signal to noise ratio must be improved by adding detectors.

Detectors may be added by preserving the single element scan pattern and stringing elements together horizontally, by making the scan one dimensional and stringing elements vertically, or by a combination of the two. Some of the possibilities are shown in Figure 8.1. Each added element requires as a minimum its own preamplifier, so the price of sensitivity improvement is increased system complexity. The increased complexity may be reduced somewhat by the techniques of serial processing or of multiplexing, both of which reduce an n-channel system to a single video channel system.

Conventional thermal imaging system designs differ from each other in two fundamental ways: the means by which the scene is dissected and the means by which preamplified detector signals are processed for video presentation. The particular implementations of the optics, scan mechanism, cooler, and display are not fundamentally significant if they meet at least minimum requirements. For example, whether a cooler uses a Stirling or a Vuilleumier cycle may influence cost, weight, or reliability, but it does not categorize a system. As we shall see, however, variations in scanning and in video processing may radically alter the functioning of the system.

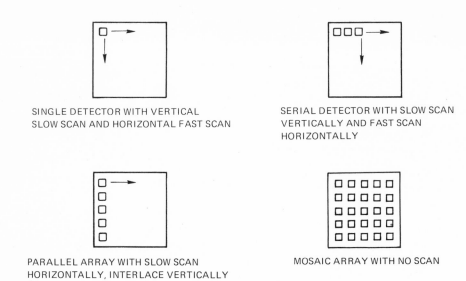

SINGLE DETECTOR WITH VERTICAL
SLOW SCAN AND HORIZONTAL FAST SCAN

SERIAL DETECTOR WITH SLOW SCAN
VERTICALLY AND FAST SCAN
HORIZONTALLY

PARALLEL ARRAY WITH SLOW SCAN
HORIZONTALLY, INTERLACE VERTICALLY

MOSAIC ARRAY WITH NO SCAN

Figure 8.1 Possible uses of discrete detectors.

8.2 Serial and Parallel Scene Dissection

In addition to distinctions between scanning mechanism types such as convergent and parallel beam scanning, and one- or two-dimensional scanning, there is a distinction between serial and parallel scene dissection and detector signal processing. The differences between the two scan types are shown in Figure 8.2. In parallel scene dissection, an array of detectors is oriented perpendicular to the primary scan axis, as in a unidimensional detector array used with an azimuth scanner. All of the detector outputs are amplified, processed and displayed simultaneously or in parallel. In serial scene dissection[13], an array of detectors is oriented parallel to the primary scan axis and each point of the image is scanned by all detectors. The detector outputs are appropriately delayed and summed by an integrating delay line which superimposes the outputs, thereby simulating a single scanning detector, or they may be read out one-for-one on a similar array of scanning display elements.

One type of scanning may be more advantageous than the other depending on such factors as overall allowed sensor size, allowed power consumption, and performance-to-cost ratio. The primary advantage of parallel processing is that an extremely compact special-purpose sensor can be built using converging beam scanning with both detector and display scanning off of the same mirror. This advantage may be lost when general-purpose parallel beam scanning is used. The major disadvantages of parallel processing are that

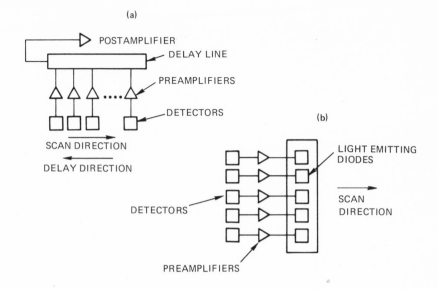

Figure 8.2 (a) Serial and (b) parallel scanning.

detector D* and responsivity variations in the array cause image non-uniformities, and that ac coupling artifacts due to the existence of different average scene values on different detectors cause nonuniformities.

The primary advantages of serial scanning are that nonuniformities are eliminated because the scene is dissected in effect by only one detector, and that the number of detectors required is dictated by sensitivity requirements rather than by the number of scan lines, as in parallel scanning. The primary disadvantage of serial scanning is that it must be done in a parallel beam, so the optics tend to be more complex. The SNR equation for a serial FLIR is not intuitively obvious. The following derivation for the effective D* and SNR in a serial scanner demonstrates how serial processing optimizes the performance of an array regardless of D* and responsivity uniformities.

The integrating delay line shown in Figure 8.3 linearly sums the detector signals and takes the root sum square value of the detector noises. The effective D* and the effective signal-to-noise ratio of this simulated single element detector are combinations of the characteristics of the individual detectors. The performance can be maximized by appropriately selecting channel gains going into the delay line as follows.

Assume that each detector has the same radiant power P incident on it, and that there are n detectors, each having responsivity R_i and detectivity $D*_i$. The signal outputs S_i and the noise outputs N_i when each detector is preamplied with an adjustable gain G_i are

$$S_i = P R_i G_i \tag{8.1}$$

and

$$N_i = R_i G_i \frac{\sqrt{A_d \, \Delta f}}{D_i^*} = \frac{R_i G_i}{D_i} . \tag{8.2}$$

Each detector is preamplified with an adjustable gain G_i, and each output is delayed by a time interval equivalent to the geometric distance of the electrical center of each detector from the final detector in the scan. These assumptions are represented by Figure 8.3. The output of the delay line is a signal S_o and a noise N_o which appears to have been produced by a detector with detectivity D_o^* and responsivity R_o. The outputs are given by

$$S_o = P \sum_{i=1}^{n} R_i G_i \tag{8.3}$$

and

$$N_o = \left[\sum_{i=1}^{n} \left(\frac{R_i G_i}{D_i} \right)^2 \right]^{1/2} . \tag{8.4}$$

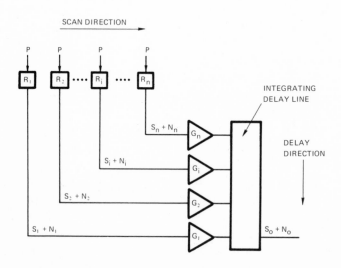

Figure 8.3 Integrating delay line technique for serial scan.

The output signal-to-noise ratio is

$$\frac{S_o}{N_o} = \frac{P \sum\limits_{i=1}^{n} R_i G_i}{\left[\sum\limits_{i=1}^{n} \left(\frac{R_i G_i}{D_i}\right)^2\right]^{1/2}} .$$

(8.5)

Defining a relative gain

$$g_i = \frac{G_i}{G_1}$$

(8.6)

such that $g_1 = 1$,

$$\frac{S_o}{N_o} = \frac{P \sum\limits_{i} R_i g_i}{\left[\sum\limits_{i} \left(\frac{R_i G_i}{D_i}\right)^2\right]^{1/2}} .$$

(8.7)

It is the quantity S_o/N_o which must be maximized by appropriate selection of relative gains g_i. To maximize S_o/N_o with respect to a particular g_k, we take the partial derivative $\partial(S_o/N_o)/\partial g_k$ and set it equal to zero. Then

$$\frac{\partial\left(\frac{S_o}{N_o}\right)}{\partial g_k} = \frac{P R_k}{\left[\sum\limits_{i} \left(\frac{R_i g_i}{D_i}\right)^2\right]^{1/2}} - \frac{P R_k^2 g_k}{D_k^2} \frac{\sum\limits_{i} R_i g_i}{\left[\sum\limits_{i} \left(\frac{R_i g_i}{D_i}\right)^2\right]^{3/2}}$$

(8.8)

or

$$\sum\limits_{i} \left(\frac{R_i g_i}{D_i}\right)^2 = \frac{g_k R_k}{D_k^2} \sum\limits_{i} R_i g_i .$$

(8.9)

Solving for g_k we get

$$\left(\frac{R_k g_k}{D_k}\right)^2 + \sum\limits_{i \neq k} \left(\frac{R_i g_i}{D_i}\right)^2 = \frac{R_k g_k}{D_k^2} \left[R_k g_k + \sum\limits_{i \neq k} R_i g_i\right]$$

(8.10)

or

$$g_K = \frac{D_K^2 \sum\limits_{i \neq K} \left(R_i \frac{g_i}{D_i} \right)^2}{R_K \sum\limits_{i \neq K} g_i R_i} . \tag{8.11}$$

Thus we must solve K equations for $K = n$ unknowns. For the simple case where $n = 2$, the solution is

$$g_2 = \frac{R_1}{R_2} \left(\frac{D_2}{D_1} \right)^2 \quad \text{for } g_1 \equiv 1. \tag{8.12}$$

This suggests a general solution

$$g_i = \frac{R_1}{R_i} \left(\frac{D_i}{D_1} \right)^2 . \tag{8.13}$$

Substituting this assumed solution into the set of equations 8.11 yields an identity, so the assumed solution is correct. Substituting the solution into equation 8.7 for S_o/N_o, we find that

$$\frac{S_o}{N_o} = \frac{P \sum\limits_i D_i^2}{\left[\sum\limits_i D_i^2 \right]^{1/2}} = P \left[\sum\limits_i D_i^2 \right]^{1/2}. \tag{8.14}$$

Thus the maximum theoretical signal-to-noise ratio corresponding to the optimum gain settings is the rss value of the individual ratios. If the responsivities are all equal, this implies that the effective specific detectivity D_o^* is

$$D_o^* = \left[\sum\limits_i D_i^{*2} \right]^{1/2}, \tag{8.15}$$

and by the mean value theorem

$$D_o^* = \sqrt{n}\, D_{rms}^*, \tag{8.16}$$

where

$$D^*_{rms} = \sqrt{\frac{\Sigma D_i^{*2}}{n}} .$$

(8.17)

8.3 Performance Comparison of Parallel and Serial Scan FLIR's

The author is a firm believer in the superiority of serial scan techniques for most FLIR applications. This section is an undisguised apologia for serial scanning, and should be recognized primarily as an expression of opinion. The assertions made here are based on present and foreseeable technology, but could easily be disproved by radical improvements in detector fabrication technology.

Two assertions are made regarding the superiority of serial scan:

1. A given level of objective performance can be achieved with a serial scan system using, at most, one-tenth as many detectors as a parallel scan system.

2. If a serial scan and a parallel scan system both exhibit the same objective performance, the serial scan system will induce the superior subjective impression of image quality.

These two assertions are made primarily on the basis of practical FLIR experiences with competing serial and parallel scan FLIR's having similar objective performance. The assertions are supported by two technical arguments. The first argument proceeds as follows. Postulate two FLIR's, one serial and one parallel, which are identical in the following respects:

1. clear aperture and optical transmission;
2. field of view;
3. detector angular subtense;
4. frame rate;
5. number of active scan lines per picture height;
6. spectral region of operation;
7. dependence of D* and responsivity on wavelength and frequency.

From the discussions of Chapter Four, we recall that the effective thermal sensitivity of a device improves as the ratio $NETD/\sqrt{F}$ decreases. An appropriate NETD equation to use for this comparison is

$$NETD = \frac{\pi \sqrt{ab\Delta f_R} \sqrt{n}}{\alpha\beta A_o \tau_o D^*(\lambda_p) \frac{\Delta W}{\Delta T}} ,$$

(8.18)

where n is the number of integrated detectors.

If we compare the two systems on the basis of $NETD/\sqrt{F}$, and for the above seven conditions of equivalence, then the comparison reduces to a consideration of a quality factor Q defined by

$$Q = \frac{1}{D^*(\lambda_p)} \sqrt{\Delta f_R} \sqrt{n} \; . \tag{8.19}$$

The reference noise equivalent bandwidth Δf_R depends on the dwelltime and the combined detector/preamplifier noise characteristics. The $D^*(\lambda_p)$ depends on the detector bias which can be used, whether or not the detectors are BLIP and the cold shielding efficiency, and the economical manufacturing yield.

At the time of this writing, useful detectors still have significant $1/f$ noise within the bandpass of parallel scanners, but insignificant $1/f$ noise within typical serial scan bandpasses. Yields of BLIP-limited detectors in large (200 element) arrays are poorer than in small (20 element) arrays. Practical cold shields tend to take simple shapes such as slots rather than arrays, so that parallel scan cold shields are longer and therefore less efficient than serial scan cold shields. To preserve system compatibility with low-power cryogenic coolers, parallel scan arrays of 200 elements must be operated at low bias to keep heat loads low, so flexibility in bias to achieve optimum $D^*(\lambda_p)$ is lost. These four considerations at the present time allow the Q for a serial scanner with $n = 20$ to equal the Q for a parallel scanner with 200 parallel-scanned elements.

The second argument regarding the subjective superiority of serial scan image quality also is primarily due to the difficulty of producing parallel scan arrays with uniform detectivities and responsivities. If a parallel scan system does not have some form of automatic responsivity control (ARC), responsivity nonuniformities will cause large spurious signal differences from line to line because of uneven responses to the high thermal background. On the other hand, if an ARC is used, it will exaggerate the line-to-line detector noise differences. Serial scanners avoid both problems by smoothing responsivity and detectivity nonuniformities in such a way that the maximum possible SNR is achieved.

8.4 Signal Processing

8.4.1 Low Frequency Response and DC Restoration

Thermal imaging systems couple the detector signal to the amplifying electronics using a dc-blocking (or ac-coupling) circuit for three reasons.

First, good contrast rendition requires background subtraction, and this function is approximated by a dc blocking circuit. Second, any dc biasing potential on the detector must be removed before signal processing. And third, the interferring effects of detector $1/f$ noise must be minimized. The simplest circuit which will provide these functions is the RC high-pass circuit shown in Figure 8.4.

The sine wave response is trivial to find, but what we are interested in is the response of this circuit to any input. The input and output voltages are

$$e_1(t) = \frac{1}{C} \int_0^t i(t)dt + \frac{Q_0}{C} + i(t)R \tag{8.20}$$

and

$$e_2(t) = i(t)R, \tag{8.21}$$

where Q_0 is the capacitor charge at time equals zero. Laplace transformation of equations 8.20 and 8.21 yields

$$E_1(s) = \frac{1}{Cs} I(s) + \frac{Q_0}{Cs} + I(s)R, \tag{8.22}$$

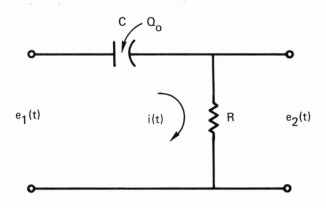

Figure 8.4 AC-coupling circuit.

and

$$E_2(s) = I(s)R. \tag{8.23}$$

Solving equation 8.22 for $I(s)$ and substituting into equation 8.23 gives

$$E_2(s) = R \left[\frac{E_1(s) - \dfrac{Q_o}{Cs}}{R + \dfrac{1}{Cs}} \right]$$

$$= \frac{RCs}{RCs + 1} E_1(s) - \frac{RQ_o}{RCs + 1}$$

$$= \frac{s}{s + \dfrac{1}{RC}} E_1(s) - \frac{Q_o}{C} \frac{1}{s + 1/RC}. \tag{8.24}$$

Expanding the first term by partial fractions,

$$E_2(s) = E_1(s) \left[-\frac{1}{RC} \frac{1}{s + 1/RC} + 1 \right] - \frac{Q_o}{C} \frac{1}{s + 1/RC}. \tag{8.25}$$

Inverse Laplace transformation gives[†]

$$e_2(t) = e_1(t) * \left[-\frac{1}{RC} e^{-t/RC} + \delta(0) \right] - \frac{Q_o}{C} e^{-t/RC}$$

$$= e_1(t) - \frac{Q_o}{C} e^{-t/RC} - \frac{1}{RC} e_1(t) * e^{-t/RC}. \tag{8.26}$$

Equation 8.26 gives the response $e_2(t)$ of the circuit to any input $e_1(t)$.

Assume that Q_o is zero for the moment, and let $e_1(t)$ be $\delta(t-\tau)$, a unit area impulse at time $t = \tau$. Then

$$e_2(t) = \delta(t - \tau) - \frac{1}{RC} e^{-(t-\tau)/RC}, \tag{8.27}$$

which is shown in Figure 8.5.

[†]where the symbol * denotes convolution

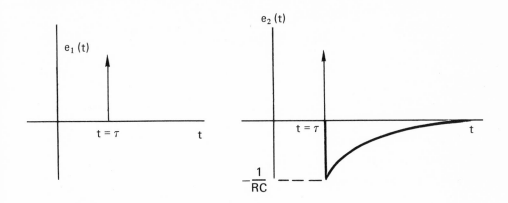

Figure 8.5 An impulse and the impulse response of an ac-coupling circuit.

Integrating $e_2(t)$ over all time gives the interesting result

$$\int_0^\infty e_2(t)dt = \int_0^\infty \delta(t-\tau)dt - \frac{1}{RC} \int_\tau^\infty e^{-(t-\tau)/RC} \, dt$$

$$= 1 - e^{-t/RC} \Big|_0^\infty$$

$$= 0. \tag{8.28}$$

This same result can be proved for any $e_2(t)$, so this circuit has the property that the average value of its signal output is zero, as one might expect since the circuit blocks dc. When Q_0 is not zero,

$$\int_0^\infty e_2(t) \, dt = - \frac{Q_0}{RC^2} . \tag{8.29}$$

It is instructive to consider the response to step and pulse inputs of unit amplitude. For a step input

$$e_1(t) = U(t - \tau) e^{-(t-\tau)/RC} \tag{8.30}$$

as shown in Figure 8.6. From equation 8.26, the step response as shown in Figure 8.6 is

$$e_2(t) = U(t - \tau) e^{-(t-\tau)/RC}, \tag{8.31}$$

For a pulse with pulse width t_p,

$$e_1(t) = \text{Rect}\left(\frac{t - \tau}{t_p}\right) \tag{8.32}$$

and

$$e_2(t) = \text{Rect}\left(\frac{t - \tau}{t_p}\right) e^{-(t-\tau)/RC} - U(t-t_p-\tau)(e^{-(t_p-\tau)/RC} - 1)$$

$$e^{-(t-t_p-\tau)/RC}. \tag{8.33}$$

This function is shown in Figure 8.7, where the areas above and below the time axis are equal.

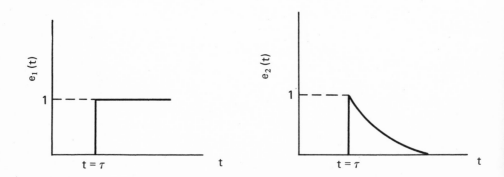

Figure 8.6 Step function and step response.

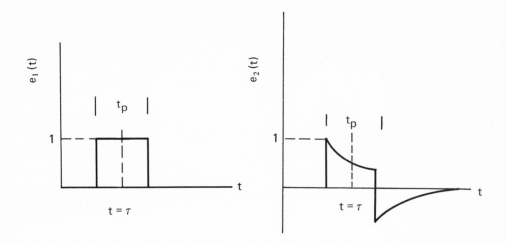

Figure 8.7 Rect function and response.

Clearly any low frequency elements of the scene will suffer from dc droop and cause undershoot whenever this circuit is used. This causes three kinds of image defects. The first type occurs with large moderate temperature difference objects, and is not very serious. An object such as that shown in Figure 8.8 will when imaged by a unidirectional ac-coupled scanner produce the droop and undershoot shown. In some cases this behavior may mask the presence of other targets. Far more serious is the presence of a very hot source, as shown in Figure 8.9. Since the average value of the circuit output is zero, the positive signal response to the hot target will be accompanied by a negative signal response of lower amplitude but of longer duration. This undershoot can easily extend across the display, as shown in Figure 8.9. The third kind of image defect arises in a multi-element scanner and is related to the charge Q_{oj} on the circuits of each of the j channels, given by

$$Q_{oj} = \int_{-\infty}^{0} i_j dt \tag{8.34}$$

where t = 0 is taken to be the observation time. Thus this charge represents the past history of the channel. It is likely that each channel has not had the same average signal on it, so the Q_{oj} will in general not be identical. Referring

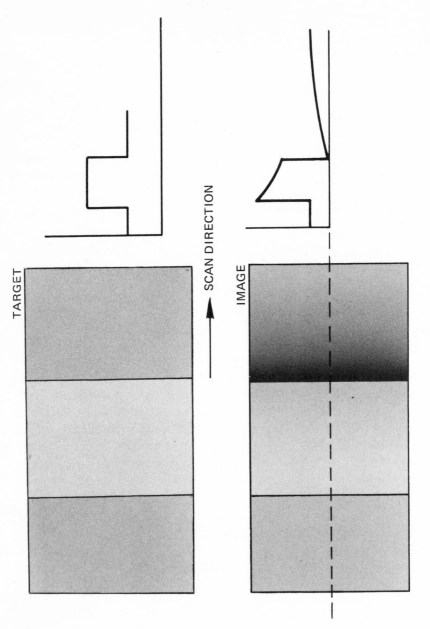

Figure 8.8   Appearance of extreme dc droop.

TARGET   IMAGE WITH UNIDIRECTIONAL   IMAGE WITH BIDIRECTIONAL
                          SCAN   SCAN

Figure 8.9 Non-dc restored effects with a small hot source.

back to equation 8.26, we see that this charge determines the level of $e_2(t)$ on each channel. Thus two channels viewing the same object against different backgrounds may give different outputs. This is clear from consideration of the target shown in Figure 8.10 and the channel responses shown in Figure 8.11.

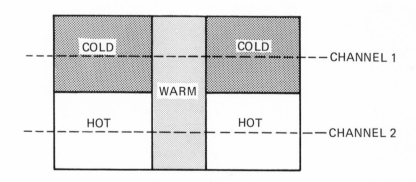

Figure 8.10 A target which demonstrates the utility of dc-restoration.

(a)

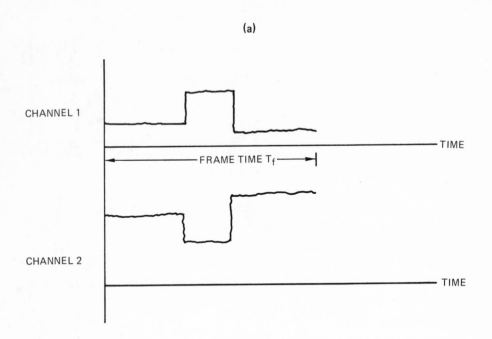

CHANNEL 1

CHANNEL 2

(b)

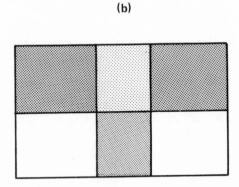

Figure 8.11 (a) Channel responses in a non-dc-restored system to the target of Figure 8.10 assuming $RC \ll T_f$ and (b) the resulting imagery

There is no satisfactory simultaneous solution to all three image defects. DC droop can be eliminated by reducing the time constant, but this increases the duration of the undershoot. Undershoot can be cured by increasing the time constant, but this increases droop. Channel responses can be made more uniform by dc coupling, but this complicates detector biasing and reduces contrasts for small delta-T targets. Droop and undershoot may be reduced by greatly increasing the signal frequency passband relative to the low frequency cut, as in serial scanning, but this does not eliminate the third defect.

A reasonably satisfactory solution is serial scanning combined with the the technique of artificial dc restoration. One version of dc restoration is shown in Figure 8.12, where the detector scan includes a view of a thermal reference source during an inactive portion of the scan cycle. This source might be a passive source such as an optical stop, or an active source such as a heated strip. When the detector sees this source, the detector signal coupling capacitor output is shorted to ground through a resistor, allowing the capacitor to charge to a dc value characteristic of the detector signal due to the source. When the detector reaches the active portion of the scan, the circuit resumes normal operation and only the signal variations around the reference capacitor voltage are passed. Restored and unrestored video are compared in Figure 8.13.

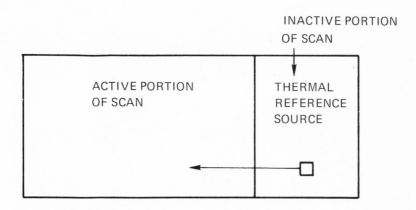

INACTIVE PORTION
OF SCAN

ACTIVE PORTION
OF SCAN

THERMAL
REFERENCE
SOURCE

Figure 8.12 DC restoration by thermal referencing.

8.4.2 Multiplexing

In systems which use parallel-processed n-element detector arrays, it is often desirable to reduce the n video channels to a single video channel. This might be necessary in order to reduce electronics power consumption, to avoid ganged multichannel video processing, to interface with a simple

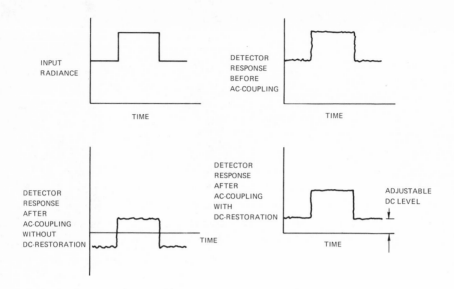

Figure 8.13 Non-restored and restored signals.

single element display such as a cathode ray tube, or to transmit video data over a single communication channel. This reduction of n channels to one channel is called multiplexing and may be accomplished either electronically or electro-optically.

Consider the simple case of a linear contiguous detector array scanning in azimuth as shown schematically in Figure 8.14. In electronic multiplexing, the preamplified detector outputs are fed to an electronic switch which samples each element in sequence and repeats the process periodically. This switch is called a time division multiplexer because it time-shares a single output channel among the input channels. The switch then feeds a video processor and a display, here shown as a CRT, whose scanning element traces in space the path in time taken by the switch, placing the multiplexed signals in the appropriate places on the screen. If the switch is operated at a sufficiently high rate, each resolvable element of the image will be sampled frequently enough so that structure-free imagery will be presented. A typical electronic multiplexer is analyzed from a sampling viewpoint in Chapter Nine.

The fabrication of an electronic switch which is sufficiently noiseless, transient-free, and fast for high performance applications has been a difficult problem in the past. The complexity of this type of system tends to make it more expensive and less reliable than some other types. Karizhenskiy and Miroshnikov[14] have described the principles of a generalized multiplexed and interlaced system.

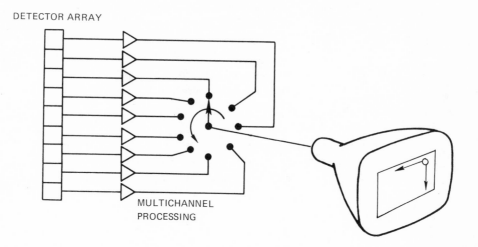

DETECTOR ARRAY

MULTICHANNEL
PROCESSING

Figure 8.14 Electronic multiplexing.

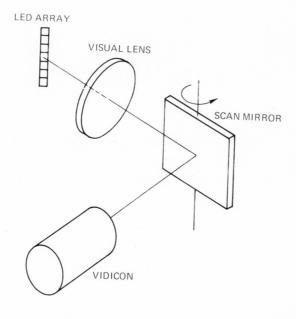

LED ARRAY

VISUAL LENS

SCAN MIRROR

VIDICON

Figure 8.15 Electro-optical multiplexer.

The electro-optical multiplexer is a particularly easy way to reduce electronic processing complexity. It operates by generating a complete low signal level video display, typically by a scanned electro-luminescent diode array, which is then viewed by a television camera to produce a single video channel as shown in Figure 8.15. There are three possible disadvantages of this scheme. First, the camera's spatial sampling rate (number of vertical lines) must considerably exceed the thermal scanner sampling rate to prevent aliasing. Second, the synchronization between the thermal and television framing action must be perfect so that the bright line which is the locus of points at which the thermal and camera scans intersect simultaneously will be constant in position on the display. This causes the bright line to be inconspicuous and allows it to be blanked electronically if desired. Third, the vidicon usually does not completely erase the image on its target surface at the end of a field, so that moving hot sources will tend to produce ghost images.

Both multiplexing techniques have offsetting advantages and disadvantages. The electronic multiplexer usually has no MTF loss, while the electro-optical cannot avoid TV camera MTF losses. The electronic may introduce noise and transients, whereas the electro-optical usually does not. Both exhibit sampling defects, the electronic by introducing a second raster perpendicular to the first, and the electro-otpical by the overlaid TV and thermal rasters.

8.5 Scanning FLIR System Types

Scanning FLIR systems are characterized by their scan patterns and by their video processing of preamplified detector signals. Video processing typically takes one of three forms: multichannel serial or parallel processing, standard television format processing, or pseudo-TV processing. In multi-channel processing there is a one-to-one throughput of detector elements to display elements. In standard video the final electronic output of the system is a single video signal whose characteristics conform to one of the commercial standards and which is suitable for use with standard composite video equipment. In pseudo-standard video, a TV-like video format and monitor are used.

Recent systems have tended to fall into one of the four categories which represent the most reasonable combinations of scan and of signal processing types, which are:

1. Parallel scan – parallel video;
2. Parallel scan – standard video;
3. Serial scan – parallel video;
4. Serial scan – standard video.

Pseudo-TV-compatible systems tend not to be as cost effective as other types, and will not be considered here. We will consider the typical features of each of the other categories and discuss their advantages and disadvantages.

The simplest conceivable multidetector system is the parallel scan parallel video system shown in Figure 8.16. This system requires only a collecting converging optic, a two-sided scanner, a detector array, amplifying electronics, display drivers, and an eyepiece. The system shown is a convergent beam scanner which uses one side of an oscillating thin mirror to generate the thermal scan and the other side to generate the visible scan. The detector signals are amplified·and shaped to appropriately drive the visible light emitters, and a visible optic magnifies and focuses the emitter scan for the observer.

Its advantages are that no scan synchronizer is needed, the scanner requires little power, and the display is compact. Its disadvantages are that only one person may observe the display, loss of any part of a channel causes loss of one line of video, all channels must be balanced individually, all channels must be controlled simultaneously (ganged), the channels may require dc restoration, and video waveform shaping must be performed in each channel. In spite of its disadvantages, the parallel-parallel system is a very cost effective way to achieve a small viewer.

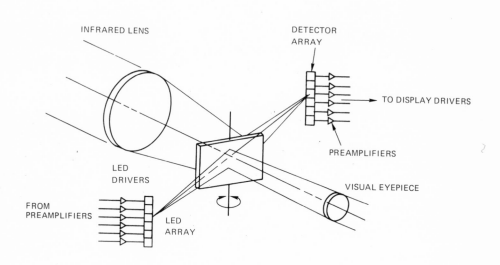

Figure 8.16 Direct view parallel scan — parallel video system.

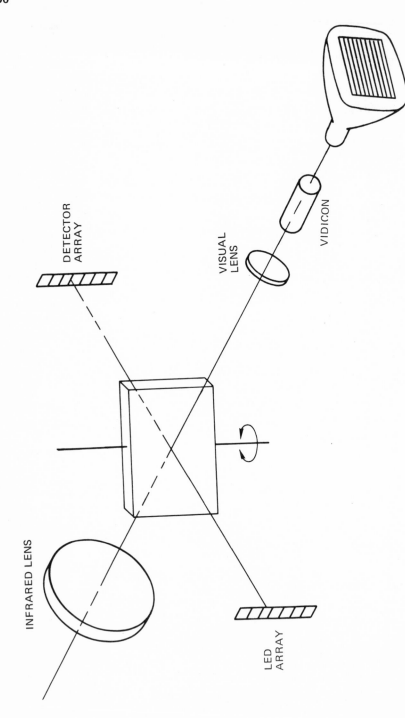

Figure 8.17 An electro-optically multiplexed parallel scan-standard video system.

The parallel scan-standard video system with signals converted to standard TV video may have a front end identical to that of the parallel-parallel system, with video conversion to a standard format being implemented by electro-optical multiplexing, as shown in Figure 8.17, and as described in in Section 8.4.2. This system views the display of a parallel-parallel system with a television camera, thereby producing standard video. Perfect synchronization of the frame rates of the thermal and of the television scans is necessary, and the effective matching of the dynamic ranges of the two devices is delicate.

The serial scan-parallel video system may be implemented as shown in Figure 8.18, where a two-dimensional array of detectors is coupled one-for-one to a similar array of light emitters. This approach uses the observer's inherent temporal integration to produce the image-smoothing effect associated with serial scanning, and combines this with the inherent simplicity of a directly viewed light-emitting diode array.

The serial scan-standard video system shown in Figure 8.19 uses the most direct method for generating standard video. It avoids schemes using multiplexers and scan converters and closely models the television process. This system combines the advantages of serial scan discussed in the previous chapter with the advantage of low cost for displays and for accessories such as video recorders, symbol generators, data retransmitters, and automatic target trackers.

The ultimate performance limits of thermal imaging systems are not yet clear because the state-of-the-art is more constrictive than theoretical limitations such as those due to diffraction and to quantum noise. The papers by Edgar[15] and by Williams[16] are interesting in this regard. Edgar's basic conclusion was that the detector state-of-the-art limits may require use of larger detectors than desired by resolution criteria, so that a larger than desired sensitivity results. Williams' conclusion was that wide field, multiline systems may require F/#'s so small or focal lengths so short that they are impractical.

8.6 Non-Scanning Thermal Imaging Systems

Conventional mechanically-scanning FLIR is considered by some to be an inelegant, inefficient, and unacceptably costly technique for thermal imaging. Raster, framing, detector cryogenic cooling, electronic signal processing, and moving optics are the often-cited defects of conventional FLIR's which stimulate continuing interest in mechanically and electronically simpler devices. Although many devices with more desirable features have been conceived, none has as yet achieved the thermal sensitivity, the resolution, and the response time necessary to displace scanners.

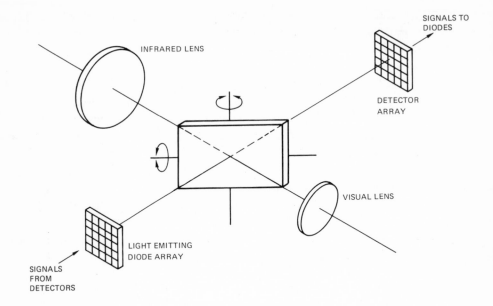

Figure 8.18 A serial scan-parallel video system using a dual-axis scanner.

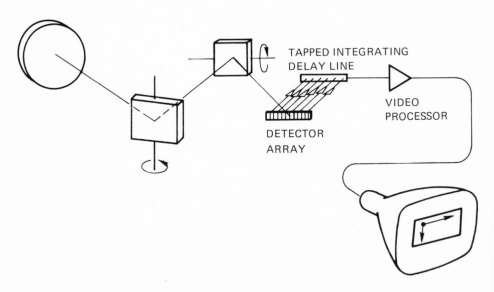

Figure 8.19 A serial scan-standard video system.

The problem is not that fundamental physical limitations exists which prevent non-scanning devices from achieving the necessary performance. It is rather that the poor radiation contrasts in the infrared make severe uniformity demands and necessitate that a linear energy conversion be followed by a background subtraction. Thus all the non-scanning imagers such as semiconductor vidicons, image converter tubes, and self-scanned charge-coupled imagers which work well in the visible will not function in the infrared without improved uniformity and a complicated electronic or electro-optical background subtraction scheme.

Since detection or energy conversion in an incoherent device necessarily occurs before background subtraction, the responsivities of the individual non-scanning detecting elements must be extremely uniform to prevent spurious image modulation. For example, consider a two dimensional matrix of detecting elements which fills an image plane and which linearly converts power in the 8 to 14 micrometer waveband to visible light. For a 300°K background, the integrated spectral radiant emittance over the band, per Chapter Two is

$$W_B = 1.72 \times 10^{-2} \text{ watts/cm}^2,$$

the change of W with T is

$$\frac{\partial W}{\partial T} = 2.62 \times 10^{-4} \text{ watt/cm}^2{}^\circ\text{K},$$

and the radiation contrast is

$$C_R = 0.74 \, \%/^\circ\text{K}.$$

Let us suppose that the individual elements of the matrix are resolvable by the eye, so that the spatial frequency composition of the matrix falls within the eye's bandpass. Further assume that the element responsivities expressed in foot-Lamberts/watt in the array have a fractional rms deviation of σ_R relative to the average element responsivity $\overline{R}$. Then before dc-subtraction, the image of a uniform target with temperature difference ΔT against a 300°K background W_B will be embedded in a fixed pattern noise induced by the responsivity variations and having an rms value of

$$\sigma = \sigma_R \, W_B \, \overline{R}. \tag{8.35}$$

The image signal will have an average value

$$\overline{S} = \left[\Delta T \left(\frac{\partial W}{\partial T} \right) \right] \overline{R}. \tag{8.36}$$

If detection is now followed by subtraction of a background level equivalent to a spectral radiant emittance $W_B' < W_B$ such that the subtraction does not clip the fixed pattern noise, (i.e., $W_B - W_B' > 3\sigma_R W_B$) then the point signal-to-noise ratio in the target image is

$$SNR = \frac{\overline{S}}{\sigma} = \frac{\Delta T \left(\dfrac{\partial W}{\partial T} \right)}{\sigma_R W_B}. \tag{8.37}$$

As an example, consider a background subtraction which removes all signals up to a level equivalent to 290°K, or

$$W_B' = 1.56 \times 10^{-2} \text{ watt/cm}^2.$$

For a ΔT of 1°K, the above condition is satisfied for small σ_R, and

$$SNR = \frac{2.62 \times 10^{-4}}{\sigma_R (1.72 \times 10^{-2})}$$

$$= 0.0152/\sigma_R.$$

Consequently if such a device is to generate a SNR of 5 which is a typical per frame SNR for a scanner viewing a 1°K ΔT, the rms responsivity variations must not exceed 0.3%.

This analysis has not accounted for the fact that spatial noise in a FLIR is not fixed-pattern in nature on a per frame basis, but random from frame to frame so that an SNR improvement by $\sqrt{T_e \dot{F}}$ occurs. Since in the hypothetical device we are discussing, the responsivity "noise" is fixed in time, the uniformity must be improved by a factor of $\sqrt{6}$ to compare favorably with a 30 frames per second FLIR, so that we now require a 0.12% uniformity. Thus even if the device is essentially noiseless on a per-element basis, responsivity variations introduce a fixed-pattern noise which interferes with performance.

The uniformity requirement derived above is likely to remain beyond the state of the art for the near term for various types of two-dimensional arrays, so the prospects for staring sensors for terrestrial thermal imaging are not good. By substituting Q's for W's in the above derivation, the reader may verify for himself that a device which converts photons rather than power has even more stringent uniformity requirements. Framing sensors suffer from the uniformity problem too, but the use of one-dimensional arrays permits better detector uniformity to be achieved. Unidimensionality of responsivity variations on the display also permits the observer to "look through" the noise easier than if it were two-dimensional.

If the uniformity problem can be solved, continuously staring systems should have one very powerful advantage over framing sensors. This is that the eye integrates signals and rms's noise at a point over a time period of T_e, so that a framing system's noise is rms'ed only a few times within T_e, whereas a staring system's noise is rms'ed continuously. Thus a staring system allows a reduction of thermal sensitivity requirements from those of a framing sensor by about one to two orders of magnitude.

Perhaps the best way to appreciate this effect is follow Budrikis[17] by modeling the eye's temporal response in the frequency domain by some approximate frequency response such as that of the low pass filter

$$\Upsilon = [1 + (2 \pi T_e f)^2]^{-1/2}. \tag{8.38}$$

Consider the images of a point source in a framing and in a staring system, both of which have the same spread function and signal transfer function, and have no fixed pattern noise, so that the signal components in each image are identical. Then the images produced by the two systems differ only in their noise components. The eye will spatially integrate signals and noise in the same fashion over both images. If we further assume that the displayed spatial noise power spectral densities differ only by an amplitude scale factor A, then the spatial integration phenomenon will have no effect on our comparison of the two systems.

Now we consider the apparent or perceived temporal equivalent noise bandwidths for each system. Let $g^2(f)$ be the displayed temporal noise power spectral density for the framing sensor, and $A^2 g^2(f)$ be that for the staring system. Then the perceived noise equivalent bandwidth for the framing sensor is

$$\Delta f_F = \frac{1}{[T_e \dot{F}]} \int_0^\infty g^2(f) \, df. \tag{8.39}$$

and for the staring sensor it is

$$\Delta f_S = \int_0^\infty A^2 \, g^2 \, (f) \, [1 + (2 \, \pi \, T_e \, f)^2 \,]^{-1} \, df. \tag{8.40}$$

If we desire that both systems have the same sensitivity, we may set their bandwidths equal and solve for A^2, yielding

$$A^2 = \frac{\dfrac{1}{[T_e \dot{F}]} \displaystyle\int_0^\infty g^2 \, (f) \, df}{\displaystyle\int_0^\infty g^2 \, (f) \, [1 + (2\pi T_e \, f)^2 \,]^{-1} \, df} \tag{8.41}$$

Since thermal sensitivity scales approximately with the square root of the bandwidth for well-behaved noise in otherwise identical systems, we may use the quantity A as a noise equivalent comparison parameter,

$$A = (T_e \dot{F})^{-1/2} \left[\frac{\displaystyle\int_0^\infty g^2 \, (f) \, df}{\displaystyle\int_0^\infty g^2 \, (f)/[1 + (2 \, \pi \, T_e \, f)^2 \,] \, df} \right]^{1/2}. \tag{8.42}$$

The significance of A is that a staring system's objective thermal sensitivity can be worse than that of a similar framing system by the factor A.

As a simple example, consider noise which is white up to an abrupt cutoff f_1, with

$$f_1 > \frac{1}{2 \, \pi \, T_e}. \tag{8.43}$$

Then

$$A = (T_e\dot{F})^{-1/2} \left[\frac{f_1}{\frac{\pi}{2} \frac{1}{2 \pi T_e}} \right]^{1/2}$$

$$= 2\left(\frac{f_1}{\dot{F}}\right)^{1/2} .$$

(8.44)

For a typical multielement framing sensor, f_1 is 10^4 Hz, and $\dot{F}$ is 30 Hz, so that

$$A = \left(\frac{10^4}{30}\right)^{1/2} = 36.4.$$

Thus some critical element of the system can have a performance which is 36 times worse than normally required. For example, the collecting aperture could be smaller, the material D* lower, or the electronic processor noise higher. It must be emphasized, however, that this advantage is meaningless if responsivity uniformity is not achieved.

The history of attempts to produce a satisfactory non-scanning thermal imager date back at least to 1929. Since that time not a single device has exhibited the combination of update rate, thermal sensitivity, and spatial resolution necessary for useful television-quality terrestrial imaging, although at least thirteen different concepts have been implemented. The following sections briefly describe the principles of operation of seven classes of non-scanning devices.

8.7 Oil Film Deformation Devices

The first thermal imaging system used the evaporagraphic principle described by Czerny[18], Czerny and Mollet[19], Robinson, et al.[20], McDaniel and Robinson[21], Ovrebo, et al.[22] and Sintsov[23]. The construction details vary, but the basic principle of all evaporagraphs is the same. A thin membrane is immersed in a supersaturated oil vapor in a chamber in thermal equilibrium. At equilibrium the rates of condensation and vaporization of the oil on the membrane are equal, but if a thermal radiation pattern is imposed on the membrane using an optical system, disequilibrium will result. If the oil film and the membrane are sufficiently thin, interference colors will be observed when the film is illuminated by white light. These variations due to differential evaporation of the volatile liquid correspond to thermal radiation absorption by the membrane.

In the early 1950's the evaporagraph principle suggested to Foshee[24] and his associates that thin film properties might be used in other ways to image thermal radiation patterns. They used a low-viscosity fluid film on the back of a thin blackened membrane, as in the evaporagraph, but now the film was illuminated and viewed through Schlieren optics. Surface deformations resulting from surface tension variations with local film temperature were thus made visible. This device did not produce satisfactory imagery because of excessive image spread due to thermal conduction in the film. The Gretag Panicon[25] overcomes this problem by allowing the film to heat up only discrete points of two dimensional pattern by using a mesh mask to isolate small film areas.

All of these devices suffer from the fact that they are thermal sensors; that is, image formation depends on scene-radiation-induced differential heating of a sensing surface. There are four serious problems with thermal sensing surfaces:

1. The necessity for good thermal sensitivity makes them susceptible to the influence of housing radiation and requires thermal isolation of the surface from all sources but the scene.

2. The necessity to image scene motion requires a short thermal time constant, which implies a rapid conduction of heat inconsistent with limiting image spread to achieve good resolution.

3. There is no simple means for background subtraction, so contrasts tend to be poor.

4. Uniformity of response is difficult to achieve over a relatively large sensing surface, so spurious modulations frequently are present.

8.8 Infrared Semiconductor Vidicons

The eclipse of mechanical television scanners by television camera tubes has inspired many researchers to attempt the same with FLIR, but thus far the record has not been good. The most overwhelming problem has been that the high background fluxes in the far infrared in terrestrial scenes tend to saturate the charge generation surfaces and to deplete electron readout beams, so that background subtraction is a necessity. This means that most vidicon schemes will work only against reduced backgrounds, such as the cold of space. Another serious problem is that vidicons which linearly convert either thermal power or quanta to electrical signals require extraordinary uniformity of response to avoid spurious modulation, as discussed in Section 8.6. Thus far four types of infrared semiconductor vidicons have been developed.

The earliest attempts were reported by Redington and van Heerden[26] in 1959. They investigated the use of camera targets composed of photoconductive doped silicon and germanium operated at cryogenic temperatures. Gold-doped silicon targets produced the best results, but spectral response was limited to about 2 μm, and satisfactory sensitivity was not conclusively demonstrated. In 1962, Heimann and Kunze[27] reported on the Resistron, at 2 μm-cutoff IR vidicon which used a lead sulfide target. Augmentation by a 1.5 kilowatt illuminator was necessary to achieve useful imagery. Berth and Brissot[28] in 1969 described two 2 μm-limited camera tubes. The first used a lead-oxysulfide photoconductive target, and the second used a mosiac of germanium photodiodes as a target. Both tubes required cryogenic cooling, and illumination was required for satisfactory imaging. In 1971, Kim and Davern[29] demonstrated the feasibility of using one- and two-dimensional arrays of indium-arsenide photodiode targets with spectral responses extending from 2.5 to 3.4 μm.

Dimmock[30] has reviewed much of this work, and has analyzed the tradeoffs between some hypothetical infrared camera tubes and conventional FLIR for varying background fluxes. Dimmock considered both mechanically-scanned linear-array vidicons and two-dimensional vidicon targets and reached the following theoretical conclusions:

1. A linear-array vidicon should perform satisfactorily in either the 3-5 or the 8-14 μm band against terrestrial backgrounds.

2. A mosaic or continuous surface vidicon with responsivity non-uniformities of less than 0.04% should perform well in the 3-5 but not in the 8-14 against terrestrial backgrounds.

3. In cold space, a two-dimensional vidicon should outperform either a linear array vidicon or a conventional scanner.

Dimmock proposed to subtract background by using an electron flood beam to subtract charge uniformly from the target, or by using photoconductive-photoemissive targets.

8.9 Pyroelectric Vidicons

A pyroelectric material has the property that a change in the material temperature produces a change in its electrical polarization. Thompsett[31] describes the use of this pyroelectric property in tri-glycine sulfate (TGS) to create an infrared vidicon. In this scheme a single crystal of TGS is electrically polarized by the application of a constant electric field. The front surface of the crystal is covered by a thin conductive layer held at a constant voltage. Incident thermal radiation produces polarization changes which induce localized potential differences between the conducting surface and the rear nonconducting surface. The rear surface is scanned by an electron beam which deposits charge to neutralize the potential differences. The beam current as a function of beam scan position is the analog of the thermal radiation pattern. Since the device is essentially thermal, the temperature of the sensing layer must be restored to uniformity prior to readout of the next frame if image smear is to be avoided. Thompsett's pyroelectric vidicon therefore uses a mechanical shutter to return the TGS crystal to a neutral temperature between frames. Thompsett concluded that a sensing surface resolution of 10^4 resolvable points per cm^2 and a thermal sensitivity of $1°C$ are possible at a frame rate of 10 Hertz.

8.10 Image Converter Tubes

A few attempts have been made to duplicate the simplicity of operation of image intensifiers by developing infrared-sensitive image tubes. Garbuny, et al.[32] described a tube called a phothermionic image converter. This device was thermal and used a scanned light spot to stimulate temperature-dependent photoemission from a thin photocathode in the focal plane of an infrared lens. It was operated at television frame rates and had good resolution, but the minimum detectable temperature difference for large targets was only about $10°C$. The paper presents an analysis of the limitations of thermal detectors.

Auphan, et al.[33] discussed a thermal image tube called le Serval which used ultraviolet-stimulated electron emission from a photoconductive cathode to produce a visible scene on a fluorescent layer. Ulmer[34] described a thermal device in which temperature changes in a thin oil film induced reflectivity changes which were made visible by illumination. Choisser and Wysoczanski[35] developed a thermal image tube called the Bolocon. The device operated by UV-stimulated photoelectron emission from a photocathode deposited on a thin semiconducting glass. Morton and Forgue[36] described a photoconductive lead sulfide sensing layer operated in a converter tube.

8.11 Laser Parametric Image Upconversion

Upconversion is a technique whereby infrared radiation is mixed with a short-wavelength collimated coherent local oscillator beam in an optically nonlinear material to produce visible light with the same spatial modulation as the infrared light. The electromagnetic theory of upconversion is described concisely by Yariv[37], and will not be reproduced here. Laser parametric upconversion can occur only in an optical crystal which does not exhibit crystal coordinate inversion symmetry so that the electric susceptibility of the crystal is nonlinear. In that case an electric field inside the crystal will produce an electric polarization which is proportional to the square of the field. An incident wave with a frequency ω_{IR} in the presence of a pump frequency ω_p will be upconverted to a visible wave with frequency ω_V expressed by

$$\omega_V = \omega_{IR} + \omega_p. \tag{8.45}$$

in such a crystal only if numerous conditions are satisfied.

The first and most important requirement is that the wave vectors must be matched over the crystal interaction length to satisfy conservation of momentum,

$$\vec{K}_V = \vec{K}_{IR} + \vec{K}_p. \tag{8.46}$$

Since the refractive indices for the pump and IR frequencies are necessarily different, this can only be satisfied if the crystal is birefringent so that the refractive indices for ordinary and extraordinary waves are different. Depending on the ratio of the ordinary index to the extraordinary index, this allows the pump to be polarized in either the ordinary or extraordinary direction, so that it will be phase matched to the component of the infrared polarized in the other direction.

The second requirement is that the crystal must be transparent to all three wavelengths involved so that good conversion efficiency is possible. The third requirement is that an image amplifier must be available which can noiselessly amplify the visible light produced. The visible photon flux cannot exceed the incident IR photon flux, and in fact the conversion efficiencies are small, so that the visible light produced is faint. The fourth requirement is that an optical filter must be available which can filter the high power pump beam from the low power visible beam when the two are separated in wavelength only by a fraction of a micrometer,

$$\lambda_p - \lambda_V = \frac{\lambda_p^2}{\lambda_p + \lambda_{IR}}. \tag{8.47}$$

Yariv derives the upconversion power efficiency P_V/P_{IR} neglecting crystal reflection and absorption losses. It is given by:

$$\frac{P_V}{P_{IR}} \cong \frac{\omega_V^2 \, \ell^2 \, d^2}{2n_{IR} \, n_p \, n_V} \left(\frac{\mu_0}{\epsilon_0} \right)^{3/2} \frac{P_p}{A} \tag{8.48}$$

where

ω_V = visible frequency

ℓ = interaction length

n   = refractive index

μ_o = free-space permeability

ϵ_o = free-space permittivity

P_p = pump power

A   = interaction cross-sectional area

and d   = nonlinear optical coefficient defined as given in Yariv.

Obviously efficiency increases as pump power density and crystal interaction length increase.

Numerous problems arise when the upconversion technique is used to produce broad band thermal imagery. The worst is probably that the angular magnification of the upconversion process is wavelength dependent and is given by

$$M = \left(\frac{n_{IR}}{\lambda_{IR}}\right)\bigg/\left(\frac{n_V}{\lambda_V}\right) \qquad\qquad (8.49)$$

If one selects the material to make this ratio constant for all λ_{IR} of interest, (i.e., uses dispersion matching) then the image components at all wavelengths image in the same place. Otherwise, one wavelength will have one angular magnification and another will have a different magnification, which is equivalent to lateral chromatic aberration. However, if dispersion matching is used, a field of view limitation is introduced by the phase-matching condition.

The material which has been used most successfully for 10 μm to visible upconversion is Proustite ($Ag_3 AsS_3$). The basic elements of an upconverter system operating in the spatial domain are shown in Figure 8.20. Advocates of parametric upconversion as an alternative to conventional scanning thermal imaging system propose that image upconversion has the following advantages:

1.   non-sampled (spatially-continuous) imagery;
2.   elimination of the need for a focal plane cryocooler;
3.   the potential for noiseless imagery;
4.   the potential for elimination of IR optics;
5.   mechanical simplicity.

The primary drawback of the technique, however, seems to be the necessity for high pump power. For example, Warner[38] concludes that a thermal sensitivity of about 1°C can be obtained in a 100 by 100 resolution element image upconverter whose pump power times the visible detector integration time is 1000 watt-seconds. This clearly is too poor a performance for such a large expenditure of power.

Milton's[39] analyses of image upconversion take into account both practical and theoretical limitations, with the conclusion that upconversion is unlikely to be competitive with conventional FLIR. The reasons are that:

1. detective quantum efficiency is too low;
2. required pump power is too high;
3. inherent field of view limitations are too restrictive;
4. a background flux subtraction device is required;
5. inherent resolution limits exist.

Milton also concluded that parametric upconversion is most applicable to laser infrared radar and to laser pulse-gated active imaging. References 40 through 42 provide good summaries of this technology.

8.12 Infrared Quantum Counters

The infrared quantum counter (IRQC) is an infrared-to-visible converter which utilizes radiative energy exchange with electronic energy levels to achieve a solid-state imager. The IRQC concept was suggested by Bloembergen[43] in 1959, and was analyzed in detail by Esterowitz, et al.[44] The IRQC is based on the premise that ionic energy levels exist in doped rare earth compounds such that transitions corresponding to both visible and infrared wavelengths exist. The principle is that an ionic energy level transition resulting from the absorption of an infrared photon can be pumped to a higher energy level by a laser local oscillator, followed by a spontaneous fluorescence of visible light from that level. This simple three level view of the IRQC is shown in Figure 8.21.

The three-level IRQC is impractical to implement for the reasons explained by Esterowitz, et al.[44], so the five level IRQC scheme shown in Figure 8.22 is used. The material most successfully used to demonstrate IRQC action is praseodymium-doped lanthanum trichloride (Pr^{3+}:$LaCl_3$). A sufficient number of transitions exist in this material to allow a broadband response in the 3 to 5 micrometer band.

The quantum efficiency of an IRQC and the output light level are so low that an image intensifier or similar device must be used to produce a useful image luminance. An effective IRQC would be a complicated device, even when compared to a mechanical scanner. To prevent image masking by spurious signals from transitions due to the crystal temperature, the crystal may have to be cooled. To achieve high IR-to-visible conversion efficiency, relatively large pump power must be used, which is inconvenient in a spectral region where lasers are few and inefficient. Finally, the IRQC is a linear and dc-coupled photon converter, so a background subtraction device must be used to improve image contrast. The noise sources of a theoretically-perfect IRQC are confined to quantum noise in the IR signal, self-emission transition noise, pump noise, and noise in the image intensifier. Limitations due to responsive nonuniformities over the crystal face have not yet been determined. Experiments with IRQC's are described in references 45 and 46.

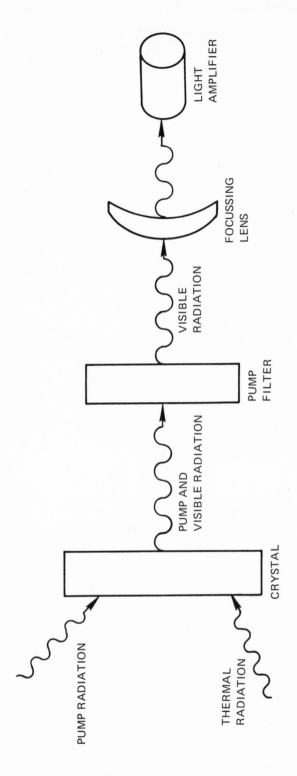

Figure 8.20 Essentials of a parametric image upconverter.

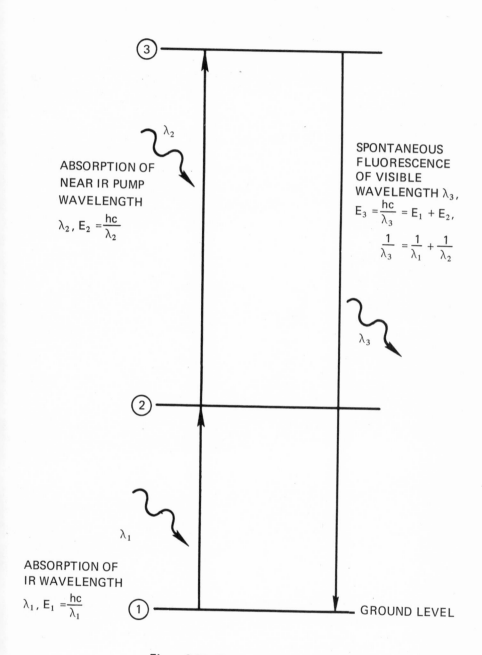

Figure 8.21 Simple three-level IRQC model.

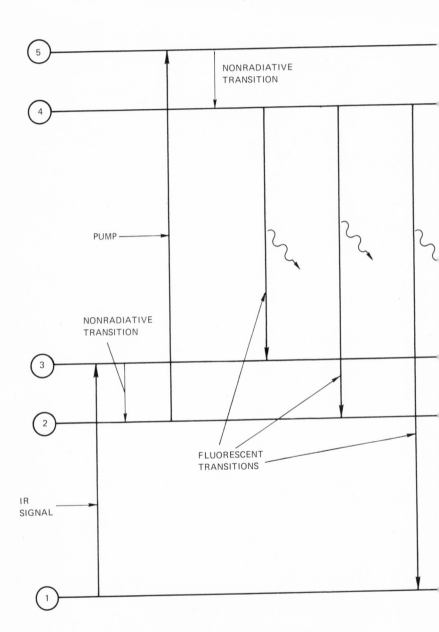

Figure 8.22 Five-level infrared quantum counter.

REFERENCES

[1] R.W. Astheimer and E.M. Wormser, "Instrument for Thermal Photography", JOSA, *49*, pp 184-187, February 1959.

[2] R.W. Astheimer and F. Schwarz, "Thermal Imaging Using Pyroelectric Detectors", Appl. Opt., *7*, pp 1687-1695, September 1968.

[3] L.W. Nichols and J. Lamar, "Conversion of Infrared Images to Visible in Color", Appl. Opt., *7*, pp 1757-1762, September 1968.

[4] E.W. Kutzcher and K.H. Zimmermann, "A Scanning Infrared Inspection System Applied to Nondestructive Testing of Bonded Aerospace Structures", Appl. Opt., *7*, pp 1715-1719, September 1968.

[5] R.W. Astheimer, "Infrared to Visible Conversion Devices", Phot. Sci. and Eng., *13*, pp 127-133, May-June 1969.

[6] M. Jatteau, "Infrared Thermography Equipment for Medical Applications", Philips Tech. Rev., *30*, pp 278-291, 1969.

[7] J. Gershon-Cohen, "Medical Thermography", Scientific American, pp 94-102, February 1967.

[8] S.B. Borg, "Thermal Imaging with Real-Time Picture Presentation", Appl. Opt., *7*, pp 1697-1703, September 1968.

[9] E. Sundstrom, "Wide-Angle Infrared Camera for Industry and Medicine", Appl. Opt., *7*, pp 1763-1768, September 1968.

[10] C.F. Gramm, "Infrared Equipment", Chapter 9 of Volume Two of *Applied Optics and Optical Engineering*, R. Kingslake editor, Academic, 1965.

[11] C.R. Borely and L.H. Guildford, "A 100 Line Thermal Viewer", Infrared Physics, *8*, pp 131-134, 1968.

[12] P.J. Lindberg, "A Prism Line-Scanner for High-Speed Thermography", Optica Acta, *15*, pp 305-316, July-August 1968.

[13] P. Laakmann, United States Patent Number 3723642, "Thermal Imaging System", 28 May 1971.

[14] E. Ga. Karizhenskiy and M.M. Miroshnikov, "Scanning Systems for Thermal Imaging Devices", Soviet Journal of Optical Technology, *37*, pp 600-603, September 1970.

[15] R.F. Edgar, "Some Design Considerations for Infrared Image Scanning Systems", Infrared Physics, *8*, pp 183-187, 1968.

[16] C.S. Williams, "Limitations on Optical Systems for Images of Mnay Discrete Elements of Area", Appl. Opt., *6*, pp 1383-1385, August 1967.

[17] Z.L. Budrikis, "Visual Thresholds and the Visibility of Random Noise in TV", Proc. IRE (Australia), pp 751-759, December 1961.

[18] M. Czerny, Physik, *53*, p 1, 1929.

[19] M. Czerny and Mollet, Zeitschrift für technische Physik, *18*, p 582, 1937.

[20] Robinson, et al., JOSA, *47*, p 340, 1957 (abstract only).

[21] G.W. McDaniel and D.Z. Robinson, "Thermal Imaging by Means of the Evaporagraph", Appl. Opt., *1*, pp 311-324, May 1962.

[22] P.J. Ovrebo, R.R. Sawyer, R.H. Ostergren, R.W. Powell, and E.L. Woodcock, "Industrial, Technical and Medical Applications of Infrared Techniques", Proc IRE, 1959.

[23] V.N. Sintsov, "Evaporagraphic Image Quality", Appl. Opt., *6*, pp 1851-1854, November 1967.

[24]L.L. Foshee, personal communication, Hughes Aircraft Company, Culver City, California.

[25]Gretag AG, CH-8105, Regensdorf, Switzerland.

[26]R.W. Redington and P.J. van Heerden, "Doped Silicon and Germanium Photoconductors as Targets for Infrared Television Camera Tubes", JOSA, 49, pp 997-1001, October 1959.

[27]W. Heimann and C. Kunze, "Infrarot-Vidikon", Infrared Physics, 2, pp 175-181, 1962.

[28]M. Berth and J.J. Brissot, "Targets for Infrared Television Camera Tubes", Philips Technical Review, 30, pp 270-279, 1969.

[29]C.W. Kim and W.E. Davern, "InAs Charge-Storage Photodiode Infrared Vidicon Targets", IEEE Trans. Elec. Dev., ED-18, pp 1062-1069, November 1971.

[30]J.O. Dimmock, "Capabilities and Limitations of Infrared Imaging Systems", Proc. SPIE Seminars "Developments in Electronic Imaging Techniques", 32, October 16-17, 1972.

[31]M.F. Thompsett, "A Pyroelectric Thermal Imaging Camera Tube", IEEE Trans. Elec. Dev., ED-18, pp 1070-1074, November 1972.

[32]M. Garbuny, T.P. Vogel, and J.R. Hansen, "Image Converter for Thermal Radiation", JOSA, 51, pp 261-273, March 1961.

[33]M. Auphan, G.A. Boutry, J.J. Brissot, H. Dormont, J. Perilhou, and G. Pietri, "Un Tube Transformateu d'Image pour l'Infrarouge Moyen a Couche Photoconductive et Couche Photoemissive Juxtaposees: Le Serval", Infrared Physics, 3, pp 117-127, 1963.

[34]W. Ulmer, "A New Type of Optical Image Converter", Infrared Physics, 11, pp 221-224, 1971.

[35]J.P. Choisser and W. Wysocyznski, personal communications, Electronic Vision Corporation, San Diego, California.

[36]G.A. Morton and S.V. Forgue, "An Infrared Pickup Tube", Proc. IRE, 47, pp 1607-1609, September 1959.

[37]A. Yariv, Introduction to Optical Electronics, Chapter Eight, Holt, Rinehart, Winston, 1971.

[38]J. Warner, "Parametric Upconversion from the Infra-Red", Optoelectronics, 3, pp 37-48, 1971.

[39]A.F. Milton, "Upconversion − A Systems View", Appl. Opt., 11, pp 2311-2330, October 1972.

[40]J.E. Midwinter and F. Žernike, "Note on Up-Converter Noise Performance", IEEE J. Quant. Electr., QE-5, pp 130-131, February 1969.

[41]R.A. Andrews, "IR Image Parametric Up-Conversion", IEEE J. Quant. Electr., QE-6, pp 68-80, January 1970.

[42]J.E. Midwinter, "Parametric Infrared Image Converters", IEEE J. Quant. Elec., QE-4, pp 716-720, November 1968.

[43]N. Bloembergen, "Solid State Infrared Quantum Counters", Phys. Rev. Letters, 2, pp 84-85, February 1959.

[44]L. Esterowitz, A. Schnitzler, J. Noonan and J. Bahler, "Rare Earth Infrared Quantum Counter", Appl. Opt., 7, pp 2053-2070, October 1968.

[45]J.F. Porter, Jr., "Sensitivity of Pr^{3+}: $LaCl_3$ Infrared Quantum Counter", IEEE J. Quant. Elect., June 1965, pp 113-115.

[46]W.B. Gandrud and H.W. Moos, "Improved Rare-Earth Trichloride Infrared Quantum Counter Sensitivity", IEEE J. Quant.Elect., QE-4, May 1968, pp 249-252.

CHAPTER NINE — SAMPLING

9.1 Introduction

Sampling effects in imaging systems are equivalent in importance to the optical transfer function and the thermal sensitivity in determining image quality. Examples of sampling processes are the use of arrays of discrete detectors for dissecting images, use of electronic multiplexing or pulse-width modulation, use of scan conversion, and relative motion between the scene and a framing sensor. Sampling in imagery has been discussed in references one through ten.

As we noted in Chapter Three, two important properties of well-behaved imaging processes are spatial invariance of the impulse response and signal transfer linearity. When these two properties are not present in FLIR, defective imagery may result. As a broad generalization, spatial and temporal sampling occurs whenever a system exhibits discrete deviations from invariance, and amplitude sampling occurs whenever there are discrete deviations from linearity. There are two types of sampling of particular importance in imaging systems. One is transmission of an object distribution through a periodic window function. The other is sampling of the average value of an object within each aperture of a periodic aperture array. We will call these processes "window" and "averaging" sampling.

The structure shown in Figure 9.1 can be used in either type of sampling. If the image formed by the system is the product of the object distribution with this structure, the array acts as a transmission function and we have window sampling. If the image consists of samples (perhaps spatially filtered) of the average value of the object within each aperture of the array the process is averaging sampling. Window sampling occurs in electronic multiplexing and in motion between an object and a framing sensor. Averaging sampling occurs perpendicular to the scan direction in a discrete scene dissection system, as in television, laser line scanners, scan converters, and half-tone photography.

Most sampling problems are not easy to solve using any methods, but the solutions are considerably simplified by using the compact Fourier analysis notion of Chapter Three, as demonstrated in the following example of window sampling. Consider an object denoted by O(x,y) which is alternately transmitted and obscured by a window sampling function W(x,y). The image I(x,y) which results is the product of the object function with the transmission function.

Then

$$I(x,y) = O(x,y) \cdot W(x,y) \tag{9.1}$$

and the Fourier transform or image spectrum is

$$\widetilde{I}(f_x,f_y) = \widetilde{O}(f_x,f_y) * \widetilde{W}(f_x,f_y). \tag{9.2}$$

If as in Figure 9.1,

$$W(x,y) = \left[\text{Rect} \left(\frac{x}{a} \right) \cdot \text{Rect} \left(\frac{y}{a} \right) \right] * \left[\text{Comb} \left(\frac{x}{b} \right) \cdot \text{Comb} \left(\frac{y}{b} \right) \right], \tag{9.3}$$

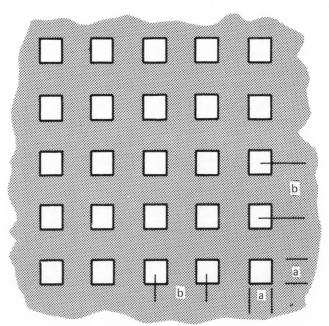

Figure 9.1 A periodically transmissive structure.

then

$$\widetilde{I}(f_x,f_y) = \widetilde{O}(f_x,f_y) * \Big[[\text{Sinc}(af_x) \cdot \text{Sinc}(af_y)] \cdot [\text{Comb}(bf_x)$$

$$\cdot \text{Comb}(bf_y)] \Big] . \qquad (9.4)$$

This image spectrum is shown in Figure 9.2, but no attempt has been made to indicate phase shift pictorially. The effect of sampling is the creation of sidebands on frequency centers of $\pm m/b$ and $\pm n/b$ which contain frequencies called aliases which are not in the original object spectrum. These aliases are merely object spatial frequencies which have been translated in the frequency domain by the sampling process. The sidebands are shown separated in Figure 9.2, but as b is necessarily greater than a, the sidebands overlap. The result is that aliased frequencies interfere with object frequencies, and worse, x-frequencies cross-alias, masquerading as y-frequencies, and *vice versa*.

Averaging sampling is not quite so simple to analyze as window sampling, and represents the more interesting sampling problems. Consider a two-dimensional object function $O(x,y)$. To demonstrate averaging sampling we will find the sampled average value of this function within three important

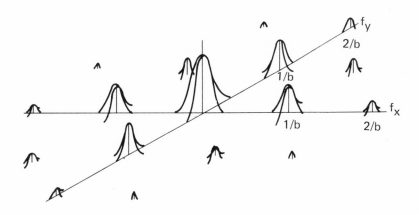

Figure 9.2 Image spectrum through a periodically transmissive structure.

aperture functions. First consider the aperture Rect (y/β) shown in Figure 9.3. Within the Rect function limits we will replace the object value at every point (x,y) with the object value averaged over y and evaluated at x. Then we will sample this new distribution with a delta function $\delta(y)$. Thus we have an image

$$I(x,y) = \left[\frac{1}{\beta} \int_{-\beta/2}^{\beta/2} O(x,y)\, dy\right] \cdot \delta(y). \tag{9.5}$$

We will prove that his expression is equivalent to convolving the object with Rect (y/β) and then sampling with $\delta(y)$. This is an important point because it considerably simplifies analysis of sampled imagery. Rewriting equation 9.5 by absorbing the limits of integration into the integrand, we get

$$I(x,y) = \left[\frac{1}{\beta} \int_{-\infty}^{\infty} O(x,y) \cdot \mathrm{Rect}\left(\frac{y}{\beta}\right) dy\right] \cdot \delta(y). \tag{9.6}$$

We may change the symbol for the y coordinate to η without affecting the value of the integral, giving

$$I(x,y) = \left[\frac{1}{\beta} \int_{-\infty}^{\infty} O(x,\eta) \cdot \mathrm{Rect}\left(\frac{\eta}{\beta}\right) d\eta\right] \delta(y). \tag{9.7}$$

Multiplication of a function by $\delta(y)$ has the effect of sifting out the value of that function at $y = 0$, so we may rewrite the factor Rect (η/β) as Rect $((\eta-y)/\beta)$ without any loss of generality. Then

$$I(x,y) = \left[\frac{1}{\beta} \int_{-\infty}^{\infty} O(x,\eta)\, \mathrm{Rect}\left(\frac{\eta-y}{\beta}\right) d\eta\right] \cdot \delta(y). \tag{9.8}$$

As the Rect is symmetrical,

$$I(x,y) = \left[\frac{1}{\beta} \int_{-\infty}^{\infty} O(x,\eta) \cdot \mathrm{Rect}\left(\frac{y-\eta}{\beta}\right) d\eta\right] \cdot \delta(y). \tag{9.9}$$

This integral is by definition a convolution integral, so we have

$$I(x,y) = \frac{1}{\beta}\left[O(x,y) * \mathrm{Rect}\left(\frac{y}{\beta}\right)\right] \cdot \delta(y). \tag{9.10}$$

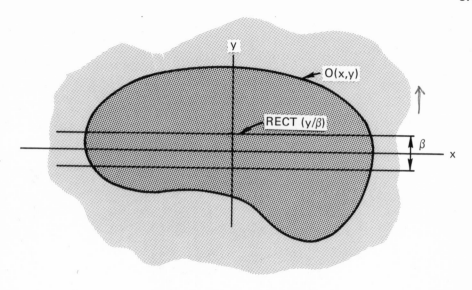

Figure 9.3 A Rect function aperture.

This proves that the image value within an averaging aperture is produced by a convolution followed by a sampling. The transform is

$$g(y) * \delta(y) = 1$$

$$\tilde{I}(f_x, f_y) = \frac{1}{\beta} \left[\tilde{O}(f_x, f_y) \cdot \text{Sinc}\,(\beta f_y) \right] * \delta(x)$$

$$= \frac{1}{\beta} \left[\tilde{O}(f_x, f_y) \cdot \text{Sinc}\,(\beta f_y) \right] \cdot \left(\delta(x) \right) \tag{9.11}$$

Sampling

Next consider the infinite array [Rect (y/β) * Comb (y/γ)] shown in Figure 9.4. If we average the object in y over each of the elementary Rect (y/β) apertures and sample with delta functions $\delta(y-n\gamma)$ with n an integer, we get

$$I(x,y) = \sum_{n=-\infty}^{\infty} \left\{ \left[\frac{1}{\beta} \int_{-\beta/2 + n\gamma}^{\beta/2 + n\gamma} O(x,y)\, dy \right] \cdot \delta(y - n\,\gamma) \right\}. \tag{9.12}$$

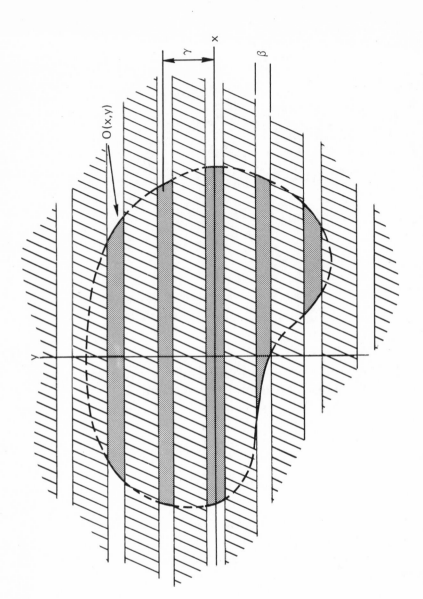

Figure 9.4 A periodic Rect function aperture sampling an object distribution.

Taking the same steps as in the single aperture case we write

$$I(x,y) = \sum_{n=-\infty}^{\infty} \left\{ \left[\frac{1}{\beta} \int_{-\infty}^{\infty} O(x,y) \cdot \text{Rect}\left(\frac{y-n\gamma}{\beta}\right) d\eta \right] \cdot \delta(y-n\gamma) \right\}$$

$$= \sum_{n=-\infty}^{\infty} \left\{ \left[\frac{1}{\beta} \int_{-\infty}^{\infty} O(x,\eta) \cdot \text{Rect}\left(\frac{\eta-n\gamma}{\beta}\right) d\eta \right] \cdot \delta(y-n\gamma) \right\}$$

$$= \sum_{n=-\infty}^{\infty} \left\{ \left[\frac{1}{\beta} \int_{-\infty}^{\infty} O(x,\eta) \cdot \text{Rect}\left(\frac{y-\eta+n\gamma}{\beta}\right) d\eta \right] \cdot \delta(y-n\gamma) \right\}$$

$$= \sum_{n=-\infty}^{\infty} \frac{1}{\beta} \left[O(x,y) * \text{Rect}\left(\frac{y+n\gamma}{\beta}\right) \right] \cdot \delta(y-n\gamma)$$

$$= \frac{1}{\beta} \left[O(x,y) * \text{Rect}\left(\frac{y}{\beta}\right) \right] \cdot \text{Comb}\left(\frac{y}{\gamma}\right) \qquad (9.13)$$

Again the effect of averaging and sampling is equivalent to convolution with the elementary averaging aperture and sampling by the delta function array. The image spectrum is

$$\tilde{I}(f_x,f_y) = \frac{1}{\beta} \left[\tilde{O}(f_x,f_y) \cdot \text{Sinc}(\beta f_y) \right] * \left[\text{Comb}(\gamma f_y) \cdot \delta(f_x) \right].$$

$$(9.14)$$

As a final example, consider the sampled average value of $O(x,y)$ within each aperture of the array of Figure 9.1, where the array is described by

$$\left[\text{Rect}\left(\frac{x}{a}\right) \cdot \text{Rect}\left(\frac{y}{a}\right) \right] * \left[\text{Comb}\left(\frac{x}{b}\right) \cdot \text{Comb}\left(\frac{y}{b}\right) \right].$$

$$(9.15)$$

The image equation 9.16 is

$$I(x,y) = \sum_{n=-\infty}^{\infty} \sum_{m=-\infty}^{\infty} \left\{ \left[\frac{1}{a^2} \int_{-a/2+nb}^{a/2+nb} \int_{-a/2+mb}^{a/2+mb} O(x,y) \, dy \right] \right.$$

$$\left. \cdot \left[\delta(y-nb) \cdot \delta(x-mb) \right] \right\} . \tag{9.16}$$

It can be shown by extension from the previous two cases that the equation can be rewritten as

$$I(x,y) = \left\{ \frac{1}{a^2} \left[O(x,y) * \left(\text{Rect}\left(\frac{x}{a}\right) \cdot \text{Rect}\left(\frac{y}{a}\right) \right) \right] \right\}$$

$$\cdot \left(\text{Comb}\left(\frac{y}{b}\right) \cdot \text{Comb}\left(\frac{x}{b}\right) \right). \tag{9.17}$$

This shows that sampling by a two-dimensional array of averaging apertures is equivalent to convolution by the basic aperture followed by sampling at the lattice points. The transform is

$$\tilde{I}(f_x,f_y) = \frac{1}{a^2} \left[\tilde{O}(f_x,f_y) \cdot \text{Sinc}(af_x) \cdot \text{Sinc}(af_y) \right]$$

$$* \left[\text{Comb}(bf_y) \cdot \text{Comb}(bf_x) \right]. \tag{9.18}$$

The simplifications made in these three examples are what make averaging sampling problems manageable. Otherwise the solutions involve explicit summations and multiple integrals. It is evident from these examples that the effect of averaging sampling by a periodic array of apertures is to prefilter the object spectrum before sampling by the delta function lattice which is the basis of the array.

The rest of this chapter is devoted to statements of four practical problems and their solutions. In all problems it is assumed that the angular magnification of the system is unity, or alternately that image plane spatial coordinates are normalized to the corresponding object plane coordinates. Scanning systems are assumed to have a linear unidirectional scan, and object functions are assumed to include the field of view limits.

9.2 Sampling in a Parallel-Processed Scanner

A common type of thermal imager uses a scanning infrared detector array coupled electrically to a scanning light-emitting diode array. As shown in Figure 9.5, this system dissects the scene by convolution in the x-direction and sampling in the y-direction. Usually the impulse response is invariant in the scan direction, and is periodic (non-stationary) in the other direction. Figure 9.6 describes this system. The left side of the sketch represents the analog channels over which the object is averaged and the right side represents the effective sampling locations centered on the raster lines.

We will determine the image $I(x,y)$ which results from the action of this system on an object $O(x,y)$. Taking the process step by step and ignoring any optics and electronics spread functions:

1. $O(x,y)$ is convolved with the detector impulse response $r_d(x,y)$ and sampled and carried by the electronics channels on optical centers $y = \pm n\gamma$,

$$I'(x,y) = [O(x,y) * r_d(x,y)] \cdot \text{Comb} \frac{y}{\gamma} . \qquad (9.19)$$

2. I' is convolved with the video monitor spread function $r_m(x,y)$.

$$I(x,y) = I'(x,y) * r_m(x,y). \qquad (9.20)$$

3. Steps (1) and (2) combine to produce

$$I(x,y) = \left[[O(x,y) * r_d(x,y)] \cdot \text{Comb} \left(\frac{y}{\gamma} \right) \right] * r_m(x,y) . \qquad (9.21)$$

Denoting the transform of an impulse response $r(x,y)$ as the optical transfer function $\tilde{r}(f_x,f_y)$, the image spectrum is

$$\tilde{I}(f_x,f_y) = \left[[\tilde{O}(f_x,f_y) \cdot \tilde{r}_d(f_x,f_y)] * \text{Comb}(\gamma f_y) \cdot \delta(f_x) \right]$$
$$\cdot \tilde{r}_m(f_x,f_y). \qquad (9.22)$$

This image spectrum is shown in Figure 9.7, where the unavoidable sideband overlap has been suppressed for clarity.

The essential features of this result are the creation of sidebands which replicate the object spectrum, and the impossibility of perfectly filtering the aliased signals from the image without drastically narrowing the image spectrum. Note that if sampling were absent, the result would be

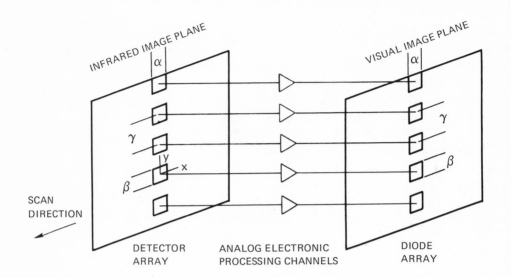

Figure 9.5 A parallel processed system.

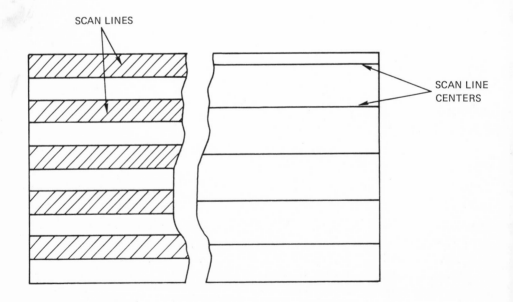

Figure 9.6 Sampling lattice of the parallel processed system.

$$\tilde{I}(f_x, f_y) = \tilde{O}(f_x, f_y) \cdot \tilde{\tau}_d(f_x, f_y) \cdot \tilde{\tau}_m(f_x, f_y), \tag{9.23}$$

where $\tilde{\tau}_d \tilde{\tau}_m$ is the assumed system MTF.

It is important to note that the sampling action may be thought of as though the detector scans in the y-direction and is sampled periodically. An interesting case of this problem occurs when the detector spread function is not the same as the diode spread function, so that the possibility of minimizing aliasing effects in the monitor arises.

9.3 Sampling in a Two-Dimensional Mosaic Staring Sensor

Image converters for thermal imaging frequently are proposed for thermal imaging which consist solely of a two-dimensional mosaic of contiguous square elements, as illustrated by Figure 9.8. Such a mosaic averages the object over each aperture at the input, samples each aperture once, and transmits this value to the output where the image is formed by the convolution of the output aperture with the sampling array. The image equation for such a device is

$$I(x,y) = \left\{ \left[O(x,y) * \left[\text{Rect}\left(\frac{x}{\alpha}\right) \cdot \text{Rect}\left(\frac{y}{\alpha}\right) \right] \right] \cdot \text{Comb}\left(\frac{x}{\alpha}\right) \cdot \text{Comb}\left(\frac{y}{\alpha}\right) \right\}$$
$$* \left[\text{Rect}\left(\frac{x}{\alpha}\right) \cdot \text{Rect}\left(\frac{y}{\alpha}\right) \right]. \tag{9.24}$$

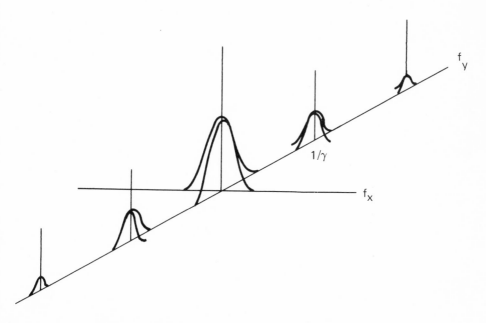

Figure 9.7 Image spectrum of the parallel processed system.

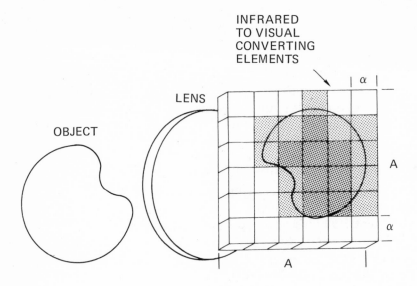

Figure 9.8 A matrix imager.

Transforming,

$$\tilde{I}(f_x, f_y) = \left\{ \left[\tilde{O}(f_x, f_y) \cdot \text{Sinc} \, (\alpha f_x) \cdot \text{Sinc} \, (\alpha f_y) \right] * \right.$$

$$* \left[\text{Comb} \, (\alpha f_x) \cdot \text{Comb} \, (\alpha f_y) \right] \right\} \cdot \left\{ \text{Sinc} \, (\alpha f_x) \right.$$

$$\left. \cdot \, \text{Sinc} \, (\alpha f_y) \right\} . \tag{9.25}$$

This spectrum is shown in Figure 9.9, where sideband overlap has been suppressed. The bandlimited object spectrum is replicated in sidebands at frequency coordinates of $(\pm n/\alpha, \pm m/\alpha)$, where m and n are integers. These sidebands are imperfectly filtered by the reconstruction filter, producing two-dimensional aliasing.

The analysis of Section 9.2 demonstrated that sampling in one direction produces aliasing in that direction. The distinguishing feature of the present problem is that aliasing occurs in both directions, as well as cross-aliasing. That is, x-frequencies masquerade as y-frequencies and *vice-versa*. These effects are evident on close examination of periodic images in color television and in half-tone pictures.

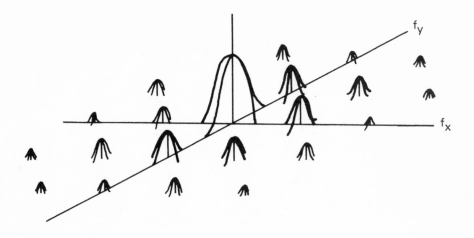

Figure 9.9 Image spectrum of a continuous mosaic imager.

9.4 Sampling in a Multiplexed System

One class of FLIR converts many channels of video to a single channel by time-division multiplexing as described in Section 8.4.2. Such a system consists of a parallel scanner followed by an electronic switch which sequentially samples the N identical bandlimited analog channels, and a display which demodulates and filters the samples. The description of a multiplexed system is identical to that of the parallel processed system prior to step two. The spatial equivalent of the time sampling pattern of the multiplexer is shown in Figure 9.10.

The horizontal raster on the left of the figure represents the analog channels which are sampled in the x-direction by the multiplexer. The slanted raster on the right represents the path of the small rectangular "window" function which successively samples each analog channel. This electronic window samples each analog channel at intervals α/s for duration α/Ns, where s is the number of samples per dwell time (assuming unity multiplexer efficiency). We will approximate the angle the sampling raster makes with the horizontal by 90°, because the exact equations are cumbersome and the exact results are not worth the added effort. The approximate sampling raster we will use is shown in Figure 9.11.

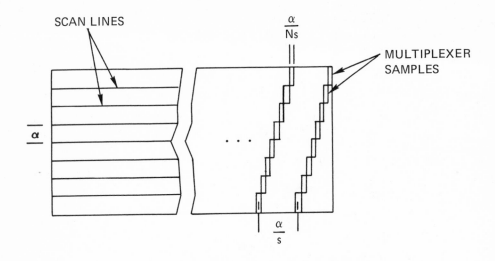

Figure 9.10 The multiplexer window function's path.

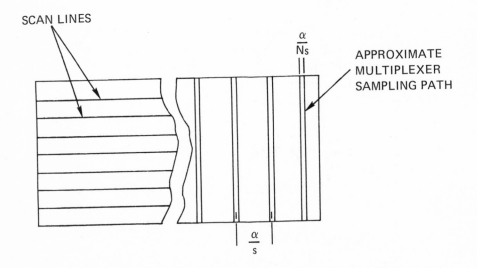

Figure 9.11 Approximate sampling raster for the electronic multiplexer.

Let the impulse response of the system prior to the multiplexer be solely due to the detector response $r_d(x,y)$, and let the impulse response of the demodulator and final filter (the monitor) be r_m. The multiplexer window function is

$$W(x) = \text{Rect}\left(\frac{x}{\alpha/Ns}\right) * \text{Comb}\left(\frac{x}{\alpha/s}\right). \tag{9.26}$$

The steps in the image construction are:

1. An image $I'(x,y)$ is formed by the usual one-dimensional raster sampling,

$$I'(x,y) = [O(x,y) * r_d(x,y)] \cdot \text{Comb}\left(\frac{y}{\beta}\right) \tag{9.27}$$

2. I' is sampled by the window function W and convolved with the multiplexer channel impulse response, which we will assume is negligible, producing

$$I''(x,y) = I'(x,y) \cdot W(x,y) . \tag{9.28}$$

3. I'' is convolved with the combined demodulator and monitor impulse response r_m to produce a final image,

$$I(x,y) = I''(x,y) * r_m(x,y). \tag{9.29}$$

Combining these,

$$I(x,y) = \left[[O(x,y) * r_d(x,y)] \cdot \text{Comb}\left(\frac{y}{\beta}\right) \cdot W(x,y)\right] * r_m(x,y). \tag{9.30}$$

Transforming,

$$\tilde{I}(f_x,f_y) = \left[[\tilde{O}(f_x,f_y) \cdot \tilde{r}_d(f_x,f_y)] * \left(\text{Comb}\left(\beta f_y\right) \cdot \delta(f_x)\right) * \tilde{W}(f_x,f_y)\right]$$

$$\cdot \tilde{r}_m(f_x,f_y). \tag{9.31}$$

The window function transform is

$$\tilde{W}(f_x, f_y) = \text{Sinc}\left(\frac{\alpha}{Ns} f_x\right) \cdot \left[\text{Comb}\left(\frac{\alpha}{s} f_x\right) \cdot \delta(f_y)\right]. \tag{9.32}$$

In the neighborhood of $f_x = 0$ and for N large,

$$\tilde{W}(f_x, f_y) \simeq \text{Comb}\left(\frac{\alpha}{s} f_x\right) \cdot \delta(f_y). \tag{9.33}$$

The image transform becomes

$$\tilde{I}(f_x, f_y) = \left[\ \tilde{O}(f_x, f_y) \cdot \tilde{r}_d(f_x, f_y) * \left(\text{Comb}\,(\beta\, f_y) \cdot \delta(f_x)\right)\right.$$

$$\left. * \left[\text{Comb}\left(\frac{\alpha}{s} f_x\right) \cdot \delta(f_y)\right]\ \cdot \tilde{r}_m(f_x, f_y) \right. \tag{9.34}$$

or

$$\tilde{I}(f_x, f_y) = \left[\left[\tilde{O}(f_x, f_y) \cdot \tilde{r}_d\,(f_x, f_y)\right] * \left[\text{comb}\,(\beta f_y) \cdot \text{comb}\left(\frac{\alpha}{s} f_x\right)\right]\right]$$
$$\cdot \tilde{r}_m\,(f_x, f_y) \tag{9.35}$$

This image spectrum is shown in Figure 9.12. The essential element of this analysis is again the existence of sidebands which replicate the pre-sampled spectrum. The analysis of these first three problems may be extended to the case of a scan converter viewing another framing sensor.

As a challenging and practical exercise in the use of this analysis, the interested reader could describe the operation of a pulsewidth modulator. This device samples each analog channel periodically for a duration briefer than a dwell-time, and converts the sampled values to constant amplitude pulses with the same period as the sampler and with pulse widths proportional to the sampled values. The pulse width modulator then drives a scanning diode display.

9.5 Practical Consequences of Sampling

The most deleterious sampling effects commonly found in thermal imagers are time sampling of image motion due to the framing action and spatial sampling due to the raster scanning process. Motion sampling causes severe corruption of the shapes of small targets moving rapidly perpendicular to the scan direction, but spatial sampling produces by far the more objectionable results. Biberman[11] presents an extensive discussion of raster sampling effects, and cites a study by Thompson[12], who determined the distribution

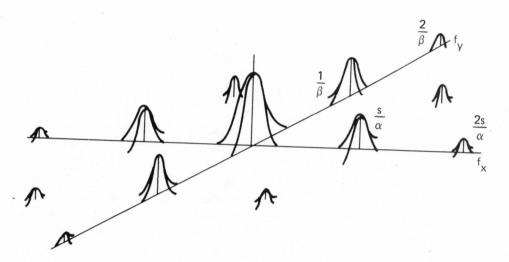

Figure 9.12 Image spectrum of a multiplexed system.

of preferred viewing distances for rasterless and rastered television imagery. Thompson found (see also Section 4.8) that the screen angle subtended at the mean preferred viewing distance with raster was about 8°, and that without raster, observers tend to halve their viewing distance to get a screen angle of about 16°. With about 480 active lines on standard television the 8° field gives about 1 arcminute per raster line, which is consistent with the maximum raster line subtense found to be tolerable by many other sources.

The significance of this is that when viewing raster-degraded imagery, viewers tend to increase the post-sampling spatial filtering effect of the visual system in order to attenuate the spurious high frequencies introduced by the raster. Consequently, viewers reduce the visibility of desired signal frequencies as well as that of undesirable ones, and the overall performance is degraded from what is achievable using a rasterless system.

Another consequence of raster sampling is loss of limiting resolution in the direction of the raster. This is expressed by the Kell factor K, which for standard television relates the vertical number of resolvable lines R_V per picture height to the number of active scan lines N_a by $R_V = K \, N_a$. According to Luxenburg and Kuehn[13], experimental values of K have a range of $0.53 \leqslant K \leqslant 0.85$, with a value of 0.7 being assumed in American television practice.

Another problem is that raster tends to catch the eye in free search, and to slow it down by distracting it from targets. The net result of sampling is that objects embedded in the fixed image structure of sampled system are harder to detect, recognize, and identify than in a nonsampled system with the same nominal resolution and sensitivity. This is difficult to believe for many of us because we are so conditioned to interpreting a television set as "good" when we see a clear, sharp raster.

A common raster defect in multi-element systems is an inoperative channel due to a detector, an electronics channel, or a display failure. Such a failure is usually dealt with either by leaving the affected raster line blank, or if possible by driving that line with the signal from either adjacent line. When several lines are missing, these solutions are not satisfactory, as they result in image features which continually distract the attention of the visual process away from the defect-free parts of the image. It has often been demonstrated that generating an artificial signal consisting of the average of the two channels adjacent to the dead channel produces imagery so good that it is difficult to find the defects, so long as the percentage of dead channels is low.

From the definition of OTF in Chapter Four, and from the present discussions, it is clear that the OTF concept applies only for image directions for which there is no sampling. Consequently some measure of sampled image quality such as the relative percentage of image power attributable to aliased frequencies must be used to describe sampling defects. Another possible measure is the resolving power for bars oriented parallel to the scan direction. At the time of this writing, no satisfactory summary measure of sampled image quality has been identified.

REFERENCES

[1] P. Mertz and F. Gray, "A Theory of Scanning and Its Relation to the Characteristics of the Transmitted Signal in Telephotography and Television," BSTJ, *13*, pp 464-515, July 1934.

[2] J.R. Jenness, Jr., W.A. Eliot, and J.A. Ake, "Intensity Ripple in a Raster Generated by a Gaussian Scanning Spot," JSMPTE, pp 549-550, June 1967.

[3] L.G. Callahan and W.B. Brown, "One- and Two-Dimensional Processing in Line-Scanning Systems," Appl. Opt., *2*, pp 401-407, April 1963.

[4] D.P. Peterson and D. Middleton, "Sampling and Reconstruction of Wave-Number-Limited Functions in N-Dimensional Euclidean Spaces," Information and Control, *5*, pp 279-323, 1962.

[5] O.H. Schade, Sr., "Image Reproduction by a Line Raster Process," Chapter 6 of *Perception of Displayed Information*, L.M. Biberman, editor, Plenum, 1973.

[6] W.D. Montgomery, "Some Consequences of Sampling in Image Transmission Systems," Research Paper P-543, Institute for Defense Analyses, December 1969.

[7] H.A. Wheeler and A.V. Loughren, "The Fine Structure of Television Images," Proc. IRE, pp 540-575, May 1938.

[8] R.M. Scott, et al., "A Symposium on Sampled Images," Perkin-Elmer Corporation publication IS 10763, Norwalk, Connecticut, 1971.

[9] S.J. Katzberg, F.O. Huck, and S.D. Wall, "Photosensor Aperture Shaping to Reduce Aliasing in Opto-Mechanical Line-Scan Imaging Systems," Appl. Opt., *12*, pp 1054-1060, May 1973.

[10] L.D. Harmon and B. Julesz, "Masking in Visual Recognition: Effects of Two-Dimensional Filtered Noise," Science, *180*, pp 1194-1197, 15 June 1973.

[11] L.M. Biberman, editor, *Perception of Displayed Information*, Plenum, 1973.

[12] F.T. Thompson, "Television Line Structure Suppression," JSMPTE, *66*, pp 602-606, October 1957.

[13] H.R. Luxenberg and R.L. Kuehn, *Display Systems Engineering*, McGraw-Hill, 1968.

CHAPTER TEN — VISUAL TARGET ACQUISITION

10.1 Fundamentals of Target Acquisition

The process of searching the display of an electro-optical imaging system for a target consists of four interrelated processes: detection, classification, recognition, and identification. Detection is the discrimination of an object from its background and its assignment to the class of potentially interesting objects. Classification is the assignment of the detected object to a gross class of objects such as vehicles or combatant vessels. Recognition is the assignment of the classified object to a specific subclass such as tanks or destroyers. Identification is the assignment of the recognized object to an even more specific category such as M-60 tanks or Spruance class destroyers.

The user of a thermal imager typically expresses the performance of the system by a single acquisition probability P[Acq] which gives the probability of accomplishing an assigned search task. We may write P[Acq] as a product of conditional probabilities if we denote the various conditions as follows:

NOTATION	MEANING
In	Target appears in the search field
Look	Target is looked at by observer
Det	Target is detected by observer
Clas	Target is classified by observer
Rec	Target is recognized by observer
Iden	Target is identified by observer

Then

$$P[Acq] = P[Iden/Rec, Clas, Det, Look, In] \cdot P[Rec/Clas, Det, Look, In]$$

$$\cdot P[Clas/Det, Look, In] \cdot P[Det/Look, In]$$

$$\cdot P[Look/In] \cdot P[In]. \tag{10.1}$$

The probability that the target is in the field is a complicated function of cueing, foreknowledge, navigation, and the search mode chosen. Thus one usually assumes $P[In] = 1$. One also usually assumes that each search task is independent of the other so that the conditional probabilities simplify to products of unconditional probabilities, giving

$$P[Acq] = P[Iden]\ P[Rec]\ P[Clas]\ P[Det]\ P[Look]. \qquad (10.2)$$

The complexity of the search process may be readily appreciated by considering the following lists of pertinent factors culled from the literature. There are fourteen significant displayed target characteristics:

> signal-to-noise ratio (SNR)
> contrast against the background (C)
> critical angular subtense (θ_c)
> edge gradients
> contour complexity
> context in the scene
> location on the display
> shape
> orientation
> perspective
> size relative to the display size
> velocity of motion through the scene (V)
> luminance
> state of operation.

There are four significant displayed scene characteristics:

> background luminance (L_B)
> clutter or false-target density
> rate of motion through the display
> scene aspect ratio.

There are six significant system characteristics:

> optical transfer function and related resolution measures
> sampling rate in time ($\dot{F}$)
> sampling rate in space (scan lines L per critical target angular subtense)
> video monitor screen angular subtense (A')
> grey scale rendition
> dynamic range.

There are eleven significant observer states:

> training
> motivation

        fatigue
        prebriefing
        age
        IQ
        personality
        workload
        search technique
        number of observers and the degree of communication between
            them
        peripheral visual acuity.

There are at least five other tactical and miscellaneous factors:

        required search area
        cockpit illumination
        vehicle noise and vibration
        image or observer vibration
        allowed search time.

These factors total forty, and there may be many others. All of these factors have been investigated to some extent and subsequent sections summarize some of the more useful experimental findings. Every visual search process paper in the literature obtains an experiment of manageable complexity by isolating a few factors from the others. Yet though there are scores of papers in this field, no two results of similar experiments are ever quite the same. The best one can hope to obtain from this tangle of data are rules of thumb which guide one in the proper direction.

Most of the experiments we will consider here select an image quality variable V, fix all of the other possible image quality variables, and vary V to determine its effect on the probability of success of a particular task T as a function of the variable, $P_V[T]$. These experiments generally assume that the effects of such parameters as noise, resolution, sampling, and target type are approximately separable. In that case, the maximum value of a particular $P_V[T]$ may not be unity because the maximum achievable probability may be limited by the fixed variables.

It is the author's opinion that field performance predictions are useful for understanding in a general way how variables affect $P_V[T]$, but that attempting to predict field test results accurately is a waste of time. The best way to approach the problem of performance prediction is to conduct intelligently designed field tests and simulations, and to extrapolate to unmeasured conditions using general principles.

It is in that spirit that the following sections are offered. They outline some reliable experimental results which illustrate general target search principles. There are few published studies concerning thermal systems, but there are many published photographic and television experiments using imagery sufficiently similar to thermal imagery to be of interest. We will examine the results of these visible-spectrum studies before considering the existing thermal imagery studies. The best single source of information on visual target acquisition is the text by Biberman[1].

10.2 Eye Motion During Visual Search

The process of searching a visual field for a target is characterized to a great extent by the limitations on the rapidity and frequency of eye movements. In free search, the center of the visual field stops for a brief period called a saccade, rapidly moves to another position, fixates again, and continues the process until detection is achieved. The rate at which a visual field can be searched is severely limited by the number of fixations possible per second and by the search pattern, which is usually selected sub-consciously. The following discussions explore these problems, considering first the mechanism of search, and then the equations describing it. The key papers describing the mechanism of search are those by Enoch[2]; Ford, et al.[3]; and Baker, et al.[4].

Enoch[2] used circular aerial maps with angular subtenses of 3°, 6°, 9°, 18°, and 24° at a viewing distance of 21.6 inches, and of 51° 18' at 13.25 inches. The maximum scene luminance was 60 fL. While observers searched these displays for a Landolt C, Enoch tracked their eye movements and found the following:

1. most eye fixations tend to fall in the center of the display

2. more are directed to the right half and to the bottom half of the display than to the rest of the display

3. fixation time decreases and the interfixation distance increases as display angle increases, as shown in the following table.

Display Angle	Average Fixation Time	Average Interfixation Distance
3°	0.578	0.87°
6°	0.468	1.82°
9°	0.384	2.13°
18°	0.361	3.72°
24°	0.355	4.33°
51° 18'	0.307	6.30°

4.   As the display angle falls below 9°, the search efficiency drops
      drastically because the number of fixations off of the display in-
      creases, as shown below:

Display Angle	Percent of Fixations Falling Off of the Display
>9°	10
6°	50
3°	75

5.   Displays larger than 9° induce an increase in central fixations at
      the expense of peripheral fixations, with a consequent reduction in
      search efficiency.

6.   The upper left hand display corner is the least often inspected area.

Enoch concluded that the optimum display angle is 9° due to the efficiency losses of observations (4) and (5). The moral of this investigation is
that a display subtense greater than 9° is justifiable only if the excess is used
primarily for orientation and navigation.

Ford, et al.[3], conducted a visual search experiment to determine the
rate of eye fixations, the fixation duration, and the interfixation distance.
Searching a 30° field twelve inches in diameter at 2.5 fL in 5 seconds for a
near-threshold 1/8 inch dot, they found a range of 2.2 to 4.4 fixations per
second, with a mean of 3.1 per second. The mean fixation duration was
0.28 second so that fixation occupied 85% of the time. The mean time in
motion was 0.04 second giving 15% of the time in motion. The mean interfixation distance was 8.6°. They also found that the central and outer por-
tions of the field tended to be neglected in search. A summary of eye
movements during search is given by White[5], who asserts that fixation time
increases as the performance requirements or the scene complexities increase.

Baker, et al.[4] conducted experiments with static displays containing
arbitrary forms. The clutter density, or the relative number of irrelevant
forms, was constant so that the number of irrelevant forms was proportional
to the display area. They found that the search time behaves as shown in
Figure 10.1. Conversely, when the number of competing objects is independent of display size, the search time is unchanged as the display size is increased. They also divided a circular static display into five contiguous
annular regions of equal width as shown in Figure 10.2, and found that the
targets were found the fastest in region 3, as shown in Figure 10.3.

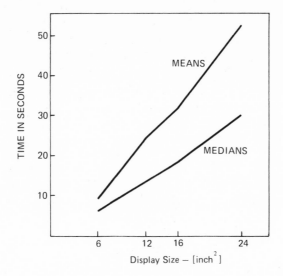

Figure 10.1 Search time as a function of the area of the display
to be searched (adapted from reference 4).

Figure 10.2 Classification of target locations on the display used
in the study of reference 4.

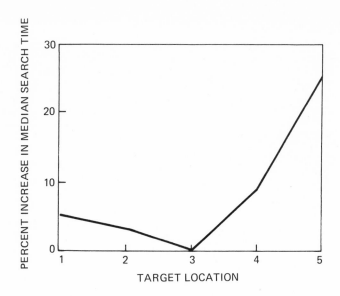

Figure 10.3 Search time as a function of target location, adapted from reference 4.

10.3 Free Search Theories

We will now apply the concept of discrete glimpses or fixations to the problem of calculating the probability of looking in the correct direction on a display for a single target.

The most efficient possible search procedure would be structured, but not necessarily ordered, so that successive glimpses at the display were non-overlapping and completely covered the display periodically without repetition. The search equation for that case is derived as follows. Let the solid angle subtended by a single glimpse be Ω_g and the total search solid angle subtended by the display be Ω_s. Then in one complete structured search of a field, there will be Ω_s/Ω_g fixations. The total possible number of glimpses in which the target is looked at is 1, and the possible number of glimpses in which the target is missed is $(\Omega_s/\Omega_g) - 1$.

The problem as stated is thus equivalent to finding the probability of selecting a black bean in n removals without replacement of beans from a jar containing "a" black beans and b white beans. The possible number of unfavorable events is b. For this binomial distribution problem, the probability of n - r favorable events occurring in n trials or selections is:

$$P(n\text{-}r) = \binom{n}{r} p^{n\text{-}r} q^r .$$
(10.3)

The probability of a favorable event in one trial is p, and q is the probability of an unfavorable event in one trial. These are related to a and b by:

$$p = \frac{a}{a+b}$$

(10.4)

and

$$q = \frac{b}{a+b} \cdot$$

(10.5)

In this simple search model, the total number of trials n is the ratio of the total allowed search time t to the glimpse time t_g,

$$n = \frac{t}{t_g} \cdot$$

(10.6)

For the case that the target is detected when it is looked at, the desired number of favorable events n-r is 1, and

$$a = 1 \text{ and } b = \frac{\Omega_s}{\Omega_g} - 1.$$

Then,

$$p = \frac{\Omega_g}{\Omega_s}$$

(10.7)

and

$$q = 1 - \frac{\Omega_g}{\Omega_s},$$

(10.8)

and

$$P(1) = \binom{t/t_g}{\frac{t}{t_g} - 1} \left(\frac{\Omega_g}{\Omega_s}\right)^1 \left(1 - \frac{\Omega_g}{\Omega_s}\right)^{t/t_g - 1}$$

(10.9)

Since there is only one possible favorable outcome, instead of using the above equation we may evaluate P(1) more simply by:

$$P(1) = 1 - P(0)$$

$$= 1 - \binom{t/t_g}{t/t_g} \left(\frac{\Omega_g}{\Omega_s}\right)^0 \left(1 - \frac{\Omega_g}{\Omega_s}\right)^{t/t_g}$$

$$= 1 - \left(1 - \frac{\Omega_g}{\Omega_s}\right)^{t/t_g}. \qquad (10.10)$$

The defects of this simple model are that free search is neither structured nor nonrepetitive so that the search efficiency is sub-optimal, and that detection does not necessarily occur when the target is looked at. Better models have been discovered or derived in the three primary papers on the theory of search by Krendel and Wodinsky[6], Williams and Borow[7], and Williams[8].

Krendel and Wodinsky[6] postulated a random search model and validated it by three experiments. They assumed that search is random and that if the probability of detection in a single glimpse is P_{sg}, the probability of detection after n glimpses is:

$$P = 1 - (1 - P_{sg})^n \qquad (10.11)$$

$$P \cong 1 - \exp\left[n \ln(1 - P_{sg})\right] \qquad (10.12)$$

for n large and P_{sg} small.

Letting the glimpse time be t_g and the total search time be t, then $n = t/t_g$ and:

$$P(t) = 1 - \exp[t \ln(1 - P_{sg})/t_g] = 1 - \exp(-mt). \qquad (10.13)$$

Their first experiment used a 72 x 63-inch screen of luminance 5 fL viewed at 10 feet, giving a display subtense of 34.92 by 30.44 degrees. The target was a near-threshold 3.87 milliradian circle which was located in one of 48 possible target locations. This experiment strongly supported the simple search model.

The second experiment used a 1.2 mrad^2 target against a large 0.01 fL background. The search areas were 0.22 or 0.098 sr. Dark-adapted observers searched for 44 randomly located targets, and the results also substantiated the model.

The third experiment was a thirty second forced choice search by dark-adapted observers of fields 6.8°, 18°, 32°, and 43° in diameter and at background luminances of 12.4, 1.03, 0.1, and 0.01 fL. The targets were circles of angular diameters 1.41, 3.81, 7.03, and 13.48 milliradians, and all target contrasts were at least twice the 95% threshold of detection for long viewing time. They concluded that the exponential search model is satisfactory under three conditions:

1.   the search time does not exceed 30 seconds

2.   the image parameters such as contrast or size are constant during search

3.   the observer does not consciously use a nonrandom search pattern.

At the time of their investigations the pattern of eye fixations was not known, so their model could not account for this. Their experiments also indicated that training (practice at the task) had no statistically significant effect on the test results.

Williams and Borow[7] verified for different clutter densities and different rates of scene motion through the display that the recognition probability takes the simple form $P = 1 - \exp(-mt)$. Williams[8] developed a somewhat different search theory based on the fact that successive fixations within an interval of a few seconds tend to be clustered, resulting in a semistructured search process. Defining N as the number of complete scans of the field completed during a given search experience and F as the fraction of the last scan completed before the desired detection occurs, Williams proposed that the probability of detection in (N+F) scans is:

$$P(N+F) = 1 - (1-P_{ss})^N + F\,P_{ss}(1-P_{ss})^N, \tag{10.14}$$

where P_{ss} is the probability of detection in a single complete scan of the field.

When the total search time t is an integral multiple of the single scan search time t_{ss}, Williams has:

$$P(t) = 1 - \exp(-mt)$$

where

$$m = -\ln(1-P_{ss})/t_{ss}. \tag{10.15}$$

In all of the above equations, the time t allowed for search can be related to physical search parameters such as sensor slewing rate or aircraft ground speed, sensor depression angle, sensor field-of-view, and sensor altitude as demonstrated in section 10.4.

Bloomfield[9] showed that search in a densely cluttered scene is facilitated when target-to-clutter contrast is increased, or when target-to-clutter size is increased. Petersen and Dugas[10] similarly found that search times are reduced when target contrast or rate of motion relative to the background are increased. In the case of target motion, they found that the free search equation must be modified by the addition of a factor to the exponential. For target velocity V through the scene of less than 5 degrees/second, this factor is $C(1 + 0.45 V^2)$ where $0.3 \leqslant C \leqslant 0.5$.

10.4 Search Geometry

Parametric tradeoff analyses cannot be performed without reference to some assumed search pattern or procedure. A simple example is that of terrain search from a moving aerial platform. To simplify the equations, we will assume that the sensor altitude is constant and that motion relative to the ground is in a straight line.

Let A be the sensor azimuth field of view, B be the sensor elevation field of view, θ_d be the sensor declination angle measured from the horizontal to the center of the sensor field of view, H be the sensor altitude and V_g be the sensor ground speed. These quantities are shown in Figure 10.5, together with the ground dimensions intercepted.

The quantities are related by:

$$H/R_2 = \sin (\theta_d - B/2) \tag{10.16}$$

$$H/R_1 = \sin (\theta_d + B/2) \tag{10.17}$$

$$H/D_1 = \cot [90° - (\theta_d + B/2)] \tag{10.18}$$

$$H/D_2 = \cot [90° - (\theta_d - B/2)] \tag{10.19}$$

$$W_1/R_1 = 2 \tan (A/2) \tag{10.20}$$

$$W_2/R_2 = 2 \tan (A/2) \tag{10.21}$$

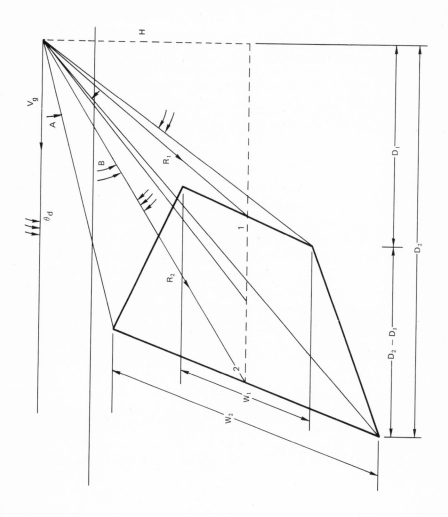

Figure 10.5 Air-to-ground search geometry.

The ground area within the field is

$$G = (D_2 - D_1)/[(W_2 + W_1)/2].$$ (10.22)

If a target enters the field at point 1 shown in Figure 10.5, and exits at point 2, the time t it is in the field is:

$$t = \frac{D_2 - D_1}{V_g}.$$ (10.23)

10.5 Equivalent Bar Chart Resolvability

Numerous investigators have demonstrated that the single most important factor in determining the level of target discrimination possible with a system is the resolvability of various bar chart equivalents of the target. This appears to be the case regardless of the nature and combination of the image defects involved. Figure 10.6 depicts this equivalent bar chart concept, where it is assumed that a target is characterizable by a critical target dimension which contains the target detail essential to discrimination. This dimension typically is the minimum projected dimension of the target, as shown in Figure 10.7. The bar chart equivalents of the target are the members of the set of bar charts whose total width is the critical target dimension and whose length spans the target in the direction normal to the critical dimension.

Figure 10.8 shows some examples of how various image degradations affect performance. The target is the same in all cases and is characterized as having an average apparent blackbody temperature difference ΔT against the background. The target has a critical angular subtense θ_C so that it subtends a fraction of the system field of view A given by θ_C/A. In Figure 10.8, system 1 is random-noise-limited, system 2 is magnification-limited, system 3 is MTF-limited, and system 4 is raster-limited. The way the degradations have been selected, 4 is better than 3, 3 is better than 2, and 2 is better than 1, as evidenced by the target imagery. These degradations allow detection with system 1, classification with system 2, recognition with system 3, and identification with system 4.

The theory relating equivalent bar target resolvability to target discrimination is that the level of discrimination can be predicted by determining the maximum resolvable frequency of an equivalent chart having the same apparent ΔT as the target and viewed under the same conditions. This theory in a somewhat different form was first proposed by Johnson[11], who sought a way to relate in-the-field performance of image intensifiers to objective laboratory measures.

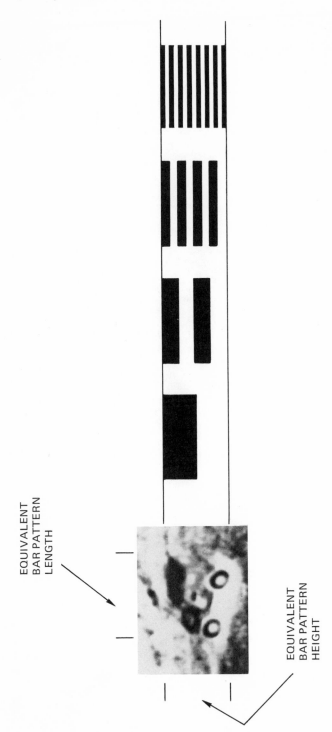

EQUIVALENT
BAR PATTERN
LENGTH

EQUIVALENT
BAR PATTERN
HEIGHT

Figure 10.6 Equivalent bar patterns.

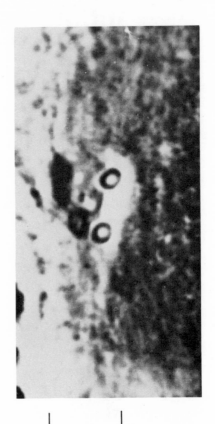

CRITICAL TARGET
ANGULAR SUBTENSE θ_C

Figure 10.7 Critical target dimension.

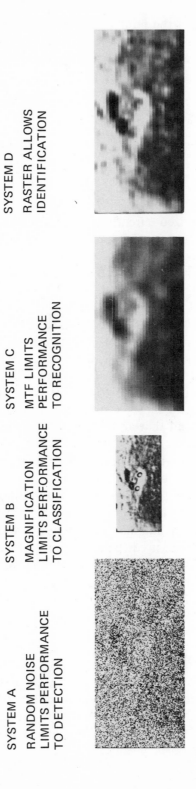

SYSTEM A
RANDOM NOISE
LIMITS PERFORMANCE
TO DETECTION

SYSTEM B
MAGNIFICATION
LIMITS PERFORMANCE
TO CLASSIFICATION

SYSTEM C
MTF LIMITS
PERFORMANCE
TO RECOGNITION

SYSTEM D
RASTER ALLOWS
IDENTIFICATION

Figure 10.8 Image quality limitations to performance.

Johnson determined the number of resolvable cycles subtended by the critical target dimension for eight military vehicles and a standing man. The now well-known Johnson criteria averaged over all target classes are:

Level of discrimination	Number of resolvable cycles required for 50% probability of correctness
detection	1.0 ± 0.25
orientation	1.4 ± 0.35
recognition	4.0 ± 0.8
identification	6.4 ± 1.5

These criteria were derived directly from image interpretation data, and their validity is in the strictest sense limited to rasterless image-intensifier-type imagery.

Rosell and Willson[12] refined the theory by applying the perceived signal-to-noise ratio concept described in Chapter Four. They derived expressions for the resolvability of bar targets embedded in additive gaussian noise on a television monitor, and performed tests to verify the theory. They then added noise to television images of vehicle targets, conducted tests with observers to determine recognition and identification probabilities as functions of perceived SNR, and plotted these probabilities as functions of the predicted perceived SNR's required to resolve the equivalent bar patterns of each target. They assumed that a seven-cycle equivalent bar pattern is required for recognition and that an eleven-cycle equivalent bar pattern is required for identification.

The results support the simple theory that target recognition and identification are predictable from equivalent bar chart resolvability. The complete experiments and results are described in Biberman[1]. These data are strictly applicable only for targets embedded in noisy but otherwise perfect 525-line television imagery and may or may not be valid for the wide variety of FLIR formats which are used. Williams[13] performed similar experiments with computer-processed FLIR imagery and found that recognition accuracy correlates well with the number of resolvable cycles on the critical target dimension. These results apply regardless of the combination of raster, resolution, contrast, and noise used, and are described in detail in section 10.9.

10.6 Probability of Detection

The probability of detection of simple geometrical targets featured against uniform backgrounds and embedded in random noise was discussed in Chapter Four. The conclusion there is that the visual system operates as if it calculates a signal-to-noise ratio and compares it with a threshold value of

SNR as a test of significance. There is considerable data to support this theory in various regimes of viewing. The theory is supported in the quantum-noise-limited or contrast limited mode by the data of Blackwell[14], and in the additive noise mode by the data of Coltman and Anderson[15], Schade[16], and Rosell and Willson[17]. Virtually no data have been published relating detectability of non-simple targets in non-uniform backgrounds to signal-to-noise ratio.

Chapter Four asserted that the probability of detection increases as target viewing time, angular subtense, and contrast increase. The tests of Bernstein[17]; Coluccio, et al.[18]; Hollanda and Harabedian[19]; and Greening and Wyman[20] with real targets and situations showed that in practical tasks, the percentage of targets detected does indeed increase as target contrast increases. Bernstein[17], for example, found that CRT images of vehicles and personnel required 90% contrast ($C \triangleq (L_T - L_B)/L_B$) to reach the maximum attainable recognition probability.

Bernstein[17] found that resolution affects detection probability only insofar as it influences signal-to-noise ratio or target contrast. However, Coluccio, et al.,[18] found that detection completeness improves with limiting resolution R [resolvable cycles/target dimension] by:

$$\text{completeness} = K_1 \log K_2 R, \qquad\qquad (10.24)$$

where K_1 and K_2 are empirical constants.

Bailey[21] used the search theory discussed earlier as a point of departure, and cited many references which indicate that the cumulative probability of detection P[Det] is a function of the clutter. When a cluttered scene of displayed solid angular subtense Ω_s must be searched for a target of displayed angular subtense Ω_T in time t, Bailey asserts that the cumulative probability of detection is given by:

$$\text{cumulative } P[\text{Det}] = P[\text{Det/Look}]\, [1 - \exp(-6.9\, t\, \Omega_T/\Omega_s\, K)].$$

$$(10.25)$$

Here P[Det/Look] is the probability that if an observer looks directly at the target, he will detect it, and K is an empirical clutter factor proportional to the density of false targets in the scene. Typical values for K in the literature surveyed by Bailey range from about 0.01 to 0.1. The experiments of Baker, et al.[4], and of Williams and Borow[8] support this theory.

10.7 Probabilities of Classification and of Recognition

To the author's knowledge, no data has been published describing the image quality necessary to convert a detection to a classification. This neglect is not, however, caused by the unimportance of classification as an imagery interpretation task. Classification is particularly important in a military context. For example, it is common that a certain type of target such as a vehicle is not supposed to be in a secured area or that a bogey is known to be in a certain sector. In such cases, it is necessary only to detect the target and to classify it simply as being a vehicle or an aircraft before firing on it.

The probability of recognition of simple and of practical targets has been exhaustively investigated for the effects of variations of a single image quality variable. Barnard[22] found the probability of recognition of randomly oriented Landolt C's, and of asterisks with one missing spoke, when masked by additive Gaussian noise. He found that the results are accurately described by assuming the eye-brain system operates as an optimum filter, which supports the contention of some visual psychophysicists that the eye-brain system consists of an assembly of stochastically independent narrow-band tuned filters.

There is agreement among all investigators that improving the resolution of the viewing device improves recognition and identification performance. Bailey[21] suggests that probability of recognition P [Rec] is related to the smallest number of resolution cells (90% detectable spots) contained in the narrowest target dimension. Johnson[11] reported that for high confidence the number of resolvable cycles per critical target dimension should be 4 ± .8 cycles for recognition.

Bennett et al.,[23] found that recognition performance improves at 15 inches viewing distance as the displayed resolution improves to approximately .25 mrad, and that no significant improvement occurs beyond that level. Greening and Wyman[20] recommend

$$ P[\text{Rec}] = \exp\left[-\left(\frac{r_s}{r}\right)^m\right], \tag{10.26} $$

where r_s = displayed solid angular resolution of the sensor including the eye in mrad^2,

r = empirically required resolution [mrad^2] for a particular target,

m = 1 for $r_s/r > 1$,

m = 2 for $r_s/r \leqslant 1$,

$$r = \frac{\Omega_T [\text{mrad}^2]}{N_R{}^2} \; ,$$

Ω_T = target solid angular subtense,

N_R = number of resolution elements on target required empirically.

Their experiments with real scenes indicated that the number of resolution elements required for 90% confidence varied from 3 to 20 depending on target complexity. Boeing Company investigators[24] found that photographic information extraction improves approximately linearly with the area under the MTF curve and with N_e for the typical types of MTFs they investigated.

Some of the most consistent results in search theory are obtained in investigations of the number of spatial samples or raster lines across a target required for recognition and identification. Johnston's[25] results for TV on the probability of recognition P[Rec] of vehicles as a function of the number of raster lines L through the target are fitted well by

$$P_L [\text{Rec}] = 1 - \exp\left\{-.018\,(L{+}1)^2\right\} \tag{10.27}$$

for $7 \leqslant L \leqslant 13$ lines.

Numerous field experiments using thermal systems yield a range of 4 ± 1 lines on target for 50% P[Rec] and 6 ± 1 lines on target for 90% P[Rec]. Scott, et al.,[26] generated vertically sampled imagery of model military vehicles with 4, 6, 9, 13.5, 20, and 30 scan lines per vehicle height. Observers were asked to match the vehicle images with the unsampled images, and the probabilities of correct vehicle recognition were computed. Their results are summarized by Figure 10.9. Hollanda and Harabedian[19] performed similar experiments and obtained similar results.

Gaven, et al.,[27] investigated information extraction using photographs sampled at equal intervals in both directions with a Gaussian spread function sigma of one-half of the lattice spacing. They used values of the number of scan lines per vehicle diagonal of 22.1, 33.1, and 49.6. The image intensities were quantized into approximately equiprobable levels of 1 to 7 bits (2 to 256 levels). The probabilities of recognition using the processed images were determined as a function of L and the number of bits in intensity, and are shown in Figures 10.10 and 10.11. They found no significant improvement in performance beyond 3 bits. Johnston[25], on the other hand, using closed circuit TV, found no significant dependence of recognition on shades of gray on a TV monitor for shades of gray between five and nine.

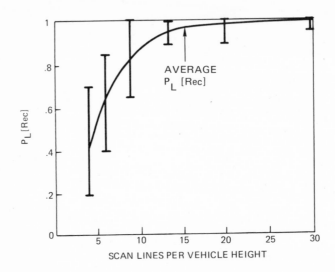

Figure 10.9 Variational bounds on probability of recognition versus lines subtended for all classes of vehicles. (Adapted from reference 26)

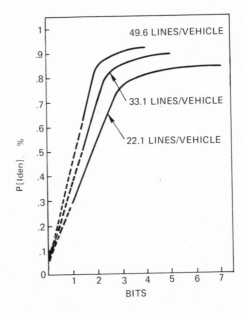

Figure 10.10 Probability of identification versus quantization level and number of scan lines per target. (Adapted from reference 27)

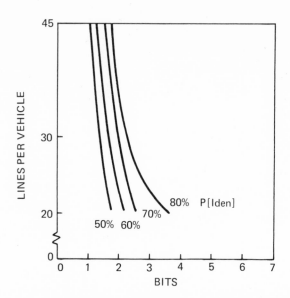

Figure 10.11 Constant probability of identification versus quantization level and lines per target. (Adapted from reference 27)

Steedman and Baker[28] investigated the dependence of target recognition on target angular subtense using circular static photographic displays of angular diameters 4.6, 9.3, 13.9, and 17.7 degrees when viewed at 24 inches. Each display contained 184 arbitrary complex forms selected from a universe of 557 forms. Twenty-four forms were designated as targets to be searched for, and an illumination of 20 foot-candles was used. All forms had high contrast and were only slightly blurred. Their results indicated that the percent recognition error and the search time are relatively constant for target angular subtenses larger than 12 arcminutes, and increase sharply below 12 arcminutes. Figure 10.18 shows the percentage change in search time and the percentage of observations in error as functions of the maximum target angular subtense.

10.8 Probability of Identification

Hollanda, et al.,[29] investigated the dependence of target identification on the number of line scans per target and the point signal-to-noise ratio. They used twenty model armored vehicles and generated simulated electro-optical imagery. The number of scan lines L per vehicle height was varied over 16, 32, and 48 scan lines per vehicle, and Gaussian noise was added to get image SNRs of 3, 5, 10, 20, and 30. Among other things, they plotted percentage correct identifications versus SNR for different L's and plotted

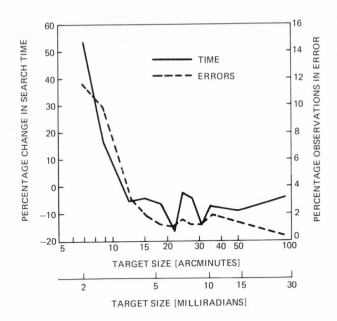

Figure 10.12 Percentage change in search time and percent change in identification errors versus target angular subtense. (Adapted from reference 28).

L versus SNR for constant identification performance. Their identification probabilities P[Iden] are shown in Figures 10.13 through 10.16 as functions of L and SNR. Their basic conclusions were that:

1. P[Iden] for vehicles increases as a strong function of SNR values up to 15.
2. The P[Iden] for noiseless images were only 5% better than those for images with SNRs of 30.
3. P[Iden] increased at most 10% when SNR increased from 20 to infinity.
4. The differences in image quality among images with SNRs of 2, 3, and 4 were negligible.

The previously described experiment of Scott et al.[26] with line scan imagery produced the dependence of P[Iden] on scan lines per target shown in Figure 10.17. They found only slight improvement in percent correct identifications beyond L = 20, and the 50% point occurred at approximately L = 7. Variations in P[Iden] with oblique 45° viewing and with nadir viewing are shown in Figure 10.18.

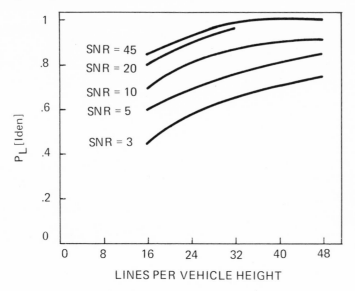

Figure 10.13 Identification versus lines for point signal to noise ratio variations with miscellaneous military vehicles. (Adapted from reference 29).

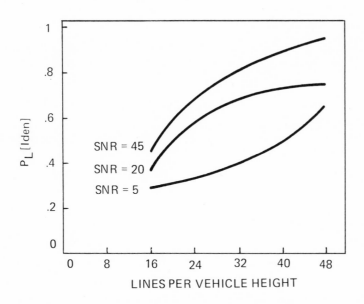

Figure 10.14 Identification versus lines for point signal to noise ratio variations with tanks. (Adapted from reference 29).

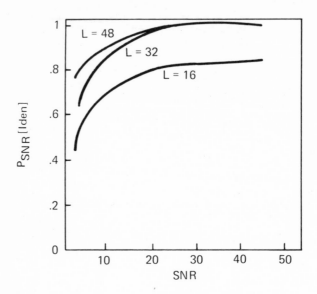

**Figure 10.15 Identification versus point signal to noise ratio for line
variations for miscellaneous military vehicles.
(Adapted from reference 29).**

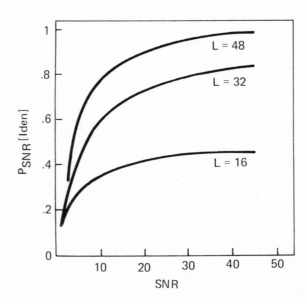

**Figure 10.16 Identification versus point signal to noise ratio for line
variations for tanks. (Adapted from reference 29).**

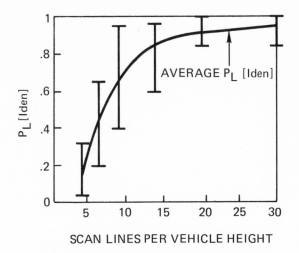

Figure 10.17 Variations in identification for all vehicles versus lines per vehicle height. (Adapted from reference 26).

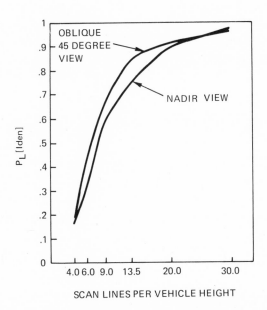

Figure 10.18 Identification versus lines per vehicle height for all vehicles. (Adapted from reference 26)

Hemingway and Erickson's[30] data relating probability of identification of arbitrary typographic symbols on a television monitor to the symbol angular subtense θ_c in milliradians and to the number of TV scanlines L through the symbol are fitted well by P[Iden] = $1 - \exp(6.4 \times 10^{-3} \, L \, \theta_c)^2$.

10.9 Experiments with Computer-Processed FLIR Imagery

The economic impracticality of constructing FLIR's which exhibit wide variations in performance parameters suitable for use in image interpretation studies has led to simulation of these effects with processed imagery. One such study by Williams[13] determined the effects on recognition accuracy of lines per target, magnification, MTF, and SNR.

Williams and his associates constructed a digital image synthesizer and used it with thermographic imagery to produce controlled image quality degradations. The purpose of these efforts was to simulate the image characteristics of conventional raster-scan FLIRs, to use the resultant imagery to quantitatively ascertain the effects of parametric changes on observers' recognition performance, and to identify summary measures of performance.

The original thermographs were essentially noiseless, had a 2-to-1 aspect ratio and 250 scan lines per picture height, and exhibited a gaussian MTF with an LSF standard deviation of one-half the scan line spacing. These photographs were processed so that the degradations in the originals were insignificant compared to the processing degradations. Each photograph was reduced to a matrix of 250 by 500 sample points recorded in six bits per point or 64 levels. The data were sampled as by a line scan process, spread in two dimensions as by a gaussian spread function, degraded by controlled noise, and printed out on film by a raster having a gaussian spread. Sixteen examples of each of ten target types were processed in this way. The following variations in image parameters were produced:

Raster lines per picture height:	11, 22, 44, 88
Raster lines per target height:	3, 6, 12, 24
Gaussian line spread function sigma as a fraction of the line spacing:	0.38, 0.48, 0.55, 0.61, 0.67, 0.72
Corresponding gaussian MTF's at a frequency of one-half of the line frequency:	0.47, 0.33, 0.23, 0.16, 0.11, 0.08
Target image SNRs:	25, 12.2, 6.58, 3.50, 2.04, 1.35

These processed pictures were then viewed at varying distances to produce the following variations in magnification:

Picture heights in degrees:	0.5, 1, 2, 4
Raster line angular subtense in milliradians:	19.8, 9.9, 4.95, 2.48
Target angular subtenses in milliradians:	23.3, 11.7, 5.82, 2.91

Fifteen thousand trials with eighteen observers were conducted using the 160 images degraded as discussed above, and the experiment was concluded by statistical analyses of the data.

Williams's results are startling, for they contradict the conventional wisdom in FLIR that small variations in MTF and SNR have significant effects on recognition performance. Williams found that the raster structure apparently so thoroughly disrupts the image that gross SNR and MTF changes have little effect on recognition performance.

The correlation of image quality variables with recognition was determined by a linear regression analysis. The best single measure of recognition accuracy for this experiment is the resolving power for tribar targets with the long bar dimension parallel to the raster. The next best parameter was the number of raster lines on the minimum target dimension.

The fact that tribar resolving power is a good predictor is consistent with the results of Johnson[11] for image intensifiers and of Rosell and Willson[12] for television. This is a useful result because vertical resolving power combines sampling rate, vertical spread function, SNR, and magnification in a single observer response measure. Thus resolving power remains a good quality measure when sampling is present.

Figure 10.19 shows the experimental limiting resolutions for bar targets with the bar length oriented in the scan direction. The limiting resolutions are plotted as functions of the overall picture height and of the number of scan lines. This clearly shows the effect of increasing magnification at constant line rate. The same data are replotted in Figure 10.20 to show the effect of increasing line rate at constant magnification. The pertinent results of the study are shown in Figures 10.21 through 10.24.

One strong conclusion of the study is that raster sampling dominates performance, so much so that noise and MTF have minor roles in recognition. This does not contradict many earlier photointerpretation studies which demonstrated a strong dependence of performance on MTF in photographs. Rather, this study shows that aliasing in sampled imagery so completely disrupts the image at low values of lines per target that MTF variations have no significant effect.

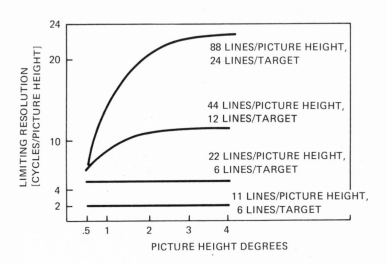

Figure 10.19 Vertical tribar limiting resolution.

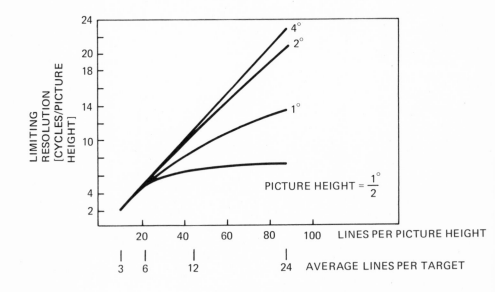

Figure 10.20 Vertical tribar limiting resolution.

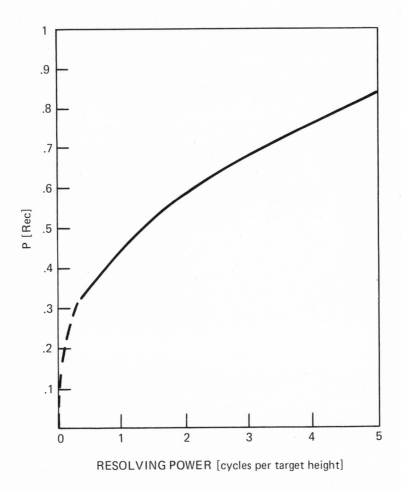

RESOLVING POWER [cycles per target height]

Figure 10.21 Probability of recognition as a function of resolving power.

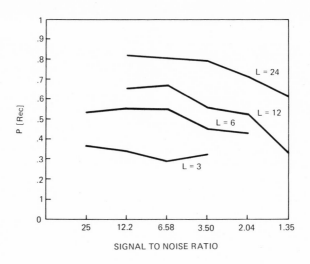

Figure 10.22 Probability of recognition as a function of lines per target and noise.

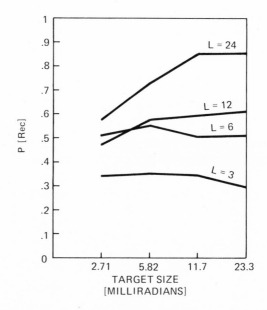

Figure 10.23 Probability of recognication as a function of lines per target and size.

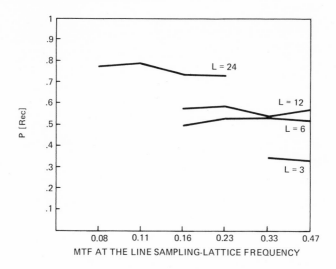

Figure 10.24 Probability of recognition as a function of lines per target and MTF at a frequency equal to the reciprocal of the line center to line center spacing.

Several rules of thumb are evident from the study results:

1. 90% recognition accuracy requires on the average 24 raster lines per target height, or 12 resolvable cycles per target height.
2. 50% recognition accuracy requires on the average 15 raster lines per target height, or 7.5 resolvable cycles per target.
3. One raster line need not subtend more than 1 mrad at the eye, but cannot subtend less than about 0.5 mrad without causing substantial loss in performance.

These values are consistent with current FLIR practice, *viz* a 350 line, 0.25 mrad sensor with a display height of six inches viewed at a maximum distance of 30 inches to give 0.57 mrad per line.

Some interesting examples of computer-processed FLIR imagery are shown in Figures 10.25 and 10.26. The imagery was processed[13] to simulate a parallel-scanned system whose MTF is dominated by a detector (read-in) aperture and by a light-emitting diode (write-out) aperture. Figure 10.25 shows imagery processed to 125 scan lines with a constant write-out aperture and with varied read-in apertures and point signal-to-noise ratios. Figure 10.26 shows 31-line imagery with the same relative variations. The original thermograph from which these variations were produced is shown in Figure 10.27.

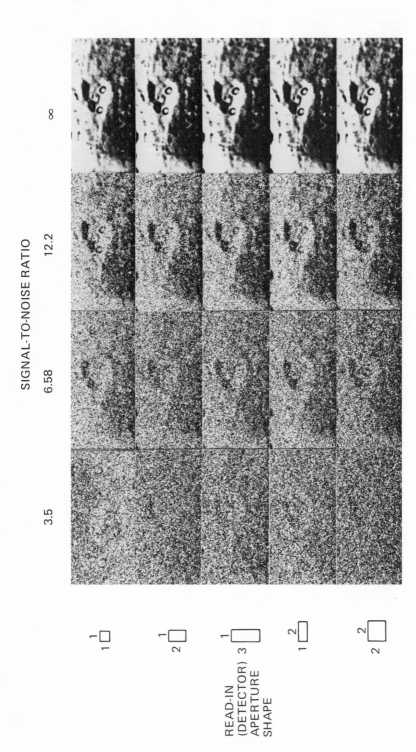

Figure 10.25 Processed thermal imagery with 125 lines per picture height and a square write-out (monitor) aperture 1 line wide.

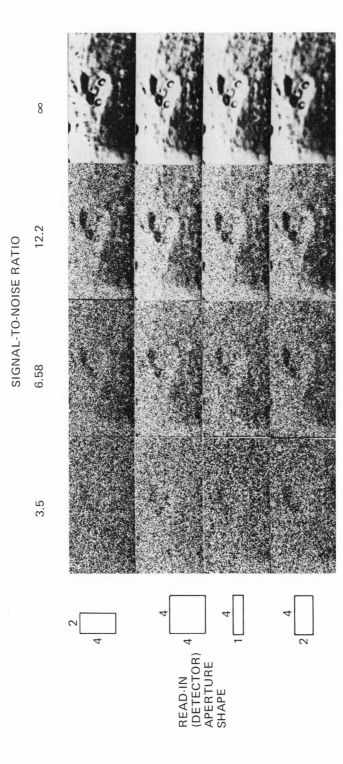

Figure 10.25 (Cont'd.) Processed thermal imagery with 125 lines per picture height and a square write-out (monitor) aperture 1 line wide.

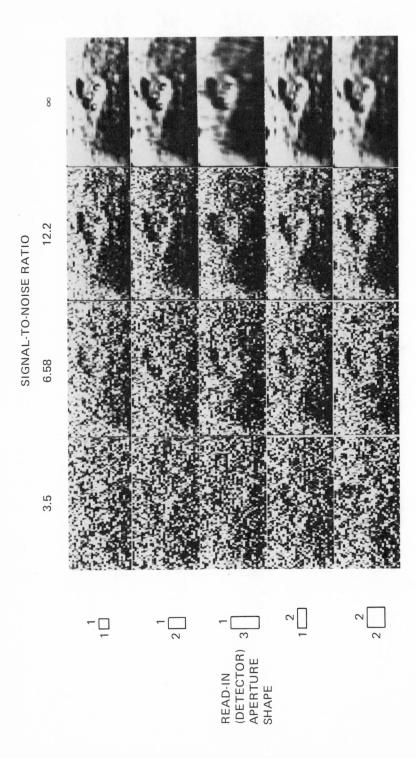

SIGNAL-TO-NOISE RATIO

∞   12.2   6.58   3.5

READ-IN
(DETECTOR)
APERTURE
SHAPE

Figure 10.26 Processed thermal imagery with 31 lines per picture height and a square write-out (monitor) aperture 1 line wide.

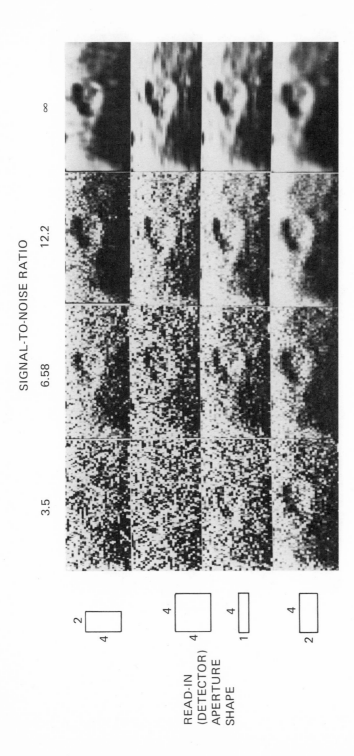

Figure 10.26 (Cont'd.) Processed thermal imagery with 31 lines per picture height and a square write-out (monitor) aperture 1 line wide.

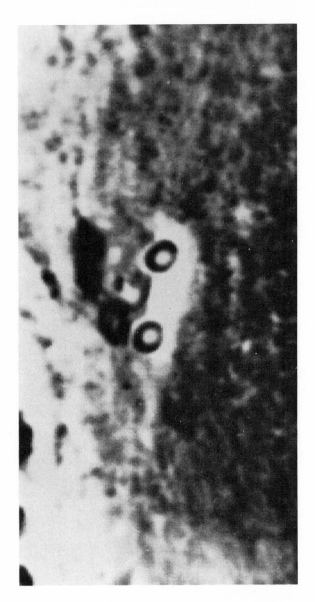

Figure 10.27 The original thermal image used for processing.

10.10 Other Observer Limitations

Bennett, et al.[23] found that one resolution element should not be magnified so that it exceeds more than 3 arcminutes (approximately 1 mrad) at the eye. This result is substantiated by Hemingway and Erickson[31], and is consistent with the investigations discussed in Section 4.7.

Williams and Borow[7] found no degradation in search performance with displayed scene angular motion of from 0 to 8 deg/s. At 16 deg/s, search time increased approximately 25%, and at 31 deg/s search time increased approximately 100%. They found also that horizontal motion produces less degradation than vertical motion.

Williams and Borow[7] also found that individual observers exhibited variations of four-to-one in information extraction capability. Similarly, Thornton, et al.[32] found that different observers exhibited variations of about five-to-one in photointerpreter accuracy, completeness, and search time. Erickson[36] found that search time decreases as observer peripheral visual acuity increases. Bennett, et al.[23] found that they could triple photointerpreter performance simply by informing observers that a particular target (one of many present) was in the scene.

Sziklai[33] generated rapidly changing sequences of television pictures containing only well-defined objects, and found that about 3 to 5 symbols or objects were recognizable per second under ideal conditions.

10.11 Optimization of System Parameters

Cost effective system design requires that only as much performance as is required be incorporated into the design. Therefore it is necessary to understand how various observer limitations interact in a system. As an example, consider a simple system having a square field of view and a contiguous raster with no overscan. Assume that for search efficiency, for display size limitations, for flicker perceptibility, for optimum surround factor, or for other reasons, that we have decided on acceptable display angles of A' by A' [milliradians]. Assume that from general image quality considerations, we have chosen an angular magnification M which couples the eye and the system efficiency. Then the sensor field-of-view size is $A = A'/M$ [milliradians] square.

Now assume that a specific recognition probability is required for a target with a critical target dimension of CTD in meters located at a slant range of R [kilometers]. The critical angular subtense of this target is θ_c [milliradians] and equals CTD/R. The required recognition probability dictates an approximate number of raster lines (or resolution elements in a nonsampled system) subtended by θ_c, say L lines. Then one line must subtend θ_c/L [milliradians]. This recognition probability also dictates in a rather loose way some minimum critical target angle subtended at the eye, say θ'_c, which is consistent with the eye's sine wave response and the target spatial frequency spectrum. Now θ_c and θ'_c are related by:

$$M = \theta'_c/\theta_c, \tag{10.28}$$

but we also have

$$M = A'/A, \tag{10.29}$$

so

$$A'/A = \theta'_c/\theta_c \text{ or } A = A' \theta_c/\theta'_c. \tag{10.30}$$

Thus the sensor field-of-view is determined by the choice of an efficient display angle for search and by a recognition requirement. For example, some reasonable numbers are:

A' = 164 mrad (approximately 9°)

θ_c = 1 mrad (3 meters at 3 kilometers)

θ'_c = 4 mrad (approximately 12 arcminutes).

These values give a field of view of A = 164 (1)/4 = 40 mrad = 2.3°, and a magnification of approximately 4.1. If we pick P[Rec] = 50%, a reasonable value of L is 4 lines. Then one line subtends θ_c/L = 0.25 mrad, and the total number of channels is set at 40 mrad/0.25 mrad = 160 channels. Finally, for an effective observer distance D from the display in meters, and for small A', the effective display width W in millimeters is $W \cong D A'_m$. A cockpit display might have D = 0.6m (approximately two feet), so that for our example W would be 0.6 (164) mm = 98 mm $\cong$ 4 inches.

Many thermal imaging systems now in use are suboptimal in at least one respect because the display size, the line subtense, the field-of-view, and the magnification are usually selected independently and for unrelated reasons. For example, the field-of-view may be chosen for ground coverage at a particular altitude, while the magnification is chosen independently for resolution efficiency. Then a large part of the display may be rendered unusable because the observer cannot absorb all its contents.

The temptation to overspecify or overdesign a system is always strong because the state-of-the-art permits a higher field-of-view to resolution ratio than a human observer can possibly use efficiently. In the numerical example given above, the figure of 160 lines per field-of-view would be low for current high performance FLIR's. The result is that FLIR performance tends to be operator-limited rather than equipment limited.

REFERENCES

[1] L.M. Biberman, editor, *Perception of Displayed Information,* Plenum, 1973.

[2] J.M. Enoch, "Effect of the Size of a Complex Display Upon Visual Search," JOSA, *49*, pp 280-286, March 1959.

[3] A. Ford, C.T. White, and M. Lichtenstein, "Analysis of Eye Movements During Free Search," JOSA, *49*, pp 287-292, March 1959.

[4] C.A. Baker, D.F. Morris, and W.C. Steedman, "Target Recognition on Complex Displays," Human Factors, *2*, pp 51-61, May 1960.

[5] C.T. White, "Ocular Behavior in Visual Search," Applied Optics, *3*, pp 569-570, May 1964.

[6] E.S. Krendel and J. Wodinsky, "Search in an Unstructured Visual Field," JOSA, *50*, pp 562-568, June 1960.

[7] L.G. Williams and M.S. Borow, "The Effect of Rate and Direction of Display Movement Upon Visual Search," Human Factors, *5*, pp 139-146, April 1963.

[8] L.G. Williams, "Target Conspicuity and Visual Search," Human Factors, *8*, pp 80-92, February 1966.

[9] J.R. Bloomfield, "Visual Search in Complex Fields," Human Factors, *14*, pp 139-148, 1972.

[10] H.E. Peterson and D.J. Dugas, "The Relative Importance of Contrast and Motion in Visual Detection," Human Factors, *14*, pp 207-216, June 1972.

[11] J. Johnson, "Analysis of Image Forming Systems," Proceedings of the Image Intensifier Symposium, U.S. Army Engineering Research Development Laboratories, Fort Belvoir, Virginia, October 1958; summarized in Biberman (Reference 1).

[12] F.A. Rosell and R.H. Willson, "Recent Psychophysical Experiments and the Display Signal-to-Noise Ratio Concept," Chapter Five of reference 1.

[13] L.G. Williams, personal communications, Honeywell, Inc., Systems and Research Division, Minneapolis, Minnesota.

[14] H.R. Blackwell, "Contrast Thresholds of the Human Eye," JOSA, *36*, pp 624-643, November 1946.

[15] J.W. Coltman and A.E. Anderson, "Noise Limitations to Resolving Power in Electronic Imaging", Proc. IRE, pp 858-865, May 1960.

[16] Otto H. Schade, Sr., "An Evaluation of Photographic Image Quality and Resolving Power," JSMPTE, *73*, pp 81-119, February 1964.

[17] B.R. Bernstein, "Detection Performance in a Simulated Real-Time Airborne Reconnaissance Mission," Human Factors, *13*, pp 1-9, February 1971.

[18] T.L. Collucio, S. MacLeod, and J.J. Maier, "Effect of Image Contrast and Resolution on Photo-interpreter Target Detection and Identification," JOSA, *59*, pp 478-481, November 1969.

[19] P.A. Hollanda and A. Harabedian, "The Informative Value of Line-Scan Images as a Function of Image Contrast and the Number of Scans per Scene Object (Ground-Level, non-simulated imagery)", Perkin-Elmer Report Number 10032, Perkin-Elmer Optical Technology Division, Danbury, Connecticut, September 1970.

[20] C.P. Greening and M.J. Wyman, "Experimental Evaluation of a Visual Detection Model," Human Factors, *12*, pp 435-445, October 1970.

[21] H.H. Bailey, "Target Detection Through Visual Recognition: A Quantitative Model," Rand Corporation Memorandum RM-6158-PR, Rand Corporation, Santa Monica, California, February 1970.

[22]T.W. Barnard, "Image Evaluation by Means of Target Recognition," Phot. Sci and Eng., *16*, pp 144-150, March-April 1972, and "An Image Evaluation Method" in "A Symposium on Sampled Images," Perkin-Elmer Corporation, Norwalk, Connecticut, 1971.

[23]C.A. Bennett, S.H. Winterstein, and R.E. Kent, "Image Quality and Target Recognition," Human Factors, *9*, pp 5-32, February 1967.

[24]Results of Boeing image evaluation studies are cited in Chapter 3 of *Perception of Displayed Information*, L.M. Biberman editor, Plenum, 1973.

[25]D.M. Johnston, "Target Recognition on TV as a Function of Horizontal Resolution and Shades of Gray," Human Factors, *10*, pp 201-209, June 1968.

[26]F. Scott, P.A. Hollanda, and A. Harabedian, "The Informative Value of Sampled Images as a Function of the Number of Scans per Scene Object," Phot. Sci. and Eng., *14*, pp 21-27, January-February 1970.

[27]J.V. Gaven, Jr., J. Tavitian, and A. Harabedian, "The Informative Value of Sampled Images as a Function of the Number of Gray Levels Used in Encoding the Images," Phot. Sci. and Eng., *14*, pp 16-20, January-February 1970.

[28]W.C. Steedman and C.A. Baker, "Target Size and Visual Recognition," Human Factors, *2*, pp 120-127, August 1960.

[29]P.A. Hollanda, F. Scott, and A. Harabedian, "The Informative Value of Sampled Images as a Function of the Number of Scans per Scene Object and the Signal-to-Noise Ratio," Phot. Sci. and Eng., *14*, pp 407-412, November-December 1970.

[30]C. Hemingway and B.A. Erickson, "Relative Effects of Raster Scan Lines and Image Subtense on Symbol Legibility on Television," Human Factors, *11*, pp 331-338, August 1969.

[31]R.A. Erickson, "Relation Between Visual Search Time and Peripheral Visual Acuity," Human Factors, *6*, pp 165-177, April 1964.

[32]C.L. Thornton, G.V. Barrett, and J.A. Davis, "Field Dependence and Target Identification," Human Factors, *10*, pp 493-496, October 1968.

[33]G.C. Sziklai, "Some Studies in the Speed of Visual Perception," IRE Trans. Info. Thry., IT-2, pp 125-128, September 1956.

CHAPTER ELEVEN – SYSTEM PERFORMANCE TESTING

11.1 Signal Transfer Function and Dynamic Range

We have defined a thermal imaging system as a device which converts optical radiation in the far infrared to visible radiation in such a way that information can be extracted from the resulting image. Typically the power gain of this conversion is controlled by the "contrast" control of the video monitor, and an arbitrary dc level is added by the "brightness" control. We must be able to describe and to measure this power conversion and its dependence on control settings to assure ourselves that the observer may optimize his information extraction capability by adjusting the controls. By this we mean that satisfactory surround factors can be achieved for the anticipated environments, that noise-limited imagery may be produced, and that high contrast imagery may be produced.

For convenience, the input-to-output power conversion is measured at a point in the image of a large square target. The input power is normally expressed in units of degrees centigrade of effective blackbody temperature difference between the target and its background measured within the thermal spectral bandpass of the system. The output power is usually represented by the photopic luminance of the displayed target image, as measured by a photometer*.

The power conversion is described by what Moulton, et al.[5] call the Signal Transfer Function (SiTF). The SiTF is defined as the photopic luminance output of the system display as a function of the target-to-background temperature difference in the test pattern of Figure 11.1 for various combinations of system control settings.

---

*The subject of photometry is treated in References 1 through 4.

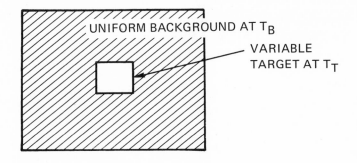

Figure 11.1 Si TF test pattern.

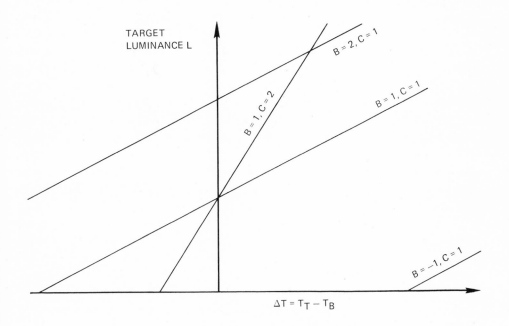

Figure 11.2 Idealized Si TF's for different settings of linear
brightness control B and contrast control C.

An SiTF test is typically performed as follows. The square and its background are set equal to the ambient temperature and the brightness and gain controls are adjusted to settings representative of some useful operating condition. A microphotometer is focused on the center of this square on the display and the luminance for a ΔT of zero degrees is recorded. The temperature of the square is subsequently changed in small increments and the luminance is plotted as a function of the temperature difference between the square and the background temperatures. The measurements are repeated for other control settings until the system's operation is sufficiently characterized. A set of idealized SiTFs for settings of linear brightness control B and contrast control C is shown in Figure 11.2.

The dynamic range of a system is described by a temperature difference dynamic range and by a luminance dynamic range, both of which are functions of control settings and are extractable from SiTF. The dynamic range of ΔT is the maximum scene ΔT which is displayed unclipped for a particular condition of control settings. The maximum dynamic range of ΔT is the maximum unclipped ΔT for the lowest gain setting. The luminance dynamic range gives the maximum luminance the system produces.

11.2 Optical Transfer Function Tests

Chapter Three defined the optical transfer function (OTF) for thermal imaging systems and presented the rationale for calculating the OTF from a measured line spread function (LSF). Figure 11.3 shows a simplified typical test scheme for generating a blackbody line source and for measuring the LSF and OTF. Two alternatives are indicated: photometric slit scanning of the image of a fixed line source, and thermal line image scanning across a fixed photometric slit.

The thermal line is generated by an approximate blackbody source such as a thermoelectric heat pump or an electrically heated plate which backlights an opaque blackbody mask with a small slit milled or chemically-etched in it. The source and the mask must have the same spectral emissivity, and both must be controllable and uniform both in contact temperature and in emissivity to within about $\pm 0.05°C$ for measurement of good-quality thermal imagers. In the scheme of Figure 11.3 the plane of the slit mask is located at the focal plane of an infrared collimator, and the source is separated from the mask sufficiently to minimize radiative heat transfer between the source and mask.

Ideally the apparent and contact temperatures of each surface should be the same, so that the source and mask temperature may be measured by embedded thermocouples, thermistors, or semiconductor diodes. This is desirable because small temperature differences may be more reliably and inexpensively measured by direct contact than by radiometry. The thermal

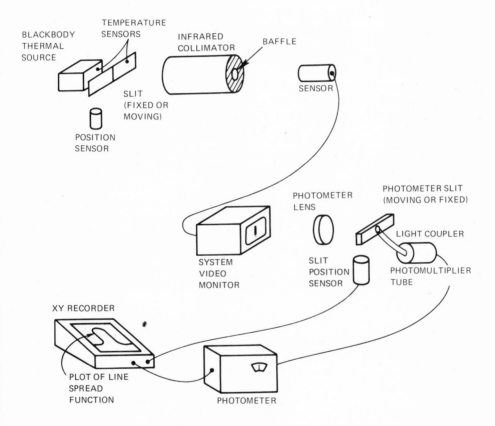

Figure 11.3 Schematic of a typical line spread function test set.

slit should be much smaller than the spatial resolution of the system, but not so small that a temperature difference sufficient to generate a good signal-to-noise ratio cannot be produced. The effective temperature difference of the slit is approximately the actual temperature difference times the ratio of the slit angular width to the system resolution.

The collimator may be either refractive or reflective, but a high-resolution, broad spectrum lens is more easily and inexpensively fabricated as a reflector. A disadvantage of reflective collimators is that baffles must be used to avoid the sensor's seeing defocused images of itself and of the laboratory retro-reflected in the lens.

Before the OTF measurement begins, the OTF's of each of the measuring and recording devices must be known, and the thermal slit spatial frequency spectrum must be known. The devices which may have non-unity OTF's are the collimator, the microphotometer, and the X-Y recorder. The non-unity spatial frequency spectrum of the line source has the effect of an OTF on the measured spread function, and this must be compensated for in the calculations.

The first step in the OTF measurement is the focusing of the sensor on the collimator focal plane image, and the alignment of the sensor so that the slit is perpendicular to the direction in which the OTF is to be measured. If the system has an eyepiece, it should be adjusted to produce a collimated output. Then the displayed slit image must be focused onto the plane of the scanning slit photometer relay optic, and must be aligned parallel with the scanning slit.

The second step is the determination of the experiment scale factor, that is, the conversion of millimeters in the scanning photometer slit plane to milliradians in object space. This allows a line spread function measured in millimeters to be converted to the desired object plane milliradian units. One procedure for measuring scale factor which minimizes error due to image spread is the introduction of a bar pattern of known angular period and the measurement and averaging of the separation between peak luminances on the display.

The third step in the OTF measurement is the establishment of the OTF existence conditions given in Chapter Three, which are: linearity, spatial invariance, and high signal-to-noise ratio. Presumably the spatial invariance requirement is satisfied by choosing a very small region on the display in which to make the measurement. However, it is exceedingly difficult to satisfy the linearity and low noise requirements simultaneously because virtually all systems are noisy, nonlinear, and have limited dynamic range. Thus a bright, high signal-to-noise ratio image is likely to have been processed nonlinearly, while a linearly processed image is likely to be noisy. It is therefore important to know the nature of the system's signal transfer function. An associated problem is that the photomultiplier used to measure the display luminance must be relatively noise-free and operated linearly. The essence of this third step is the selection of an effective temperature difference which gives a signal-to-noise ratio of around ten or better, and the selection of system control settings for which the SiTF is quasi-linear over the signal and noise range.

The fourth step is the generation of the line spread function and dissection of it with the moving photometric slit probe. This probe must be moved slowly so that the framing action of the system in combination with the probe motion does not undersample the LSF. This also allows narrowband noise suppression circuits to be used without distorting the LSF. The LSF is commonly recorded on an X-Y plotter, although the more sophisticated automated techniques described later may be used.

The LSF may be transformed approximately by fitting it to an LSF whose transform is known, such as a gaussian. Except in special cases, this is not sufficiently accurate, and the LSF must be transformed using a computer. This apparent OTF must then be compensated for measurement errors by dividing by the OTF of the measurement apparatus and by the line source spatial frequency spectrum.

The OTF test procedure outlined is sufficiently time consuming and emotionally exhausting to warrant automating the data taking. Typically the investigator will perform the measurement several times until all the experimental problems are worked out and a satisfactory result is obtained, a process that can easily consume a full working day. Moulton, et al.,[5] and the author[6] have integrated OTF test equipments with digital control computers to reduce the aggravation of manual OTF. Figure 11.4 is a schematic of the essentials of such a computer controlled test set. The analog voltages representing the scanning slit position and the corresponding LSF amplitude are relayed to a computer-controlled dual channel analog-to-digital converter. The A-to-D converter digitizes the analog data and relays them to the computer for storage in memory. The computer performs a fast-Fourier-transform of the LSF and plots the LSF, MTF, and PTF on a plotter. An example of such a measurement was given earlier in Figure 3.10.

There are many experimental errors which can cause results of an OTF test to be invalid. Sendall[7] has pointed out that the high instantaneous display element luminance required to achieve an acceptable average display luminance in a fast-framing system may saturate the photomultiplier tube used. Thus care should be taken to insure that the photomultiplier has an adequate dynamic range. Brown[8] has indicated the necessity to view the LSF through a photometer entrance pupil of eye size when measuring systems with eyepieces. This prevents lens aberrations not seen by the operator from yielding OTF's inconsistent with visual experience. Another problem

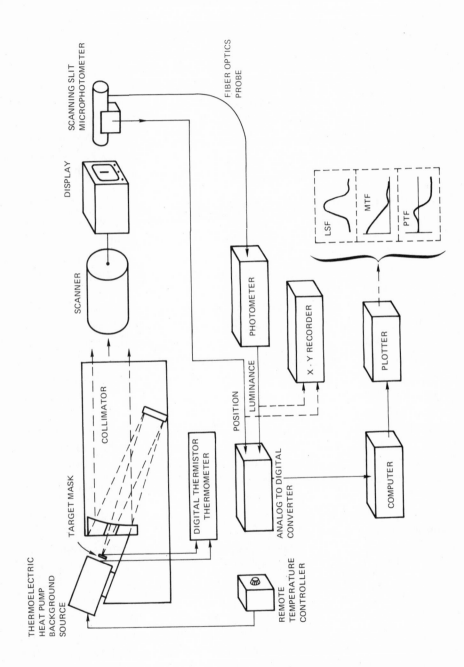

Figure 11.4 Schematic of an automated OTF test set

with the OTF measurement is that it is difficult to assess the accuracy of the result. Perhaps the best way to recognize whether a given measurement is the "true" OTF is to compare the measurement with predictions and with the nature of the MRTD measured. OTF measurement techniques employed on visible spectrum systems are described in references 9 through 16.

11.3 Minimum Resolvable Temperature Difference Test

Chapter Five defined the minimum resolvable temperature difference (MRTD) as the equivalent blackbody temperature difference between the target of Figure 5.5 at a given frequency and its background such that the individual bars are resolvable by an observer. Care must be taken in devising test equipment and a test procedure to ensure that the MRTD existence conditions are satisfied. The system controls must be set, or varied between limits, such that the system is operated quasi-linearly, and so that noise is just visible on the display. In this way the system will be "noise-limited" rather than "contrast-limited." The first condition can be verified by referring to SiTF curves, while the second condition can be verified only by the observer's subjective impression of the noise level.

The test charts may be either finite conjugate sources located at some in-focus object distance, or they may be infinite conjugate sources simulated by a collimator. Finite conjugate sources have the disadvantage that low frequency patterns must be rather large. In that case uniformity and controllability of the targets will likely be a problem. The test procedure discussed here assumes that a collimator is used.

A sufficient number of test patterns must be available to investigate the spatial frequency range of the system. The collimator focal plane must be determined and the targets located there. The target and background emissivities must be equal, uniform, and known. The temperature differential generating and sensing equipment must be calibrated. The target generator should be stable, controllable, and spatially uniform to within ±0.01°C, and the accuracy of the measuring devices should be within ±0.01°C. The collimator should be optically baffled and the sensor oriented so that back reflections of the laboratory scene into the sensor are minimized. The spectral transfer and optical transfer functions of the collimator must have no detrimental effect on the sensor.

The observer or observers conducting the test must have good visual acuity and normal color vision. They should be thoroughly familiar with thermal imagery and be able to interpret noisy imagery. They should be dark-adapted to the average luminance level of the display, and the background (laboratory) ambient luminance should approximately match the display luminance. Further, the observers should be trained to identify and to reproduce the conditions for which their confidence in recognition accuracy is high.

A typical test procedure would progress as follows. The sensor of the system to be tested is mounted in the beam of the collimator, and the sensor is focused using an appropriate technique. The lowest frequency test target is introduced into the field at a position known to the observer, and at $0°C$ ΔT. The ΔT is increased positively and at a rate slow enough to ensure target uniformity and measurement accuracy, and to allow the observer a sufficient number of decision times to make a valid decision. The observer then indicates the target ΔT at which he resolves the bars. The procedure is repeated for higher and higher frequencies until the observer is no longer able to resolve the bars at any ΔT. For target temperature differences low enough to assure linear system operation for both positive and negative contrast targets (usually only low frequencies), negative ΔT's may be used to find the negative MRT. This will indicate any temperature offset in the test set.

11.4 Noise Equivalent Temperature Difference Test

Citation of a measured NETD is virtually meaningless unless the noise-limiting electrical filter used in the measurement is also specified. It is desirable for standardization and comparison purposes to use as the noise filter a single – RC low pass filter whose 3-db frequency is the reciprocal of twice the dwelltime. Since electrical noise filtering in the system itself is usually minimal for good quality, this noise filter can frequently be external to the system and a part of the test equipment.

To assure that all noise sources prior to the video display are accounted for in the measurement, it is customary to take the test signals from the final processing stage at which an analog video signal exists. This means prior to a multiplexer or pulse-width modulator. If this point is conveniently accessible, the conduct of the NETD test is trivial. One simply generates a large square blackbody thermal target whose angular dimension exceeds the system resolution by several times and whose temperature exceeds the expected NETD by a factor of ten or so, and then measures the peak to peak signal and true RMS noise in the electronic image of that target. The NETD is then the target temperature difference divided by the peak signal to rms noise ratio. Care must be taken to assure that the system is operated linearly and that no noise sources are included which are not displayed, such as switching transients, or reference signals which appear during the display blank time. Though it is a somewhat dubious figure of merit, NETD will long endure because its measurement does not require the skill and experience necessary for OTF and MRTD tests, so that virtually anyone can do it correctly.

REFERENCES

[1] P. Moon, *The Scientific Basis of Illuminating Engineering*, Dover, 1961.

[2] R.P. Teel, "Photometry", Chapter 1, Volume 1, of *Applied Optics and Optical Engineering*, edited by R. Kingslake, Academic Press, 1965.

[3] S. Sherr, *Fundamentals of Display System Design*, Wiley-Interscience, 1970.

[4] H.R. Luxenberg and R.L. Kuehn, *Display Systems Engineering*, McGraw-Hill, 1968.

[5] J.R. Moulton and J.T. Wood, "Two-Port Evaluation Techniques Applied to Commercial Imaging Thermographs", Proceedings of the Technical Program, Electro-Optical Systems Design Conference, September 22-24, 1970.

[6] Work performed in association with S.F. Layman, B.A. Blecha, and H.J. Orlando at the United States Army Night Vision Laboratory, and published privately.

[7] R.L. Sendall, personal communication, Xerox Electro-Optical Systems, Pasadena, California.

[8] D.P. Brown, personal communication, Hughes Aircraft Company, Culver City, California.

[9] R.R. Shannon and A.H. Newman, Appl. Opt., *2*, pp 365-369, April 1963.

[10] R.A. Jones and F.C. Yeadon, Phot. Sci. and Eng., *13*, pp 200-204, August 1969.

[11] R.A. Jones, Phot. Sci. and Eng., *11*, pp 102-106, March-April 1967.

[12] O.H. Schade, Sr., JSMPTE, *73*, pp 81, 1964.

[13] F.S. Blackman, Phot. Sci. and Eng., *12*, pp 244-250, September-October 1968.

[14] A.A. Sanders, "Modulation Transfer Function Measurements for Infrared Systems", Master's Thesis, The George Washington University, Washington, D.C. 1970.

[15] B. Tatian, "Method for Obtaining the Transfer Function from the Edge Response Function", JOSA, *55*, pp 1014-1019, August 1965.

[16] E.F. Brown, "A Method for Measuring the Spatial-Frequency Response of a Television System", JSMPTE, *76*, pp 884-888, September 1967.

CHAPTER TWELVE — THERMAL IMAGERY

The quality of imagery which the current FLIR state-of-the-art can produce is exemplified by the eleven figures which follow. These photographs were produced by Sawyer[1] from a FLIR monitor using an exposure time of 0.2 second to closely approximate the visual impression of noisiness in the real-time imagery. The particular FLIR used is a serial scan television-compatible system. Some of the imagery has superimposed on it a reticle, the square gate of an automatic target tracker, and date-time symbology.

---

[1] L.H. Sawyer, personal communication, Honeywell Radiation Center, Lexington, Massachusetts.

Figure 12.1 Highway and bridge through a light industrial area.

Figure 12.2 An urban scene.

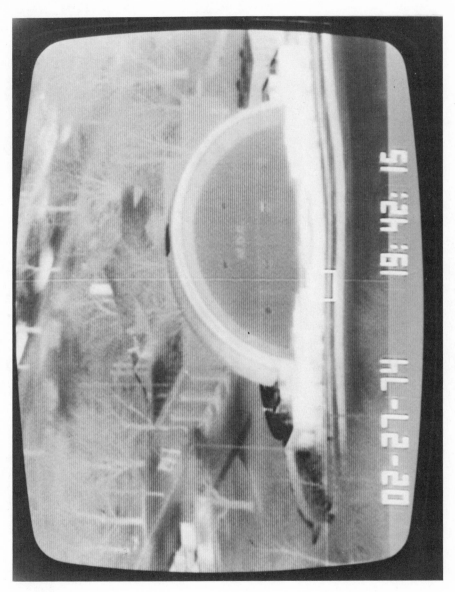

Figure 12.3 A concert shell.

Figure 12.4 A cargo helicopter.

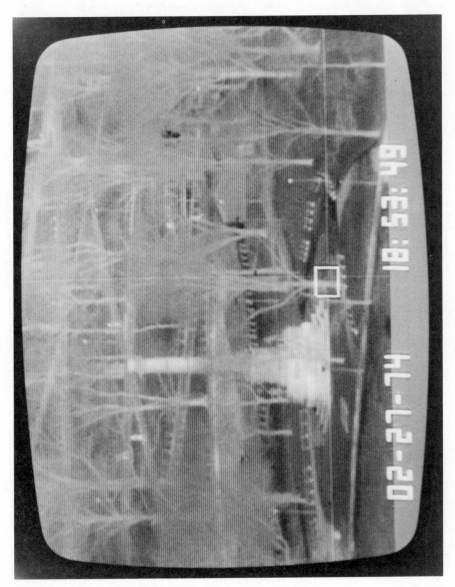

Figure 12.5 A park.

Figure 12.6 A capitol building.

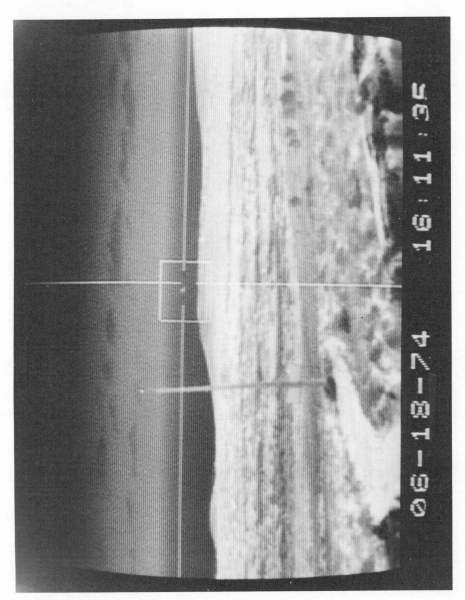

Figure 12.7 A sparsely vegetated area.

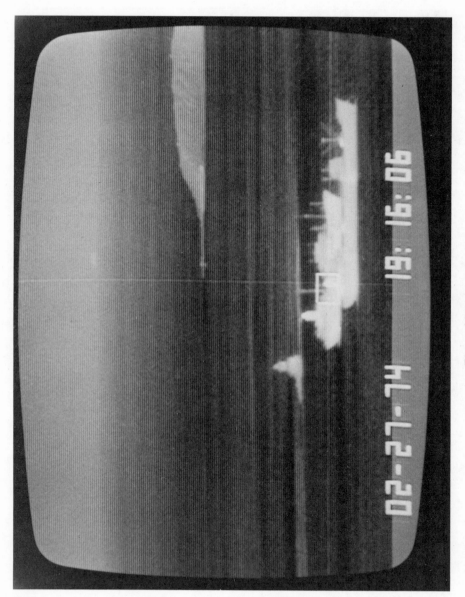

Figure 12.8 Freighters.

Figure 12.9 A power plant.

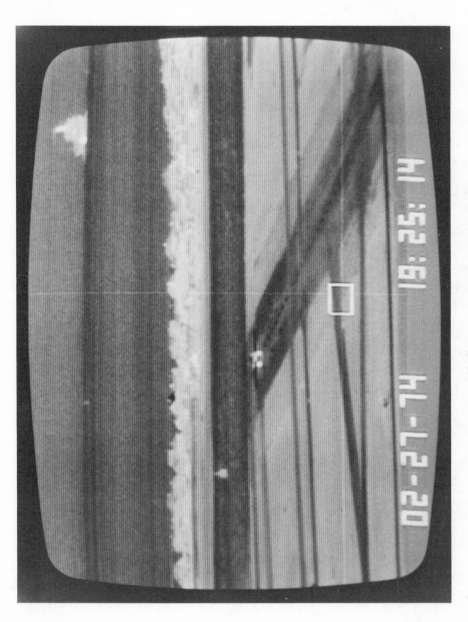

Figure 12.10 An aircraft immediately after landing.

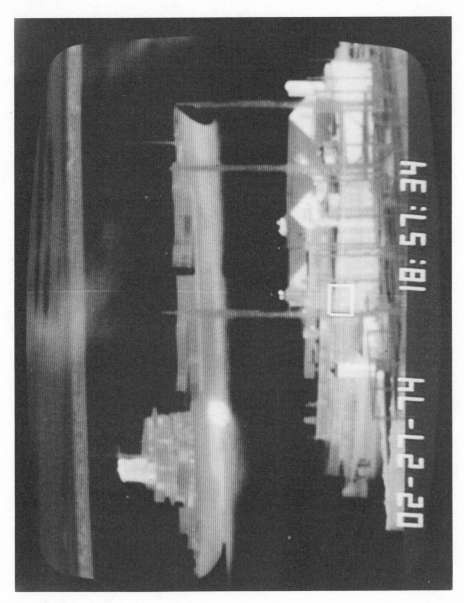

Figure 12.11 A freighter.

INDEX

V

W

U

Z